AN INTRODUCTION TO ATMOSPHERIC RADIATION

AN INTRODUCTION TO ATMOSPHERIC RADIATION

Kuo-Nan Liou

Department of Meteorology
University of Utah
Salt Lake City, Utah

ACADEMIC PRESS
A Subsidiary of Harcourt Brace Jovanovich, Publishers
New York   London   Toronto   Sydney   San Francisco   1980

ACADEMIC PRESS, INC.
111 Fifth Avenue, New York, New York 10003

United Kingdom Edition published by
ACADEMIC PRESS, INC. (LONDON) LTD.
24/28 Oval Road, London NW1 7DX

Library of Congress Cataloging in Publication Data

Liou, Kuo–Nan.
  An introduction to atmospheric radiation.

  (International geophysics series ; v.   )
  Bibliography: p.
  Includes indexes.
  1. Atmospheric radiation. I. Title. II. Series.
QC912.3.L56   551.5'273   80–769
ISBN 0–12–451450–2

PRINTED IN THE UNITED STATES OF AMERICA

80 81 82 83   9 8 7 6 5 4 3 2 1

CONTENTS

v

Chapter 4 Infrared Radiation Transfer in the Atmosphere

Chapter 5 Light Scattering by Particulates in the Atmosphere

Chapter 6 Principle of Multiple Scattering in Plane-Parallel Atmospheres

PREFACE

During recent years, the problems involved in understanding and predicting climate and climatic changes have become topics of increasing interest among both scientists and the public. This upsurge of interest has resulted from the realization that human endeavors are vulnerable to uncertainties in climate, and that human activities may be causing climatic changes. The transfer of solar and infrared radiation represents the prime physical process that drives the circulation of the atmosphere and the ocean currents. It is apparent that an understanding of climate and the mechanisms of climatic changes must begin with detailed understanding of radiative processes and the radiative balance of the earth and the atmosphere.

Moreover, since the successive launches of meteorological satellites in the sixties, applications of the principle of radiative transfer have been fruitful. Using data gathered by these satellites and the principle of radiative transfer, we now are capable of deriving profiles of the temperature and various optically active gases such as water vapor and ozone in our atmospheres. Information of such profiles significantly enhance our understanding of weather and climate of the earth. With the progressive comprehension of the physical interaction of clouds and aerosols with solar and infrared radiation, the quantitative inference of the composition and structure of globally distributed cloud systems and aerosols appears feasible. It is evident that the sounding techniques developed for the earth's atmosphere may be applied directly to other planetary atmospheres as well.

Although there have been a number of important reference books written in the field of atmospheric radiation, none of them can be adequately adopted as textbooks in atmospheric sciences. This is either because the books are oriented toward a literature survey or because they are lacking in presentations on one or several aspects of scattering and absorption processes in planetary atmospheres. Furthermore, none of the published books so far has presented applications of light scattering and radiative transfer principles to remote sensing and radiation climatology. At this time, when satellite sensing, laser applications, and radiative transfer are becoming increasingly important in conjunction with the study of weather and climate of planetary atmospheres, there is indeed an urgent need for a coherent and logical development on the subject of radiation processes in planetary atmospheres. It is the purpose of this book to present and to unify all of the topics associated with the fundamentals of atmospheric radiation. The level of presentation is in such a manner that seniors and graduate students in the atmospheric sciences, and research beginners in atmospheric radiation can follow and absorb the mathematical deductions and fundamental physical laws that govern the radiation field of planetary atmospheres.

The book is divided into eight chapters. Chapter 1 introduces concepts, definitions, various basic radiation laws, and the fundamental equations for radiative transfer. Chapter 2 describes the characteristics of the solar radiation that is available at the top of the earth's atmosphere. Chapter 3 is concerned with the absorption and scattering processes of solar radiation in molecular atmospheres. Photochemical processes involving ultraviolet radiation and ozone are discussed, and the concept of polarization and the scattering of sunlight by Rayleigh molecules are presented. Chapter 4 deals with infrared radiative transfer in the earth–atmosphere system. The fundamental theory of infrared transfer is covered, and absorption band models and the principle of radiation charts are discussed. Chapter 5 presents the single scattering processes involving aerosols and cloud particles in the atmosphere. The Maxwell equations are first introduced, and the solution of the vector wave equation, which leads to the Mie theory, is derived. The geometrical ray optics approach to light scattering by spherical water drops and hexagonal ice crystals is outlined. In Chapter 6, the principles of multiple scattering in plane–parallel atmospheres are introduced. This chapter includes the presentation of the basic equations, some approximations for radiative transfer problems, the principles of invariance, and various methods for solving the fundamental transfer equation. Applications of the basic radiative transfer theory to remote sensing of the atmosphere are given in Chapter 7. Discussions are made on the inversion principles used in determining temperature and gaseous profiles by means of satellite infrared sounding channels. Various inversion methods also are introduced. The uses of microwave sounders, and the reflected and transmitted sunlight as a means of remote sensing are further discussed. The basic principles of radar and lidar back-scattering techniques for cloud and precipitation detection also are described. The subject matter associated with radiation climatology is covered in the final chapter. This chapter introduces broadband radiation observations from satellites, and reports the latitudinal and global radiation budgets determined from satellite measurements. Theoretical radiation budget studies, and simple climate models based on the radiative balance are further described. Problem sets with varying degrees of difficulty are prepared in each chapter.

In writing this book, I have assumed that the readers already have had introductory courses in physics and calculus. Although the book has been written primarily for students and researchers in the field of atmospheric sciences, students and researchers in other disciplines, including planetary exploration, electromagnetic scattering, optics, and geophysics, also may find various topics in the text of some interest and use. I have used materials in Chapters 1–4 in a senior and first-year graduate course entitled "Atmospheric Radiation: Physical Meteorology I." I also have utilized the subject matter in Chapters 5–6 and Chapters 7–8 in advanced graduate courses entitled "Radiative Transfer" and "Remote Sensing from Satellites," respectively. Some of the materials presented in the text are original and have not been published elsewhere.

During the course of the writing, I have found an enormous amount of literature in the field of atmospheric radiation, resulting from the overlap of meteorology, astrophysics, planetology, electrical engineering, and applied physics. Generally, I have avoided citing the original reference on the topic discussed in the text. Interested readers who wish to further study the subject matter can find the relevant papers from the suggested references which are either published books or review papers. However, I have attempted to make reference to important contributions, which represent recent developments and significant finds in the field of atmospheric radiation and remote sensing. I have undertaken an almost impossible task of unifying diffuse notations used in fields of scattering, absorption and emission, radiative transfer, and satellite sensing. Unfortunately, I find that it is unavoidable to repeat some symbols to preserve the distinction of various content areas. Finally, a number of subject matters, which are not described in the text, are presented through exercises at the end of each chapter.

I am indebted to the following friends and colleagues who took the time to read various chapters of the manuscript and offered many helpful suggestions for improvements: P. Barber, K. L. Coulson, A. Fymat, J. F. King, C. B. Leovy, J. North, and T. Sasamori. During the course of the writing, my research programs have been continuously supported by the Atmospheric

Research Section of the National Science Foundation and the Air Force Geophysics Laboratory. Their support has made possible a number of presentations in the text. Appreciation is extended to the University of Utah for granting me a David P. Gardner Faculty Fellow Award which released my teaching duty in the winter quarter of the 1978/1979 academic year during which considerable writing was accomplished. I would also like to thank R. Coleman and K. Hutchison for independently working out most of the exercises and for assisting me in proofreading the manuscript, and Mrs. D. Plumhof for typing various versions of the manuscript.

In the northern darkness there is a fish and his name is Kun. The Kun is so huge that he measures many thousand miles. He changes and becomes a bird whose name is Peng. The back of the Peng also measures many thousand miles across and, when he rises up and flies off, his wings are like clouds all over the sky. When the sea begins to move, this bird journeys to the southern darkness, and the waters are roiled for three thousand miles. He beats the whirlwind and rises ninety thousand miles, setting off on the sixth month gale, wavering heat, bits of dust, living things blown about by the wind—the sky looks very blue. Is that its real color, or is it because it is so far away and has no end?

Chuang Tzu
~339–295 B.C.

AN INTRODUCTION TO ATMOSPHERIC RADIATION

Chapter 1
FUNDAMENTALS OF RADIATION

1.1 CONCEPTS, DEFINITIONS, AND UNITS

1.1.1 Electromagnetic Spectrum

The most important of the processes responsible for energy transfer in the atmosphere is electromagnetic radiation. Electromagnetic radiation travels in the wave form, and all electromagnetic waves travel at the same speed, which is the speed of light. This is $2.99793 \pm 1 \times 10^8$ m sec^{-1} in a vacuum and at very nearly this speed in air. Visible light together with gamma rays, x rays, ultraviolet light, infrared radiation, microwaves, television signals, and radio waves form the *electromagnetic spectrum*.

The retina of the human eye is sensitive to electromagnetic waves with frequencies between 4.3×10^{14} vibrations per second (usually written as cycles per second and abbreviated cps) and 7.5×10^{14} cps. Hence, this band of frequencies is called the *visible* region of the electromagnetic spectrum. The eye, however, does not respond to frequencies of the electromagnetic waves higher than 7.5×10^{14} cps. Such waves, lying beyond the violet edge of the spectrum, are called *ultraviolet* light. Moreover, if the waves have frequencies lower than 4.3×10^{14} cps, the eye again does not respond to them. These waves, having frequencies lower than the lowest frequency of visible light at the red end of the spectrum and higher than about 3×10^{12} cps, are called *infrared light* or *infrared radiation*. Just beyond the infrared portion of the spectrum are the *microwaves*, which cover the frequency from about 3×10^{10} to 3×10^{12} cps. The most significant spectral regions associated

Name of region	Wavelength (cm)	Frequency (cps)
Gamma rays		
	10^{-9}	3×10^{19}
x rays		
	10^{-6}	3×10^{16}
Ultraviolet	3×10^{-5}	10^{15}
Visible		
	10^{-4}	
Infrared		3×10^{11}
	10^{-1}	
Microwaves	1	3×10^{10}
Spacecraft		3×10^{8}
	10^{2}	
Television & FM	10^{3}	3×10^{7}
Shortwave	10^{4}	3×10^{6}
AM	10^{5}	3×10^{5}
Radio waves		

Violet
Purple
Blue
Green
Yellow
Orange
Red

Fig. 1.1   The electromagnetic spectrum.

with the radiative energy transfer in planetary atmospheres lie between the ultraviolet light and microwaves.

The *x ray* region of the electromagnetic spectrum consists of waves with frequencies ranging from about 3×10^{16} to 5×10^{18} cps, and is adjacent to the ultraviolet region in the spectrum. The *gamma-ray* region of the spectrum has the highest frequencies of all, ranging upward from about 3×10^{19} cps. At the other end of the spectrum beyond the microwave region is the *television* and *FM* band of frequencies, extending from about 3×10^{8} to 3×10^{5} cps. *Radio* waves have the lowest frequencies in the spectrum, extending downward from about 3×10^{5} cps.

Electromagnetic waves often are described in terms of their wavelength rather than their frequency. The following general formula connects frequency $\tilde{\nu}$ and wavelength λ:

$$\lambda = c/\tilde{\nu}, \tag{1.1}$$

where c represents the speed of light in a vacuum. The formula is valid for any type of wave, and is not restricted to electromagnetic waves. It is customary to use wave number ν to describe the characteristics of infrared radiation. It is defined by

$$\nu = \tilde{\nu}/c = 1/\lambda. \tag{1.2}$$

Thus, a 10 micrometer (μm) (1 μm $= 10^{-4}$ cm) wavelength is equal to a 1000 cm^{-1} wave number. In the microwave region, however, a frequency unit called gigahertz (GHz) is commonly used. One GHz is equal to 10^{9} cycles per second. Thus, 1 cm is equivalent to 30 GHz.

Figure 1.1 shows the complete electromagnetic spectrum with frequencies and wavelengths indicated. The names given to the various parts of the spectrum are also shown.

1.1.2   Solid Angle

The analysis of a radiation field often requires the consideration of the amount of radiant energy confined to an element of solid angle. The solid angle is defined as the ratio of the area σ of a spherical surface intercepted by the core to the square of the radius, r, as indicated in Fig. 1.2. It can be written

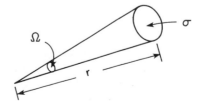

Fig. 1.2   Definition of a solid angle.

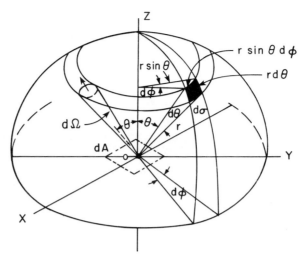

Fig. 1.3 Illustration of a solid angle and its representation in polar coordinates. Also shown is a pencil of radiation through an element of area dA in directions confined to an element of solid angle $d\Omega$.

as

$$\Omega = \sigma/r^2. \tag{1.3}$$

Units of the solid angle are expressed in terms of the steradian (sr). For a sphere whose surface area is $4\pi r^2$, its solid angle is 4π sr.

To obtain a differential elemental solid angle, we construct a sphere whose central point is denoted as O. Assuming a line through point O moving in space and intersecting an arbitrary surface located at a distance r from point O, then as evident from Fig. 1.3 the differential area in polar coordinates is given by

$$d\sigma = (r\,d\theta)(r\sin\theta\,d\phi). \tag{1.4}$$

Hence, the differential solid angle is

$$d\Omega = d\sigma/r^2 = \sin\theta\,d\theta\,d\phi, \tag{1.5}$$

where θ and ϕ denote the zenithal and azimuthal angles, respectively, in polar coordinates.

1.1.3   Basic Radiometric Quantities

Consider the differential amount of radiant energy dE_λ in a time interval dt and in a specified wavelength interval λ to $\lambda + d\lambda$, which crosses an element of area dA depicted in Fig. 1.3, and in directions confined to a differential solid angle, which is oriented at an angle θ to the normal of dA. This energy is expressed in terms of the specific intensity I_λ by

$$dE_\lambda = I_\lambda \cos \theta \, d\Omega \, dA \, d\lambda \, dt. \tag{1.6}$$

Equation (1.6) defines the *monochromatic intensity* (or *radiance*) in a general way as

$$I_\lambda = \frac{dE_\lambda}{\cos \theta \, d\Omega \, d\lambda \, dt \, dA}. \tag{1.7}$$

Thus the intensity is in units of energy per area per time per frequency and per steradian. It is evident that the intensity implies a directionality in the radiation stream. Commonly, the intensity is said to be confined in a pencil of radiation.

The *monochromatic flux density* or the *monochromatic irradiance* of radiant energy is defined by the normal component of I_λ integrated over the entire spherical solid angle and may be written as

$$F_\lambda = \int_\Omega I_\lambda \cos \theta \, d\Omega. \tag{1.8}$$

In polar coordinates, we write

$$F_\lambda = \int_0^{2\pi} \int_0^{\pi/2} I_\lambda(\theta, \phi) \cos \theta \sin \theta \, d\theta \, d\phi. \tag{1.9}$$

It can easily be shown that for isotropic radiation, i.e., if radiant intensity is independent of direction, the monochromatic flux density is

$$F_\lambda = \pi I_\lambda. \tag{1.10}$$

The total *flux density* of radiant energy, or irradiance for all wavelengths (energy per area per time), can be obtained by integrating the monochromatic flux density over the entire electromagnetic spectrum:

$$F = \int_0^\infty F_\lambda \, d\lambda. \tag{1.11}$$

Moreover, the total flux f, or radiant power W (energy per time) is defined by

$$f = \int_A F \, dA. \tag{1.12}$$

The monochromatic flux density in the frequency domain may be written in the form

$$F_{\tilde{\nu}} = \frac{dF}{d\tilde{\nu}}. \tag{1.13}$$

TABLE 1.1 *Symbols, Dimensions, and Units of Various Radiometric Quantities*

Symbol	Quantity	Dimension[a]	Unit (cgs)[b]
E	Energy	ML^2T^{-2}	Erg
f	Flux	ML^2T^{-3}	Erg per second (erg sec^{-1})
	Luminosity	MT^{-3}	Erg per second per square centimeter
F	Flux density (irradiance)		(erg cm^{-2} sec^{-1})
	Emittance	MT^{-3}	Erg per second per square centimeter
I	Intensity (radiance)		per steradian (erg cm^{-2} sec^{-1} sr^{-1})
	Brightness (luminance)		

[a] M is mass, L is length, and T is time.
[b] 1 erg = 10^{-7} joule (J), 1 watt (W) = 1 joule sec^{-1}.

From the relation between the wavelength and frequency denoted in Eq. (1.1), we shall have

$$F_{\tilde{v}} = -(\lambda^2/c)F_\lambda. \tag{1.14}$$

Likewise, the intensity quantity in wavelength and frequency domains shall be connected by

$$I_{\tilde{v}} = -(\lambda^2/c)I_\lambda. \tag{1.15}$$

A similar relation between the monochromatic flux density or intensity in wave number and wavelength (or frequency) domains may be expressed by means of Eq. (1.2).

When the flux density or the irradiance is from an emitting surface, the quantity is called the *emittance*. When expressed in terms of wavelength, it is referred to as the *monochromatic emittance*. The intensity or the radiance is also called the *brightness* or *luminance* (photometric brightness). The total flux from an emitting surface is often called *luminosity*. The basic radiometric quantities are summarized in Table 1.1, along with their symbols, dimensions, and units.

1.1.4 Concepts of Scattering and Absorption

Most of the light that reaches our eyes comes not directly from its sources but indirectly by the process of *scattering*. We see diffusely scattered sunlight when we look at clouds or at the sky. The land and water surfaces, and the objects surrounding us are visible through the light that they scatter. An electric lamp does not send us light directly from the luminous filament but usually glows with the light that has been scattered by the glass bulb. Unless we look at a source, such as the sun, a flame, or an incandescent filament with a clear bulb, we see light that has been scattered. In the atmosphere, we see many colorful examples of scattering generated by molecules, aerosols, and clouds containing droplets and ice crystals. Blue sky, white clouds, and magnificent rainbows and halos, to name a few, are all optical phenomena due to scattering. *Scattering* is a fundamental physical process associated with the light and its interaction with matter. It occurs at all wavelengths covering the entire electromagnetic spectrum.

Scattering is a physical process by which a particle in the path of an electromagnetic wave continuously abstracts energy from the incident wave and reradiates that energy in all directions. Therefore, the particle may be thought of as a point source of the scattered energy. In the atmosphere, the particles responsible for scattering cover the sizes from gas molecules ($\sim 10^{-8}$ cm) to large raindrops and hail particles (~ 1 cm). The relative intensity of the scattering pattern depends strongly on the ratio of particle size to wavelength of the incident wave. If scattering is isotropic, the scattering pattern is symmetric about the direction of the incident wave. A small anisotropic particle tends to scatter light equally into the forward and rear directions. When the particle becomes larger, the scattered energy is increasingly concentrated in the forward directions with greater complexities as evident from Fig. 1.4, where scattering patterns of three particle sizes are illustrated. Distribution of the scattered energy involving spherical and certain symmetrical particles may be quantitatively determined by means of the electromagnetic wave theory. When particles are much smaller than the incident wavelength, the scattering is called *Rayleigh scattering*, which leads to the explanation of blue sky and sky polarization as will be discussed in Chapter 3. For particles whose sizes are comparable to or larger than the wavelength, the scattering is customarily referred to as *Mie scattering*. The mathematical theory of Mie scattering for spherical particles, and the associated geometrical optics for water droplets and hexagonal crystals will be presented in Chapter 5.

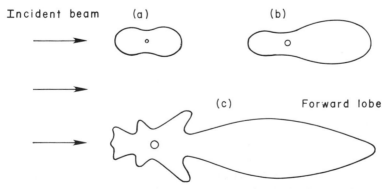

Fig. 1.4 Demonstrative angular patterns of the scattered intensity from particles of three sizes: (a) small particles, (b) large particles, and (c) larger particles.

In a scattering volume, which contains many particles, each particle is exposed to, and also scatters, the light which has already been scattered by other particles. To demonstrate this concept we refer to Fig. 1.5. A particle at position P removes the incident light by scattering just once, i.e., single scattering, in all directions. Meanwhile, a portion of this scattered light reaches the particle at position Q, where it is scattered again in all directions. This is called *secondary scattering*. Likewise, a subsequent third-order scattering involving the particle at position R takes place. Scattering more than once is called *multiple scattering*. It is apparent from Fig. 1.5 that some of the incident light that has been first scattered away from the direction **d** may reappear in this direction by means of multiple scattering. Multiple scattering is an important process for the transfer of radiant energy in the atmosphere, especially when aerosols and clouds are involved. Chapter 6 deals with the theory of multiple scattering.

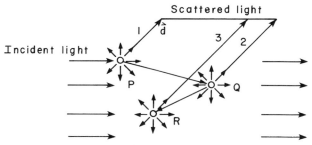

Fig. 1.5   Multiple scattering process.

Scattering is often accompanied by *absorption*. Grass looks green because it scatters green light more effectively than red and blue light. Apparently, red and blue light incident on the grass is absorbed. The absorbed energy is converted into some other form, and it is no longer present as red or blue light. In the visible spectrum, absorption of energy is nearly absent in molecular atmospheres. Clouds also absorb very little visible light. Both scattering and absorption remove energy from a beam of light traversing the medium. The beam of light is attenuated, and we call this attenuation *extinction*. Thus, extinction is a result of scattering plus absorption. In a nonabsorbing medium, scattering is the sole process of extinction.

In the field of light scattering and radiative transfer, it is customary to use a term called *cross section*, which is analogous to the geometrical area, to denote the amount of energy removed from the original beam by the particles. In the case when the cross section is referred to a particle, its units are in area (cm^2). Thus, the extinction cross section, in units of area, is the sum of the scattering and absorption cross sections. However, when the cross section is in reference to unit mass, its units are in area per mass ($cm^2\ g^{-1}$). In this case, the term mass extinction cross section is used in the transfer study. The mass extinction cross section is therefore the sum of the mass absorption and mass scattering cross sections. Furthermore, when the extinction cross section is multiplied by the particle number density (cm^{-3}) or when the mass extinction cross section is multiplied by the density ($g\ cm^{-3}$) the quantity is referred to as *extinction coefficient*, which has units of per length (cm^{-1}). In the field of infrared radiative transfer, the mass absorption cross section is simply referred to as *absorption coefficient*.

Absorption of energy by particles and molecules leads to *emission*. The concept of emission is associated with blackbody radiation, and it is to be discussed in the following section. Moreover, a number of minor atmospheric constituents exhibit complicated absorption line structures in the infrared regions. Section 1.3 and Chapter 4 will provide discussions on the fundamentals of the line formation and the transfer of infrared radiation in the atmosphere. A fundamental understanding of the scattering and absorption processes in the atmosphere is imperative for the studies of the radiation budget and climate of planetary atmospheres and for the exploration of remote sounding techniques to infer the atmospheric composition and structure.

1.2 BLACKBODY RADIATION

1.2.1 Planck's Law

In order to have a theoretical explanation for the cavity radiation, Planck in 1901 was led to make two assumptions about the atomic oscillators. First, he postulated that an oscillator cannot have any energy but only energies given by

$$E = nh\tilde{v}, \tag{1.16}$$

where $\tilde{v}$ is the oscillator frequency, h is Planck's constant, and n is called a quantum number that can take on only integral values. Equation (1.16) asserts that the oscillator energy is quantized. Although later developments revealed that the correct formula for a harmonic oscillator is $E = (n + \frac{1}{2})h\tilde{v}$, the change introduces no difference to Planck's conclusions. Secondly, he postulated that the oscillators do not radiate energy continuously, but only in jumps, or in quanta. These quanta of energy are emitted when an oscillator changes from one to another of its quantized energy states. Hence, if the quantum number changes by one unit, the amount of energy that is radiated is given by

$$\Delta E = \Delta n\, h\tilde{v} = h\tilde{v}. \tag{1.17}$$

On the basis of these two assumptions, Planck was able to derive from a theoretical point of view the so-called Planck function which is expressed by

$$B_{\tilde{v}}(T) = \frac{2h\tilde{v}^3}{c^2(e^{h\tilde{v}/KT} - 1)},$$ (1.18)

where K is Boltzmann's constant, c the velocity of light, and T the absolute temperature. The Planck and Boltzmann constants are determined from the

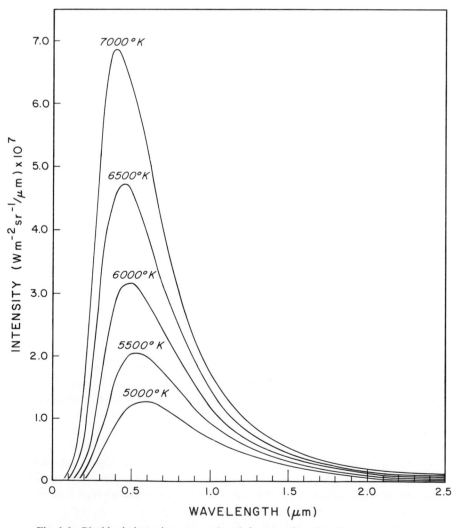

Fig. 1.6 Blackbody intensity per wavelength for a number of emitting temperatures.

experiment, and they are given by $h = 6.6262 \times 10^{-27}$ erg sec, and $K = 1.3806 \times 10^{-16}$ erg deg^{-1}. The derivation of the Planck function is given in Appendix C.

The Planck function relates the emitted monochromatic intensity with the frequency and the temperature of the emitting substance. By utilizing the relation between frequency and wavelength shown in Eq. (1.15), Eq. (1.18) can be written as

$$B_\lambda(T) = \frac{2hc^2}{\lambda^5(e^{hc/K\lambda T} - 1)}. \tag{1.19}$$

Figure 1.6 shows curves of $B_\lambda(T)$ versus wavelength for a number of emitting temperatures. It is evident that the blackbody radiant intensity increases with the temperature, and that the wavelength of the maximum intensity decreases with increasing temperature.

1.2.2   Stefan-Boltzmann Law

The total radiant intensity of a blackbody can be derived by integrating the Planck function over the entire wavelength domain from 0 to ∞. Hence,

$$B(T) = \int_0^\infty B_\lambda(T)\, d\lambda = \int_0^\infty \frac{2hc^2 \lambda^{-5}\, d\lambda}{(e^{hc/kT} - 1)}. \tag{1.20}$$

On introducing a new variable $x = hc/k\lambda T$, Eq. (1.20) becomes

$$B(T) = \frac{2k^4 T^4}{h^3 c^2} \int_0^\infty \frac{x^3\, dx}{(e^x - 1)}. \tag{1.21}$$

The integral term in Eq. (1.21) is equal to $\pi^4/15$. Thus, defining

$$b = 2\pi^4 k^4/(15c^2 h^3), \tag{1.22}$$

we then have

$$B(T) = bT^4. \tag{1.23}$$

Since blackbody radiation is isotropic, the flux density emitted by a blackbody is therefore [see Eq. (1.10)]

$$F = \pi B(T) = \sigma T^4, \tag{1.24}$$

where σ is the Stefan–Boltzmann constant and is equal to 5.67×10^{-5} erg cm^{-2} sec^{-1} deg^{-4}. Equation (1.24) states that the flux density emitted by a blackbody is proportional to the fourth power of the *absolute* temperature. This is the Stefan–Boltzmann law, which is fundamental in the field of infrared radiative transfer.

1.2.3   Wien's Displacement Law

Wien's displacement law states that the wavelength of the maximum intensity for blackbody radiation is inversely proportional to the temperature. By differentiating the Planck function with respect to wavelength, and by setting the result equal to zero, i.e.,

$$\frac{\partial B_\lambda(T)}{\partial \lambda} = 0, \tag{1.25}$$

we obtain the wavelength of the maximum

$$\lambda_m = a/T, \tag{1.26}$$

where $a = 0.2897$ cm deg. From this relation, we may determine the temperature of a blackbody from the measurement of the maximum monochromatic intensity. The dependence of the position of the maximum intensity on temperature can be seen from the blackbody curves depicted in Fig. 1.6.

1.2.4   Kirchoff's Law

The foregoing three fundamental laws are essentially concerned with radiant intensity emitted by a blackbody. The amount of radiant intensity is associated with the emitting wavelength and the temperature of the medium. A medium may absorb radiation of a particular wavelength, and at the same time also may emit radiation of the same wavelength. The rate at which emission takes place is a function of temperature and wavelength. This is the fundamental property of a medium under the condition of *thermodynamic equilibrium*. The physical statement regarding absorption and emission was first proposed by Kirchhoff in 1859.

To understand the physical meaning of Kirchhoff's law, we consider a perfectly insulated enclosure having black walls. Assume that this system has reached the state of thermodynamic equilibrium characterized by uniform temperature and isotropic radiation. Because the walls are black, radiation emitted by the system to the walls is absorbed. Moreover, because there is an equilibrium, the same amount of radiation absorbed by the walls is also emitted. Since the blackbody absorbs the maximum possible radiation, it has to emit that same amount of radiation. If it emitted more, equilibrium would not be possible, and this would violate the second law of thermodynamics. Radiation within the system is referred to as blackbody radiation, and the amount of radiant intensity is a function of temperature only.

On the basis of the preceding, for a given wavelength, the emissivity ε_λ, defined as the ratio of the emitting intensity to the Planck function, of a medium is equal to the absorptivity A_λ, defined as the ratio of the absorbed intensity to the Planck function, of that medium under thermodynamic equilibrium. Hence we may write

$$\varepsilon_\lambda = A_\lambda. \tag{1.27}$$

A medium with an absorptivity A_λ absorbs only A_λ times the blackbody radiant intensity $B_\lambda(T)$, and therefore emits ε_λ times the blackbody radiant intensity. For a *blackbody*, absorption is a maximum and so is emission. Thus, we shall have

$$A_\lambda = \varepsilon_\lambda = 1 \tag{1.28}$$

for all wavelengths. A *gray body* is characterized by incomplete absorption and emission, and may be described by

$$A_\lambda = \varepsilon_\lambda < 1. \tag{1.29}$$

Kirchhoff's law requires the condition of thermodynamic equilibrium, such that uniform temperature and isotropic radiation are achieved. Obviously, the radiation field of the earth's atmosphere as a whole is not isotropic and its temperatures are not uniform. However, in a localized volume below about 40 km, to a good approximation, it may be considered to be isotropic with a uniform temperature in which energy transitions are determined by molecular collisions. It is in the context of this *local* thermodynamic equilibrium that Kirchhoff's law is applicable to the atmosphere.

1.3   ABSORPTION (EMISSION) LINE FORMATION AND LINE SHAPE

An inspection of the high resolution spectroscopy reveals that the emission spectra of certain gases are composed of a large number of individual and characteristic spectral lines (see Fig. 4.2). In the previous section, we indicated that Planck successfully explained the nature of radiation from heated solid objects of which the cavity radiator formed the prototype. Such radiation generates continuous spectra, and is contrary to line spectra. We note, however, that Planck's quantization ideas, properly extended, lead to an understanding of line spectra also.

Investigation of the hydrogen spectrum led Bohr in 1913 to postulate that the circular orbits of the electrons were quantized, that is, their angular momentum could have only integral multiples of a basic value. He assumed that the hydrogen atoms exists, like Planck's oscillators, in certain stationary states in which it does not radiate. Radiation occurs only when the atom makes a transition from one state with energy E_k to a state with lower energy

E_j. Thus we write

$$E_k - E_j = h\tilde{v}, \qquad (1.30)$$

where $h\tilde{v}$ represents the quantum of energy carried away by the photon, which is emitted from the atom during the transition. The lowest energy state is called the *ground state* of the atom. When an electron of an atom absorbs energy due to collisions and jumps into a larger orbit, for example, the atom is said to be in an *excited state*. Then, according to Eq. (1.30), a sudden transition will take place, and the atom emits a photon of energy and collapses to a lower energy state. This is illustrated in Fig. 1.7 for a hydrogen atom. Also shown in this figure is the absorption of a photon by a stationary hydrogen atom.

Bohr further postulated that the angular momentum L can take on only discrete values given by

$$L = n(h/2\pi), \qquad n = 1, 2, 3, \dots. \qquad (1.31)$$

EMISSION

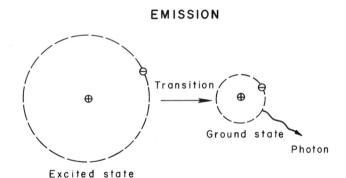

ABSORPTION

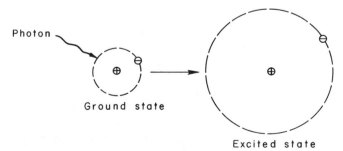

Fig. 1.7 Illustration of emission and absorption for a hydrogen atom which is composed of one proton and one electron. The radius of the circular orbit r is given by $n^2 \times 0.53$ Å, where n is the quantum number, and 1 Å $= 10^{-8}$ cm.

With this selection rule, he showed from the equation of motion for the electron that the total energy state of the system is (cgs units, see Exercise 1.9)

$$E = -(2\pi^2 m_e e^4/h^2)n^{-2}, \qquad n = 1, 2, 3, \ldots, \qquad (1.32)$$

where m_e is the mass of the electron, and e the charge carried by the electron. It follows from Eq. (1.30) that the frequency of emission or absorption lines in the hydrogen spectrum is

$$\tilde{v} = \frac{2\pi^2 m_e e^4}{h^3}\left(\frac{1}{j^2} - \frac{1}{k^2}\right), \qquad (1.33)$$

where j and k are integers describing, respectively, the lower and higher energy states. Figure 1.8 shows an energy diagram for the hydrogen. In the field of spectroscopy, energy is usually given in units of electron volts (eV) or in units of wave number (cm^{-1}). An electron volt is the energy acquired by an electron accelerated through a potential difference of one volt, and is equivalent to 1.602×10^{-12} erg.

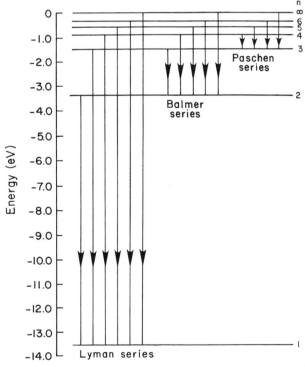

Fig. 1.8   An energy level diagram for the hydrogen showing the quantum number n for each level and some of the transitions that appear in the spectrum. An infinite number of levels is crowded in between the levels marked $n = 6$ and $n = \infty$.

Each quantum jump between fixed energy levels results in emission or absorption of characteristic frequency or wavelength. These quanta appear in the spectrum as emission or absorption lines. For the simple hydrogen atom described previously the line spectrum is relatively simple, whereas the spectra of water vapor, carbon dioxide, and ozone molecules are considerably more complex.

Monochromatic emission is practically never observed. Energy levels during energy transitions are normally changed slightly due to external influences on atoms and molecules, and due to the loss of energy in emission. As a consequence, radiation emitted during repeated energy transitions is nonmonochromatic, and spectral lines of finite widths are observed. The broadening of spectral lines is caused by (1) the damping of vibrations of oscillators resulting from the loss of energy in emission (the broadening of lines in this case is considered to be normal), (2) the perturbations due to reciprocal collisions between the absorbing molecules, and between the absorbing and nonabsorbing molecules, and (3) the Doppler effect resulting from the difference in thermal velocities of atoms and molecules. The broadening of lines due to the loss of energy in emission (natural broadening) is practically negligible as compared with that caused by collisions and the Doppler effect. In the upper atmosphere, we find a combination of collision broadening and Doppler broadening, whereas in the lower atmosphere, below about 40 km, the collision broadening prevails because of the pressure effect.

1.3.1 Pressure Broadening

The shape of spectral lines due to the pressure broadening is given by the *Lorentz profile*. It is expressed by the formula

$$k_{\tilde{\nu}} = \frac{S}{\pi} \frac{\alpha}{(\tilde{\nu} - \tilde{\nu}_0)^2 + \alpha^2} = Sf(\tilde{\nu} - \tilde{\nu}_0), \qquad (1.34)$$

where $k_{\tilde{\nu}}$ denotes the absorption coefficient, $\tilde{\nu}_0$ is the frequency of an ideal, monochromatic line, α is the half width of the line at the half maximum and is a function of pressure and to a lesser degree of the temperature, $f(\tilde{\nu} - \tilde{\nu}_0)$ represents the shape factor of a spectral line, and the line strength or line intensity S is defined by

$$\int_{-\infty}^{\infty} k_{\tilde{\nu}} d\tilde{\nu} = S. \qquad (1.35)$$

In this case, we say the absorption coefficient is normalized. Figure 1.9 depicts a plot for the Lorentz profile.

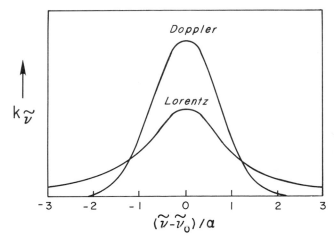

Fig. 1.9   Lorentz and Doppler line shapes for similar intensities and line widths.

The Lorentz shape of infrared lines is fundamental for the theory of infrared radiative transfer in the atmosphere. Thus it is desirable to give a brief explanation of how the formula denoted in Eq. (1.34) is derived. An isolated molecule emits or absorbs an almost purely harmonic wave given by

$$f(t) = A \cos 2\pi \tilde{\nu}_0 t, \qquad (1.36a)$$

where A is an arbitrary amplitude. During the period $-t/2$ to $t/2$, the distribution of amplitude $g(\tilde{\nu})$ of the wave in the discrete frequency domain may be obtained from the Fourier cosine transform as follows:

$$g(\tilde{\nu}) = \sqrt{\frac{2}{\pi}} \int_0^{t/2} (A \cos 2\pi \tilde{\nu}_0 t') \cos 2\pi \tilde{\nu} t' \, dt'$$

$$= \frac{A}{(2\pi)^{3/2}} \left[\frac{\sin \pi(\tilde{\nu}_0 + \tilde{\nu})t}{\tilde{\nu}_0 + \tilde{\nu}} + \frac{\sin \pi(\tilde{\nu}_0 - \tilde{\nu})t}{\tilde{\nu}_0 - \tilde{\nu}} \right]. \qquad (1.36b)$$

Generally, the widths of absorption lines are much smaller than $\tilde{\nu}_0$, i.e., $\tilde{\nu} = \tilde{\nu}_0 + \Delta\tilde{\nu}$, so that the first term in Eq. (1.36b) may be neglected when it is compared with the second.

The only deviation from purely harmonic behavior would be produced by the damping due to the loss of energy in emission. In the infrared, the spectroscopic effect of this damping is extremely small. However, a radiating molecule upon collision with another molecule would alter the radiating harmonic wave train owing to the intermolecular forces, and the frequency of the emitting molecule would be temporarily shifted by an appreciable amount. Since the collision may be considered to be instantaneous, one may assume that the principal effect of the collision is to destroy the phase coherence of the emitted wave train. That is to say, that after the collision the molecule starts emitting with another phase and the new phases are now randomly distributed. From general statistical principles, the time between collisions is distributed according to Poisson's law that the probability a collision occurs between t and $t + dt$ is e^{-t/t_0}, where t_0 is the mean time between collisions. All the initial phases of the wave trains have to be averaged. Thus, the absorption coefficient will be given by

$$k_{\tilde{v}} = A' \int_0^\infty [g(\tilde{v})]^2 e^{-t/t_0} \, dt, \qquad (1.34a)$$

where $[g(\tilde{v})]^2$ is the distribution of intensity, and A' is a constant. The integral may easily be evaluated. Further, by letting $1/t_0 = 2\pi\alpha$ and utilizing Eq. (1.35), we find Eq. (1.34a) becomes equivalent to Eq. (1.34). Here, $2\pi\alpha$ is the number of collisions per molecule per unit time. [Exercise 1.10 requires the derivation of Eq. (1.34) from Eq. (1.34a).] We note that the Lorentz line shape also can be derived from the classical theory of absorption and dispersion as shown in Appendix D.

From the kinetic theory of gases, the dependence of the half width α on the pressure and temperature is given by

$$\alpha = \alpha_0 (P/P_0)(T_0/T)^{1/2}, \qquad (1.37)$$

where α_0 is the width at the standard pressure P_0 and temperature T_0.

1.3.2   Doppler Broadening

Assuming that there is no collision broadening in a highly rarefied gas, a molecule in a given quantum state radiates at frequency $\tilde{v}_0$. If this molecule has a velocity component u in the line of sight (the line joining the molecule and the observer), and if $v \ll c$, the velocity of light, the frequency $\tilde{v}_0$ appears shifted as seen by a stationary observer to the frequency

$$\tilde{v} = \tilde{v}_0 (1 \pm v/c). \qquad (1.38)$$

Let the probability that the velocity component lies between v and $v + dv$ be $p(v) \, dv$. From the kinetic theory, if the translational states are in thermodynamic equilibrium, $p(v)$ is given by the Maxwell–Boltzmann distribution so that

$$p(v) \, dv = (m/2\pi KT)^{1/2} \exp(-mv^2/2KT) \, dv, \qquad (1.39)$$

where m is the mass of the molecule, K the Boltzmann constant, and T the absolute temperature.

To obtain the *Doppler* distribution, we insert the expression of v in Eq. (1.38) into Eq. (1.39), and perform normalization to an integrated line intensity S shown in Eq. (1.35). After these operations we find the absorption coefficient in the form

$$k_{\tilde{v}} = \frac{S}{\alpha_D \sqrt{\pi}} \exp\left[-\left(\frac{\tilde{v} - \tilde{v}_0}{\alpha_D}\right)^2 \right], \tag{1.40}$$

where

$$\alpha_D = (\tilde{v}_0/c)(2KT/m)^{1/2} \tag{1.41}$$

is a measure of the Doppler width of the line. The half width at the half maximum is $\alpha_D \sqrt{\ln 2}$.

A graphical representation of the Doppler line shape is also shown in Fig. 1.9. Since the absorption coefficient of a Doppler line is dependent on $\exp[-(\tilde{v} - \tilde{v}_0)^2]$, it is more intense at the line center and much weaker in the wings than the Lorentz shape. This implies that when a line is fully absorbed at the center, any addition of absorption will occur in the wings and will be caused by collision effects rather than Doppler effects. One final note may be in order. In the atmosphere above about 40 km where gases are at low pressures, it becomes important to incorporate the combined influence of the Lorentz and the Doppler broadening in the infrared transfer calculations. The shape factor for the combined profile is called the *Voigt profile*, which involves an infinite integral of a complicated function determined from Eqs. (1.34) and (1.40) (e.g., see Penner, 1959). Numerical calculations are therefore required to evaluate the absorption coefficient.

1.4 SIMPLE ASPECTS OF RADIATIVE TRANSFER

1.4.1 The Equation of Transfer

A pencil of radiation traversing a medium will be weakened by its interaction with matter. If the intensity of radiation I_λ becomes $I_\lambda + dI_\lambda$ after traversing a thickness ds in the direction of its propagation, then

$$dI_\lambda = -k_\lambda \rho I_\lambda \, ds, \tag{1.42}$$

where ρ is the density of the material, and k_λ denotes the mass extinction cross section (in units of area per mass) for radiation of wavelength λ. The mass extinction cross section is the sum of the mass absorption and scattering cross sections as discussed in Section 1.1.4. Thus, the reduction in intensity is caused by absorption in the material as well as scattering of radiation by the material.

On the other hand, the intensity may be strengthened by emission of the material plus multiple scattering from all other directions into the pencil under consideration at the same wavelength (see Fig. 1.5). We define the source function coefficient j_λ such that the increase of intensity due to emission and multiple scattering is given by

$$dI_\lambda = j_\lambda \rho \, ds, \tag{1.43}$$

where the source function coefficient j_λ has the same physical meaning as the mass extinction cross section. Upon combining Eqs. (1.42) and (1.43), we obtain

$$dI_\lambda = -k_\lambda \rho I_\lambda \, ds + j_\lambda \rho \, ds. \tag{1.44}$$

Moreover, it is convenient to define the source function J_λ such that

$$J_\lambda \equiv j_\lambda / k_\lambda. \tag{1.45}$$

In this manner, the source function has units of radiant intensity. It follows that Eq. (1.44) may be rearranged to yield

$$\frac{dI_\lambda}{k_\lambda \rho \, ds} = -I_\lambda + J_\lambda. \tag{1.46}$$

This is the general equation of transfer without any coordinate system imposed. It is fundamental in the discussion of any radiative transfer process.

1.4.2 Beer–Bouguer–Lambert Law

When both scattering and emission contributions may be neglected, Eq. (1.46) reduces to the form

$$\frac{dI_\lambda}{k_\lambda \rho \, ds} = -I_\lambda, \tag{1.47}$$

where k_λ now represents the mass absorption cross section (or simply absorption coefficient) only. If the incident intensity at $s = 0$ is $I_\lambda(0)$, then the emergent intensity at a distance s apart shown in Fig. 1.10 can be obtained by integrating Eq. (1.47), and is given by

$$I_\lambda(s_1) = I_\lambda(0) \exp\left(-\int_0^{s_1} k_\lambda \rho \, ds \right). \tag{1.48}$$

Assuming that the medium is homogeneous, then k_λ is independent of the distance s. Thus, by defining the path length

$$u = \int_0^{s_1} \rho \, ds, \tag{1.49}$$

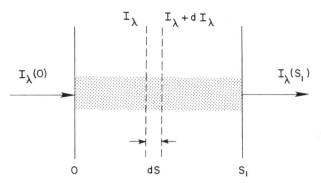

Fig. 1.10 Depletion of the radiant intensity in traversing an absorbing medium.

Eq. (1.48) becomes

$$I_\lambda(s_1) = I_\lambda(0)e^{-k_\lambda u}. \tag{1.50}$$

This is known as Beer law or Bouguer law or Lambert law, and is referred to here as the Beer–Bouguer–Lambert law, which states that the decrease of the radiant intensity traversing a homogeneous absorbing medium is according to the simple exponential function whose argument is the product of the mass absorption cross section and the path length. It should be noted that since this law involves no directional dependence, it is applicable not only to the intensity quantity but also the flux density and the flux.

By virtue of Eq. (1.50), we may define the monochromatic transmissivity $\mathscr{T}_\lambda$ as

$$\mathscr{T}_\lambda = I_\lambda(s_1)/I_\lambda(0) = e^{-k_\lambda u}. \tag{1.51}$$

Moreover, for a nonscattering medium, the monochromatic absorptivity, representing the fractional part of the incident radiation that is absorbed by the medium, is given by

$$A_\lambda = 1 - \mathscr{T}_\lambda = 1 - e^{-k_\lambda u}. \tag{1.52}$$

Equations (1.51) and (1.52) are normally expressed in the wave number domain in conjunction with the applications of infrared radiation. We note that if there is a scattering contribution from the medium, certain portions of the incident radiation may reflect back to the incident direction. Under this circumstance, we may define the monochromatic reflectivity R_λ, which is the ratio of the reflected (backscattered) intensity to the incident intensity. On the basis of the conservation of energy we must have

$$\mathscr{T}_\lambda + A_\lambda + R_\lambda = 1 \tag{1.53}$$

for the transfer of radiation through a scattering and absorbing medium.

1.4.3   Schwarzschild's Equation and Its Solution

Consider a nonscattering medium, which is a blackbody and which is in local thermodynamic equilibrium. A beam of intensity I_λ passing through it will undergo absorption process, while emission from the matter also takes place simultaneously. The source function in this case is given by the Planck function, and can be expressed by

$$J_\lambda = B_\lambda(T). \tag{1.54}$$

Hence, the equation of transfer may be written as

$$\frac{dI_\lambda}{k_\lambda\rho\, ds} = -I_\lambda + B_\lambda(T). \tag{1.55}$$

This equation is called Schwarzschild's equation. The first term in the right-hand side of Eq. (1.55) denotes the reduction of the radiant intensity due to absorption, whereas the second term represents the increase of the radiant intensity arising from blackbody emission of the material. To seek a solution for the Schwarzschild equation, we define the monochromatic optical thickness of the medium between points s and s_1 as shown in Fig. 1.11 in the form

$$\tau_\lambda(s_1, s) = \int_s^{s_1} k_\lambda\rho\, ds'. \tag{1.56}$$

By noting that

$$d\tau_\lambda(s_1, s) = -k_\lambda\rho\, ds, \tag{1.57}$$

Eq. (1.55) becomes

$$-\frac{dI_\lambda(s)}{d\tau_\lambda(s_1, s)} = -I_\lambda(s) + B_\lambda[T(s)]. \tag{1.58}$$

Upon multiplying Eq. (1.58) by a factor $e^{-\tau_\lambda(s_1,s)}$, and integrating the thickness ds from 0 to s_1, we find

$$-\int_0^{s_1} d\{I_\lambda(s)e^{-\tau_\lambda(s_1,s)}\} = \int_0^{s_1} B_\lambda[T(s)]e^{-\tau_\lambda(s_1,s)}\, d\tau_\lambda(s_1, s). \tag{1.59}$$

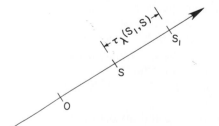

Fig. 1.11 Configuration of the optical thickness.

Consequently,

$$I_\lambda(s_1) = I_\lambda(0)e^{-\tau_\lambda(s_1,0)} + \int_0^{s_1} B_\lambda[T(s)]e^{-\tau_\lambda(s_1,s)}k_\lambda\rho\,ds. \qquad (1.60)$$

The first term in Eq. (1.60) is essentially equivalent to Eq. (1.48), representing the absorption attentuation of the radiant intensity by the medium. The second term represents the emission contribution from the medium along the path from 0 to s_1. If the temperature and density of the medium, and the associated absorption coefficient along the path of the beam are known, Eq. (1.60) can be integrated numerically to yield the intensity at the point s_1. Applications of Eq. (1.60) to infrared radiative transfer and to remote sounding of atmospheric temperature profiles and compositions from orbiting meterological satellites will be discussed in Chapters 4 and 7, respectively.

1.4.4 The Equation of Transfer for Plane–Parallel Atmospheres

In problems of radiative transfer in plane–parallel atmospheres it is convenient to measure linear distances normal to the plane of stratification (see Fig. 1.12). If z denotes this distance, then the general equation of transfer in Eq. (1.46) becomes

$$\cos\theta\,\frac{dI(z;\theta,\phi)}{k\rho\,dz} = -I(z;\theta,\phi) + J(z;\theta,\phi), \qquad (1.61)$$

where θ denotes the inclination to the upward normal, and ϕ the azimuthal angle in reference to the X axis. Here, we omit the subscript λ on various radiative quantities.

Introducing the normal optical thickness

$$\tau = \int_z^\infty k\rho\,dz' \qquad (1.62)$$

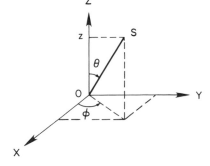

Fig. 1.12 Geometry for plane–parallel atmospheres.

measured from the outer boundary downward, we have

$$\mu \frac{dI(\tau; \mu, \phi)}{d\tau} = I(\tau; \mu, \phi) - J(\tau; \mu, \phi), \tag{1.63}$$

where $\mu = \cos\theta$. This is the basic equation for the problem of multiple scattering in plane–parallel atmospheres.

Following the same procedure as that described in Section 1.4.3, Eq. (1.63) can be solved to give the upward and downward intensities for a finite atmosphere which is bounded on two sides at $\tau = 0$ and $\tau = \tau_1$ as depicted in Fig. 1.13. To obtain the upward intensity ($\mu > 0$) at level τ, we multiply Eq. (1.63) by a factor $e^{-\tau/\mu}$ and perform integration from τ to $\tau = \tau_1$. This leads to

$$I(\tau; \mu, \phi) = I(\tau_1; \mu, \phi)e^{-(\tau_1 - \tau)/\mu}$$

$$+ \int_\tau^{\tau_1} J(\tau'; \mu, \phi)e^{-(\tau' - \tau)/\mu} \frac{d\tau'}{\mu} \qquad (1 \geq \mu > 0). \tag{1.64}$$

To get the downward intensity ($\mu < 0$) at level τ, a factor $e^{\tau/\mu}$ is used and μ is replaced by $-\mu$. After carrying out integration from $\tau = 0$ to τ, we obtain the expression

$$I(\tau; -\mu, \phi) = I(0; -\mu, \phi)e^{-\tau/\mu}$$

$$+ \int_0^\tau J(\tau'; -\mu, \phi)e^{-(\tau - \tau')/\mu} \frac{d\tau'}{\mu} \qquad (1 \geq \mu > 0). \tag{1.65}$$

In Eqs. (1.64) and (1.65), $I(\tau_1; \mu, \phi)$ and $I(0; -\mu, \phi)$ represent, respectively, the inward source intensities at the bottom and top surfaces (see Fig. 1.13).

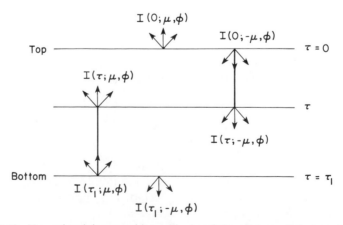

Fig 1.13   Upward and downward intensities in a finite, plane-parallel atmosphere.

For planetary applications, it is desirable to measure the emergent outward intensities at the top and bottom of the atmosphere in conjunction with remote sensing of atmospheric compositions, and radiative balance studies. Upon setting $\tau = 0$ in Eq. (1.64), we find

$$I(0; \mu, \phi) = I(\tau_1; \mu, \phi)e^{-\tau_1/\mu} + \int_0^{\tau_1} J(\tau'; \mu, \phi)e^{-\tau'/\mu} \frac{d\tau'}{\mu}, \qquad (1.66)$$

where the first and second terms represent, respectively, the bottom surface contribution (attenuated to the top) and the internal atmospheric contribution. Moreover, upon setting $\tau = \tau_1$ in Eq. (1.65), we get

$$I(\tau_1; -\mu, \phi) = I(0; -\mu, \phi)e^{-\tau_1/\mu} + \int_0^{\tau_1} J(\tau'; -\mu, \phi)e^{-(\tau_1 - \tau')/\mu} \frac{d\tau'}{\mu}, \qquad (1.67)$$

where again the first and second terms represent, respectively, the top surface contribution (attenuated to the bottom) and the internal atmospheric contribution. Detailed applications of these two equations associated with infrared transfer and multiple scattering will be discussed in Chapters 4 and 6.

EXERCISES

1.1   Show that for isotropic radiation, the monochromatic flux density is $F_\lambda = \pi I_\lambda$.

1.2   A meteorological satellite circles the earth at a height h above the earth's surface. Let the radius of the earth be a_e and show that the solid angle under which the earth is seen by the satellite sensor is

$$2\pi[1 - (2a_e h + h^2)^{1/2}/(a_e + h)].$$

1.3   Express the Planck function in the wavelength and wave number domains from that in the frequency domain.

1.4   From Eq. (1.25), show that Eq. (1.26) is true.

1.5   Show that the maximum intensity of the Planck function is proportional to the fifth power of the temperature.

1.6   An infrared scanning radiometer aboard a meteorological satellite measures the outgoing radiation emitted from the earth's surface at the 10 μm window region. Assuming that the effect of the atmosphere between the satellite and surface can be neglected, what would be the temperature of the surface if the observed radiance at 10 μm is 0.98×10^4 erg/sec/cm^2/μm/sr?

1.7   A black land surface with a temperature of 15°C emits radiation at all frequencies. What would be the emitted radiances at 0.7 μm, 1000 cm^{-1}, and 31.4 GHz? (Note: Use the appropriate Planck functions in the calculations.)

1.8   Assuming the average normal body temperature is 98°F, what would be the emittance of the body? If it is not a blackbody but absorbs only 90% of the incoming radiation averaged over all wavelengths, what would be the emittance in this case? Also, at which wavelength does the body emit the maximum energy?

1.9   (a)   From Newton's second law for motion and Coulomb's law, find the kinetic energy of the electron in the hydrogen atom moving with a velocity v in a circular orbit of radius r centered on its nucleus. Express r in terms of the quantum number n using the selection rule for the angular momentum mvr. Then find the potential energy of the proton–electron system. By combining the kinetic and potential energy, derive Eq. (1.32).

(b)   Consider only the transitions between the ground state ($n = 1$) and the excited states and let the highest quantum number be 6, compute the wavelengths of hydrogen emission lines.

1.10   Derive Eq. (1.34) from Eq. (1.34a).

1.11   Prove that the line intensity $S = \int_{-\infty}^{\infty} k_{\tilde{v}} \, d\tilde{v}$ for both Lorentz and Doppler absorption lines.

1.12   Calculate and plot the shape factor of the Lorentz and Doppler profiles for ozone whose half width is assumed to be 0.1 cm^{-1} in the wave number domain.

1.13   A He–Ne laser beam at 0.6328 μm with an output power of 5 mW (10^{-3} W) is passing through an artificial cloud layer 10 m in thickness and is directed at 30° from the normal to the layer. Neglecting the effect of multiple scattering, calculate the extinction coefficients (per length) if the measured powers are 1.57576 and 0.01554 mW. Also calculate the normal optical depths in these cases.

1.14   The contrast of the object against its surroundings is defined by

$$C \equiv (B - B_0)/B_0,$$

where B represents the brightness of the object and B_0 the brightness of the background sky. By these definitions, $B = B_0$ when $x \to \infty$, whereas $B = 0$ when $x = 0$, where x denotes the distance between the object and the observer. For the normal eye the threshold contrast has a value of ± 0.02. Assuming that the extinction coefficient β is independent of the wavelength, show that the visual range or visibility is given by $x = 3.912/\beta$.

1.15   In reference to Fig. 1.10, if the reflectivity at $s = 0$ and $s = s_1$ are R_λ, what would be the value of $I_\lambda(s_1)$?

1.16   By differentiation with respect to the optical thickness τ, show that Eqs. (1.64) and (1.65) reduce to Eq. (1.63), the equation of transfer for plane parallel atmospheres. *Hint*: Use the following Leibnitz's rule for differentiation of integrals:

$$\frac{d}{dy} \int_{\phi_1(y)}^{\phi_2(y)} F(x, y)\, dx = \int_{\phi_1(y)}^{\phi_2(y)} \frac{\partial F}{\partial y}\, dx - F(\phi_1, y)\frac{d\phi_1}{dy} + F(\phi_2, y)\frac{d\phi_2}{dy}.$$

1.17   Consider an isothermal nonscattering atmosphere with a temperature T and let the surface temperature of such an atmosphere be T_s, derive an expression for the emergent flux density at the top of an atmosphere whose optical depth is τ_1. *Hint*: Make use of Eq. (1.66).

SUGGESTED REFERENCES

Halliday, D., and Resnick, R. (1974). *Fundamentals of Physics*, Revised Printing. Wiley, New York. Chapter 39 gives elementary discussions on the production of radiation.

Jastrow, R., and Thompson, M. H. (1974). *Astronomy: Fundamentals and Frontiers*, 2nd ed., Wiley, New York. Chapter 4 contains excellent descriptions in lay terms for the absorption line formation.

Penner, S. S. (1959). *Quantitative Molecular Spectroscopy and Gas Emissivities*. Addison-Wesley, Reading, Massachusetts. Chapters 1–3 provide fundamental discussions on blackbody radiation and absorption line profiles.

Chapter 2
SOLAR RADIATION AT THE TOP OF THE ATMOSPHERE

2.1 THE SUN AS AN ENERGY SOURCE

The sun, formed about 4.6 billion years ago, is an ordinary body in the cosmic hierarchy. Among billions of stars in the universe, the sun is about average in mass but below average in size. The sun has one unique feature in that it is 300,000 times closer to the earth than the next nearest star. With a mean distance of about 1.5×10^8 km between the earth and the sun, virtually all of the energy that the earth receives and that drives the earth's atmosphere in motion comes from the sun.

The sun is a gaseous sphere whose radius is about 6.96×10^5 km with a mass of approximately 1.99×10^{35} g. Its main ingredients are primordial hydrogen and helium plus a small amount of heavier elements such as iron, silicon, neon, and carbon. Hydrogen makes up about 75% of the mass, while the remaining 25% or so is helium. The temperature of the sun decreases from a central value of about $5 \times 10^{6\circ}$K to about $5800°$K at the surface. The density within the sun falls off very rapidly with increasing distance from the center. The central density is about 150 g cm^{-3}, and at the surface, it is about 10^{-7} g cm^{-3}. The average density is about 1.4 g cm^{-3}. Approximately 90% of the sun's mass is contained in the inner half of its radius.

The source of solar energy is believed to be generated from the steady conversion of four hydrogen atoms to one helium atom in fusion reactions, which take place in the deep interior of the sun with temperatures up to many millions of degrees. The amount of energy released in nuclear fusions causes a reduction of the sun's mass. According to Einstein's law relating the mass and energy, $E = mc^2$, and converting the energy radiated by the sun, we find that almost five million tons of mass per second are radiated by the sun in the form of electromagnetic energy. In a billion years, it is believed that the sun will radiate into space about 10^{29} grams, which is less than one part in 10^4 of its total mass. Thus, only an insignificant fraction of the sun's substance has been lost by electromagnetic radiation. It is estimated that only 5% of the sun's total mass has been converted from hydrogen to helium in its lifetime thus far.

Because of the extremely high temperatures in the deep interior, collisions between atoms are sufficiently violent to eject many electrons from their orbits. Only the tightly bound inner electrons of heavy atoms will be retained. The emitted energy caused by nuclear fusions in forms of photons can pass through the inner part of the sun without being absorbed by the electrons. However, closer to the sun's surface, the temperature decreases and the heavier atoms such as iron begin to recapture their outer electrons. These outer electrons are bound to the nucleus by relatively small forces, and can be easily separated from the nucleus by the absorption of photons. It follows that the flow of photons coming from the interior is blocked by the appearance of the absorbing atoms. The blocking of these photons will cause the temperature to drop sharply at some depth below the surface. Thus, the outer region of the sun consists of a layer of relatively cool gas resting on the top of a hotter interior. As a result of this situation, the gas at the bottom of the cool outer layer is heated by the hot gas in the interior. It undergoes expansion and rises toward the surface. Once it reaches the surface, the hot gas loses its heat to space, cools, and descends into the interior. The entire outer layer breaks up into ascending columns of heated gas and descending columns of cooler gas. The region in which this large-scale upward and downward movement of gases occurs is called the *zone of convection*, which extends from a depth of about 150,000 km to the surface of the sun. Below this depth, it is believed that energy is transported within the sun by means of electromagnetic radiation, i.e., by the flow of photons. Near the surface, however, because of the substantial blocking of radiant energy by the absorption of heavier elements, energy is transferred partly by convection and partly by electromagnetic radiation. Above the surface, energy transport is again by means of electromagnetic radiation.

2.1.1   The Structure

The visible region of the sun is called the *photosphere*, where most of the electromagnetic energy reaching the earth originates. Although the sun is in a gaseous form, the photosphere is referred to as the *surface* of the sun. The photosphere is marked by relatively bright *granules* about 1500 km in diameter. The bright granules are separated by dark regions known as *faculae* and variable features called *sunspots*. They are fairly uniformly distributed over the solar disk, and are believed to be associated with ascending hot gases in the uppermost layer of the zone of convection discussed previously.

The photosphere is a comparatively thin layer about 500 km thick which constitutes the source of the sun's visible radiation. The temperature in this layer varies from 8000°K in lower layer to 4000°K in the upper layer. By matching the theoretical Planck curve versus wavelength depicted in Fig. 1.6 with the measured spectral radiant energy emitted by the sun, the best agreement was found for a temperature of approximately 6000°K. This temperature is an average over the temperature range of the photosphere. The *effective temperature* of the photosphere also may be obtained by measuring the luminosity of the sun. On the basis of the Stefan–Boltzmann Law and with the knowledge that the sun's radius assumes a sharply defined surface, the effective temperature yields a value of 5800°K. This value agrees closely with the temperature of 6000°K derived from the Planck curve. Radiation emitted from the photosphere is essentially continuous.

The region above the photosphere is called the *solar atmosphere*. It is characterized by the tenuous and transparent solar gases. The solar atmosphere is divided into two regions called the *chromosphere* and *corona*.

The chromosphere lies above the photosphere to a height of approximately 5000 km. The temperature of the chromosphere increases from a minimum of about 4000°K and stays between 4000 and 6000°K up to about 2000 km. Above this height, the temperature rises drastically reaching about 10^6°K at an altitude of about 5000 km. The layer with a minimum temperature of 4000°K extends to a few thousand kilometers, consisting of relatively cool gases lying over the hotter gases. These cool gases absorb continuous radiation emitted from the photosphere at wavelengths characteristic of the atoms in the sun, and generate the solar absorption spectrum. In accord with the discussion in Section 1.3, when an atom absorbs radiant energy, it is excited to a new energy level. The excited atom then makes a transition to a lower excited state, or to the ground state, during which a quantum of energy is emitted. Consequently, the emission spectrum of the chromosphere is formed. Since the absorption spectrum is produced by the initial transition of atoms from a low energy to a high energy state, while the emission spectrum results from the subsequent transition of the same atoms in the reverse direction, it is clear that the lines in the emission spectrum are the same as those in the sun's absorption spectrum. When the photosphere is eclipsed by the moon or by instrument, the line emission, mostly from hydrogen, helium, and calcium, can be observed. Because a bright line emission spectrum flashes into view briefly at the beginning and the end of the period of the total eclipse, it is called the *flash spectrum*. The 6563 Å line of hydrogen is one of the strongest absorption lines in the solar spectrum. Owing to the large amount of energy emitted in this line, the chromosphere becomes visible and has a characteristic reddish appearance during an eclipse.

Above the chromosphere lies the region of the solar atmosphere called *corona*. The corona layer extends out from the edge of the solar disk many millions of kilometers. It is visible as a faint white halo during total eclipses. Figure 2.1 illustrates the solar corona during the total eclipse of March 1970. It is generally believed that the corona has no outer boundary. A stream of gas called *solar wind* flows out of the corona and into the solar system continuously. An instrument called *coronagraph* has been used frequently in the past to study the chromosphere and corona in the absence of a natural eclipse. Strong emission lines of hydrogen and helium originating from the chromosphere disappear with increasing altitude, and they are replaced by the continuous spectrum of white light characteristic of the corona. The spectrum of the corona contains a number of weak emission

Fig. 2.1   The solar corona during the total eclipse of March 7, 1970. Features are visible at a distance of about 4.5 solar radii or 3 million kilometers (courtesy of G. Newkirk, Jr., High Altitude Observatory, Boulder, Colorado).

lines, among which, the most intense is the green line of ionized iron. The generation of this emission line requires an enormous amount of energy, and it is believed that the temperature in large regions of corona is close to $10^{6\circ}$ K.

2.1.2 Solar Surface Activity: Sunspots

Several observable features of the sun are particularly interesting and important because of their transient occurrence. The best known and largest observed of these variable features are the *sunspots*. Sunspots are relatively dark regions on the photosphere—the surface of the sun. The sunspots have an average size of about 10,000 km but range from barely visible to areas that cover more than 150,000 km on the sun's surface. The spots usually occur in pairs, or in complex groups, which follow a leader spot in the direction of the sun's rotation. Small sunspots persist for several days or a week, while the largest spots may last for several weeks, long enough for these spots to reappear during the course of the sun's 27-day rotation. Sunspots are almost entirely confined to the zone of latitudes between 40° and the equator, and never appear near the poles. Sunspots are cooler regions having an average temperature of about 4000°K, compared to an average temperature of 6000°K for the photosphere. Owing to the relatively low temperature, sunspots appear black. Figure 2.2 illustrates a large sunspot group photographed with the 100-inch telescope on Mount Wilson.

The number of sunspots that appear on the solar disk averaged over a period of time is highly variable. There are periods of time when the spots are relatively numerous, while a few years later spots occur hardly at all. These periods are called *sunspot maxima* and *sunspot minima*, respectively. The periodic change in the sunspot number is referred to as the *sunspot cycle*. For about 200 years, the number of spots appearing everyday and the position of these spots on the face of the sun have been recorded continuously. The average length of time between sunspot maxima is about 11 years; the so-called *11-year cycle*. Figure 2.3 depicts the variation of the sunspot number since about 1730. In the years of sunspot maxima, the sun's surface is violently disturbed, and the outbursts of particles and radiation are commonly observed. During the sunspot minima period, however, outbursts are much less frequent. These outbursts are usually observed in the vicinity of large, complex groups of sunspots, and are called *solar flares*. The burst of radiation and energetic particles from a large flare may produce interferences with radio communications and cause substantial variations in the earth's magnetic field.

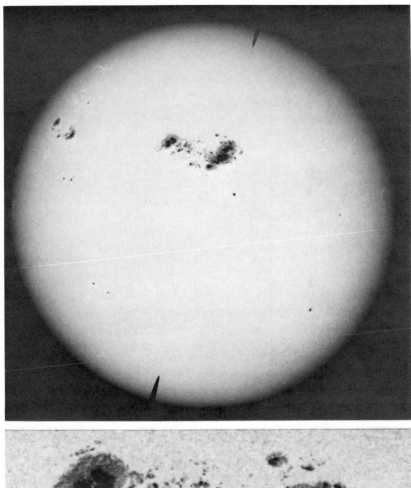

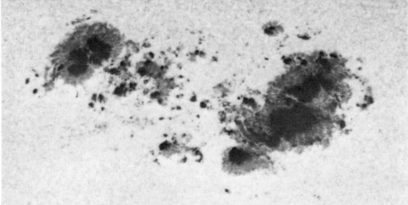

Fig. 2.2 A large cluster of sunspots photographed with the 100-inch telescope in 1947 at sunspot maximum. The lower photograph is an enlarged view (courtesy of the Hale Observatories, Pasadena, California).

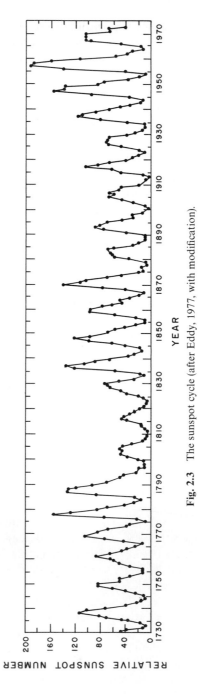

Fig. 2.3   The sunspot cycle (after Eddy, 1977, with modification).

It is believed that sunspots are associated with the very strong magnetic fields that exist in their interiors. Magnetic field measurements utilizing the Zeeman effect (the splitting of a spectral line into several separate lines) show that pairs of sunspots often have opposite magnetic polarities. For a given sunspot cycle, the polarity of the leader spot is always the same for a given hemisphere. With each new sunspot cycle, the polarities reverse. The cycle of the sunspot maximum having the same polarity is referred to as the 22-*year cycle*. The sunspot activities have been found to have a profound influence on many geophysical phenomena and on atmospheric processes.

2.2   THE EARTH'S ORBIT ABOUT THE SUN

The earth is one of the nine planets in the solar system. The four planets closest to the sun, i.e., Mercury, Venus, Earth, and Mars, are referred to as the terrestrial planets, and the remaining planets; Jupiter, Saturn, Uranus, Neptune, and Pluto, are called the major planets. All of the planets revolve around the sun in the same direction, and except Uranus, they also rotate in the same direction about their axis. Except Mercury and Pluto, all the planetary orbits lie in almost the same plane, and we call the plane of the earth's orbit the plane of the *ecliptic*.

Once every 24 hours with respect to the sun, the earth makes a complete rotation eastward about an axis through the poles. This rotation is the cause of the most obvious of all time periods involving the alternation of day and night, which comes about as the sun shines on the different parts of the earth exposed to it. Meanwhile, the earth with a mass of 6×10^{27} g, moves eastward around the sun once in approximately 365 days. The earth's orbit about the sun and the earth's rotation about its axis are the most important factors determining the amount of solar radiant energy reaching the earth, and the climate and climatic changes of the earth–atmosphere system. Owing to the rotation of the earth about its axis, the earth assumes the shape of an oblate spheroid, having equatorial and polar radii of 6378.17 and 6356.79 km, respectively. Its orbit around the sun is an ellipse, and the axis of its rotation is tilted as shown in Fig. 2.4.

There are three ways in which the earth's orbit about the sun varies. The earth orbital *eccentricity*, defined as the ratio of the distance between the two foci to the major axis of the ellipse, fluctuates within about 0.05 with a variable period of about 100,000 years. The mean eccentricity of the earth's orbit is about 0.017. In reference to Fig. 2.4, the axis of the earth's rotation is tilted at an angle of 23.5° from the normal to the plane of the ecliptic, the *inclination angle*. This angle, representing the obliquity of the

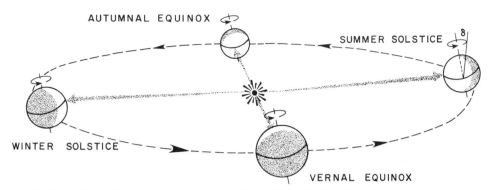

Fig. 2.4 The earth's orbit about the sun and effects of the obliquity of the ecliptic on the seasons.

ecliptic, varies cyclically over an average range of 1.5° with a period of about 41,000 years. In addition to these two factors, there is a very slow westward motion of the equinoctial points along the ecliptic, called *precession*, caused by the attraction of other planets upon the earth. Owing to the wobbling motion, the time when the earth is closest to the sun advances by about 25 minutes each year. Thus, the periodic precession index is about 21,000 years. Obviously, these three orbital changes affect the distribution of the amount of solar energy on earth–atmosphere system.

In recent years, much concern has been focused on the climate and climatic changes of the earth in view of the fact that human endeavors are particularly vulnerable to uncertainties in climate, and that human activity may be causing climatic change even now. The cause of fluctuations in the Pleistocene ice sheets has been a topic of scientific debates and speculations. A number of external factors have been speculated to be the major causes of the earth's climatic variations; variations in the output of the sun, seasonal and latitudinal distribution of incoming radiation due to the earth's orbital changes, the volcanic dust content of the atmosphere, and the distribution of carbon dioxide between the atmosphere and ocean are the most popular hypotheses.

The orbital theory of the climatic change proposed some years ago by the astronomer Milankovitch (1941) increasingly has gained scientific support in recent years. A group of climatologists (Hays *et al.*, 1976) recently reconstructed the climatic record of the earth. The reconstruction was based on measurements of the oxygen isotopic composition of planktonic foraminifera from the deep-sea sediment cores in the southern Indian ocean, and estimates of summer sea-surface temperatures at the core site derived from statistical analyses of radiolarian assemblages. These deep-sea sediment cores contain a continuous climatic record up to about 500,000 years.

On the basis of statistical analyses, these climatologists found that (1) the dominant 100,000-year climatic component has an average period close to the orbital eccentricity variations; (2) the 40,000-year climatic component has the same period as variations in the obliquity of the earth's axis; and (3) the 23,000-year climatic variation is associated with the periodic precession index. The dashed line in the center of Fig. 2.5 shows the variation of the estimated sea-surface temperature based on their investigation. The dotted line depicts a plot of orbital eccentricity variations. The upper and lower curves are the 23,000 and 40,000-year frequency components extracted from the estimated sea-surface temperature by a statistical filter method.

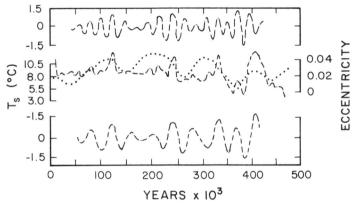

Fig. 2.5 Variation in eccentricity and climate over the past 500,000 years. Dashed line in the center shows variations in the estimated sea-surface temperature T_s. Dotted line denotes the orbital eccentricity. Upper and lower curves are the 23,000- and 40,000-year frequency components extracted from T_s based on a statistical filter method (after Hays *et al.*, 1976, with modification).

The most distinguishable feature of climatic changes is the seasons. The revolution of the earth about the sun and the tilt of the earth's axis cause the seasonal variation. At the time of the summer solstice, which occurs about June 22, the sun appears directly overhead at noon on latitude 23.5°N, called the *Tropic of Cancer*. The elevation of the sun above the horizon and the length of the day reach their maximum values in the northern hemisphere at the summer solstice, and everywhere north of the Arctic Circle (latitude 66.5°N) the sun remains above the horizon all day. In the southern hemisphere, the sun's elevation is at a minimum, the days are shortest, and everywhere south of the Antarctic Circle (latitude 66.5°S) the sun does not rise above the horizon on the June solstice. This is the beginning of the northern hemisphere summer, and the southern hemisphere summer begins at the winter solstice on about December 22.

Having reached the southermost point in its annual migration, the sun then stands directly overhead at noon on latitude 23.5°S, called the *Tropic of Capricorn*. Both the elevation of the sun above the horizon and the length of the day are then at their minimum values in the northern hemisphere, and the sun does not rise within the Arctic Circle or set within the Antarctic Circle. At the vernal (spring) and autumnal equinoxes, the days and nights everywhere are equal (12 hours), and the sun appears directly overhead on the equator at noon. The sun crosses the equator from north to south at the autumnal equinox, and from south to north at the vernal equinox.

The distances between the centers of the sun and earth vary between the extreme values 147×10^6 km at about winter solstice, and 153×10^6 km at about summer solstice. The mean distance between the sun and earth is about 150×10^6 km, denoted earlier in Section 2.1.

2.3   THE SOLAR SPECTRUM AND SOLAR CONSTANT

The distribution of electromagnetic radiation emitted by the sun as a function of the wavelength incident on the top of the atmosphere is called the *solar spectrum*. The *solar constant S* is a quantity denoting the amount of total solar energy reaching the top of the atmosphere. It is defined as the flux of solar energy (energy per time) across a surface of unit area normal to the solar beam at the mean distance between the sun and earth. The solar spectrum and solar constant have been the topics of extensive investigations for a long period of time. Abbot undertook a long series of ground-based measurements, resulting in a value of about 1350 W m^{-2} for the solar constant. Subsequent to Abbot's work and prior to more recent measurements carried out from high-altitude platforms, solar constant values of 1396 and 1380 W m^{-2} proposed by Johnson and Nicolet, respectively, were widely accepted. Recently, based on a series of measurements from high-altitude platforms, a revised value of 1353 ($\pm$21) W m^{-2} or 1.94 ($\pm$0.03) cal cm^{-2} min^{-1} issued by the National Aeronautics and Space Administration (NASA) has been accepted as a standard solar constant (Thekaekara, 1976).

The standard solar spectrum in terms of the spectral irradiance is shown in the top solid curve of Fig. 2.6. Also shown in this diagram is the spectral solar irradiance reaching the sea level in a clear atmosphere. The shaded areas represent the amount of absorption by various gases, primarily H_2O, CO_2, O_3, and O_2. Absorption and scattering of solar radiation in clear atmospheres will be discussed in Chapter 3. If one matches the solar spectral irradiance curve with theoretical blackbody values, we find that a temperature of about 6000°K fits the observed curve closely in the visible and infrared wavelengths. The reader is invited to carry out this exercise [Ex-

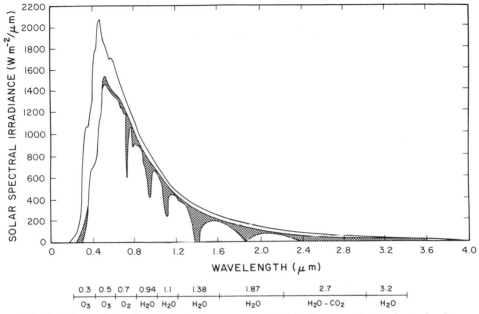

Fig. 2.6   Spectral irradiance distribution curves related to the sun; (1) the observed solar irradiance at the top of the atmosphere (after Thekaekara, 1976), and (2) solar irradiance observed at sea level. The shaded areas represent absorption due to various gases in a clear atmosphere. See Chapter 3 for further information. The outer envelope of the shaded areas denotes the reduction of solar irradiance due to scattering.

ercise (2.1)]. As has been pointed out in Section 2.1, most of the electromagnetic energy reaching the earth originates from the sun's surface—the photosphere. Of the electromagnetic energy emitted from the sun, approximately 50% lies in wavelengths longer than the visible region, about 40% in the visible region (0.4–0.7 μm), and about 10% in wavelengths shorter than the visible.

According to solar flux observations, the ultraviolet region (<0.4 μm) of the solar spectrum deviates greatly from the visible and infrared regions in terms of the equivalent blackbody temperature of the sun. Figure 2.7 illustrates a detailed observed solar spectrum from about 0.1 to 0.4 μm, along with blackbody temperatures of 4500, 5000, 5500, and 6000°K. In the ultraviolet region, wavelength units used are normally in angstroms (Å); $1Å = 10^{-4}$ μm. In the interval 2100–2600 Å, the equivalent blackbody temperature of the sun lies somewhat above 5000°K. It falls gradually to a minimum level of about 4700°K at about 1400 Å. From there toward shorter wavelengths, a large amount of energy flux is observed at the Lyman α emission line of 1216 Å associated with the transition of the first excited

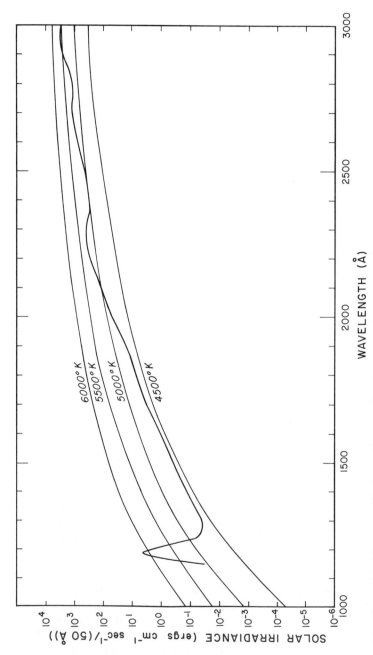

Fig. 2.7 Observed irradiance outside the earth's atmosphere in the ultraviolet region (after Thekaekara, 1974). Blackbody curves are shown for comparison.

and ground states of hydrogen atoms. The ultraviolet portion of the solar spectrum below 0.3 μm contains a relatively small amount of energy. However, because the ozone, and the molecular and atomic oxygen and nitrogen in the upper atmosphere absorb all this energy, it represents the prime source for the energetics of the atmosphere above 10 km.

The sun emits energy at the rate of 6.2×10^7 W m^{-2}, or 9.0×10^4 cal min^{-1} cm^{-2}. On the basis of the energy conservation principle, the energy emitted from the sun must remain the same at some distance away from the sun. Thus,

$$F4\pi a_s^2 = S4\pi d_m^2, \tag{2.1}$$

where F denotes the solar emittance, a_s the radius of the sun, and d_m the mean distance between the sun and earth. Hence, the solar constant may be expressed by

$$S = F(a_s/d_m)^2. \tag{2.2}$$

The total energy intercepted by the earth whose radius is a_e is given by $S\pi a_e^2$. If this energy is spread uniformly over the full surface of the earth, then the amount received per unit area and unit time at the top of the atmosphere is given by

$$\bar{Q}_s = S\pi a_e^2/(4\pi a_e^2) = S/4. \tag{2.3}$$

To estimate the effective temperature T of the sun we assume that the sun is a blackbody. Thus, by virtue of the Stefan–Boltzmann law, i.e., $F = \sigma T^4$, we find

$$T^4 = (d_m/a_s)^2(S/\sigma). \tag{2.4}$$

Inserting values of S, σ, d_m, and a_s into Eq. (2.4), we obtain an effective temperature of about 5800°K for the sun.

2.4 DETERMINATIONS OF THE SOLAR CONSTANT

There are two techniques of measuring the solar constant from the ground-based radiometer, called the *long* and *short* methods of the Smithsonian Institution. The long method is more fundamental and establishes the basis for the short method.

Observations of solar energy for the purpose of determining the solar constant require three primary instruments. These are the *pyrheliometer*, the *pyranometer*, and the *spectrobolometer*. The pyrheliometer was used to measure the direct plus some diffuse solar radiation, while the pyranometer measured only the diffuse solar radiation for arriving at a pyrheliometer correction utilizing a suitable shield to block the direct solar radiation from striking the instrument. The amount of the direct sunlight then can be evaluated by subtracting the flux density measured by the pyranometer from that by the pyrheliometer. The spectrobolometer is a combination of a spectrograph and a coelostat. The coelostat is a mirror which follows the sun and focuses its rays continuously on the entrance slit of the spectrograph, which disperses the solar radiation into different wavelengths by means of a prism or diffraction grating. In the Smithsonian solar constant measurements, about 40 standard wavelengths between 0.34 and 2.5μm are measured nearly simultaneously from the record of the spectrograph. The instrument corresponding to these measurements is called bologram.

2.4.1 Long Method

Assume that the atmosphere consists of plane parallel layers. At a given sun's position, which is denoted by the solar zenith angle θ_0, the effective path length of the air mass is $u \sec \theta_0$, where

$$u = \int_{z_1}^{\infty} \rho \, dz. \tag{2.5}$$

In this equation, z_1 is the height of the station. On the basis of the Beer–Bouguer–Lambert law, the irradiance F of the direct solar radiation of wavelength λ observed at the surface level is given by

$$F_{\lambda} = F_{\lambda 0} \exp(-k_{\lambda} u \sec \theta_0) = F_{\lambda 0} \mathcal{T}_{\lambda}^{m}, \tag{2.6}$$

where $F_{\lambda 0}$ is the monochromatic solar irradiance at the top of the atmosphere, k_{λ} denotes the monochromatic mass extinction cross section, $\mathcal{T}_{\lambda}$ is the monochromatic transmissivity defined in Eq. (1.51), and $m \, (= \sec \theta_0)$ represents the ratio of the air mass between the sun and observer and that at the local zenith distance. Upon taking the logarithm, we find

$$\ln F_{\lambda} = \ln F_{\lambda 0} + m \ln \mathcal{T}_{\lambda}. \tag{2.7}$$

Observations of F_{λ} may be made for several zenith angles during a single day. If the atmospheric properties do not change during the observational period, then the transmissivity $\mathcal{T}_{\lambda}$ is constant. A plot of F_{λ} versus m shown in Fig. 2.8 may be extrapolated to the zero point, which represents the top of the atmosphere ($m = 0$). If observations of the monochromatic irradiance are carried out for wavelengths covering the entire solar spectrum, then from Eq. (2.6) we have

$$F_0 = \int_0^{\infty} F_{\lambda 0} \, d\lambda \approx \sum_{i=1}^{N} F_{\lambda_i 0} \, \Delta\lambda_i, \tag{2.8}$$

where N is the total number of the monochromatic irradiance measured. Let d denote the actual distance between the earth and the sun, then from the

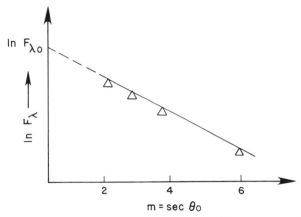

Fig. 2.8   Hypothetical observed monochromatic solar irradiances F_λ as a function of the effective path length.

energy conservation principle the solar constant is simply given by

$$S = F_0(d/d_m)^2. \tag{2.9}$$

The foregoing outlines theoretical procedures of the Smithsonian long method for the determination of the solar constant. However, as illustrated in Fig. 2.6, the atmosphere is essentially opaque for wavelengths shorter than about 0.34 μm, and for wavelengths longer than about 2.5 μm. Consequently, flux density observations cannot be made in these regions. Therefore empirical corrections are needed for the omitted ranges, which account for about 8% of the solar flux.

There are sources of error inherent in the Smithsonian long method caused by (1) empirical corrections for absorption of ultraviolet by ozone, and absorption of infrared by water vapor and carbon dioxide in the wings of the solar spectrum; (2) an unknown amount of diffuse radiation entering the aperture of the observing instrument; (3) variations of k_λ and the possible effects of aerosols during a series of measurements, and (4) measurement errors. Therefore, in spite of very careful evaluations and observations, a certain amount of error is inevitable.

Employing the Smithsonian long method, each determination requires about two to three hours of observational time plus twice that much time for the data reduction. In addition, there is no assurance that atmospheric properties and solar conditions remain unchanged in a consistent manner during the observational period. Because of this uncertainty and the burdensome, time-consuming work, a short method was devised to determine the solar constant.

2.4.2 Short Method

In the short method, the diffuse component of solar radiation (the sky brightness) has been measured for a given locality over a long period of time, so that a mean diffuse intensity has been determined. Thus, a pyranometer reading of the diffuse solar radiation will differ from the mean by an amount ε, called the *pyranometer excess*. In reference to Section 1.1.4 and Fig. 2.6, the attenuation of solar radiation in a clear day is due to scattering by molecules and aerosol particles, and absorption by various gases, primarily, the water vapor. If the total precipitable water is given by w, an empirical relationship between the attenuation of the direct solar irradiance and the scattering and absorption effects may be expressed in the form

$$F_\lambda = w + q_\lambda \varepsilon, \tag{2.10}$$

where q_λ is a constant empirically determined for each wavelength for a given locality. With q_λ known, the spectral value of the solar irradiance can be found from the observed precipitable water and a pyranometer reading.

On the basis of a long series of previous observations of F_λ, m, and $\mathscr{T}_\lambda$ at a given location where the solar constant measurement has been made, a graph of F_λ versus air mass m has been constructed for a set value of $\mathscr{T}_\lambda$. Thus, for a particular measurement of F_λ with a known air mass m, the corresponding transmissivity $\mathscr{T}_\lambda$ can be found from the graph. Once $\mathscr{T}_\lambda$ has been determined, the solar irradiance at the top of the atmosphere $F_{0\lambda}$ may be evaluated through Eq. (2.6). After this point, evaluation of the solar constant proceeds in the same manner as in the long method. In the short method, the required measurements include a bologram of the sun, an observation of the sky brightness by the pyranometer, and the air mass determined by the position of the sun from a theodolite. These three measurements take only about 10 to 15 minutes.

From thousands of observations at various locations over the world during a period of more than a half century, the best value of the solar constant determined by the Smithsonian methods is 1.94 cal cm^{-2} min^{-1} (1353 W m^{-2}). It is interesting to note that this is the same value recently accepted as standard value for the solar constant.

A number of measurements have also been made in the upper atmosphere and outer space to minimize atmospheric effects in solar constant determinations. These observations included balloons floating in the 27–35-km altitude range, jet aircraft at about 12 km, the X-15 rocket aircraft at 82 km, and the Mars Mariner VI and VII spacecrafts entirely outside the atmosphere. The solar constant derived from these experiments ranges from about 1.92 to 1.95 cm^{-2} min^{-1}. A standard value for the solar constant recently adopted by NASA is 1.94 ± 0.03 cal cm^{-2} min^{-1} as indicated previously.

More recently, measurements of the incoming solar irradiance also have been made from the satellite platform. The earth radiation budget (ERB) experiment including the solar constant determination was launched into a circular sun-synchronous earth-orbit aboard the Nimbus VI and Nimbus VII statellites in June 1975 and October 1978, respectively (see Appendix H). In the ERB experiment, the incoming solar radiation is observed with an array of 10 telescopes which measure the total solar irradiance and spectral irradiances contained within various broad and narrow subdivisions of the solar spectrum. One of the objectives of the ERB experiment has been to monitor continuously the solar radiation input to the earth–atmosphere system and to investigate the possible variability of the solar constant. Values 1391 and 1368 W m^{-2} have been cited for the solar constant during the 1975–1977 period (Jacobowitz *et al.*, 1979).

2.5 DISTRIBUTION OF ISOLATION OUTSIDE THE ATMOSPHERE

Insolation is defined as the flux of solar radiation per unit horizontal area. It depends strongly on the solar zenith angle and to some extent on the variable distance of the earth from the sun. The flux density at the top of the atmosphere may be expressed by

$$F = F_0 \cos \theta_0, \tag{2.11}$$

where F_0 represents the solar flux density at the top of the atmosphere when the instantaneous distance between the earth and sun is d, and θ_0 denotes the solar zenith angle. From Eq. (2.9), we find

$$F = S(d_m/d)^2 \cos \theta_0. \tag{2.12}$$

Define the solar heating received at the top of the atmosphere per unit area as Q, then the solar flux density may be written as

$$F = \frac{dQ}{dt}. \tag{2.13}$$

Thus, the insolation for a given period of time is

$$Q = \int_t F(t)\, dt. \tag{2.14}$$

The total solar energy received on an unit area per one day may be evaluated by integrating over the daylight hours. Upon substituting Eq. (2.12) into (2.14), we find the daily insolation as follows:

$$Q = S \left(\frac{d_m}{d}\right)^2 \int_{\text{sunrise}}^{\text{sunset}} \cos \theta_0(t)\, dt. \tag{2.15}$$

The solar zenith angle is normally determined from other angles that are known. In reference to Fig. 2.9, let P be the point of observation and OZ the zenith through this point. Assume that the sun is in the direction OS or PS and let D be the point directly under the sun. Then the plane of OZ and OS will intersect the surface of the earth in a great circle. The angle ZOS, measured by the arc PD of this circle, is equal to the sun's zenith distance θ_0. In the spherical triangle NPD, the arc ND is equal to $90°$ minus the solar inclination δ which is the angular distance of the sun north (positive) or south (negative) of the equator. The arc NP is equal to $90°$ minus the latitude λ of the observation point, and the angle h is the hour angle, or the angle through which the earth must turn to bring the meridian of P directly under the sun. From the spherical trigonometry, the cosine of the solar zenith angle as given in Appendix F is

$$\cos\theta_0 = \sin\lambda\sin\delta + \cos\lambda\cos\delta\cos h. \qquad (2.16)$$

The solar inclination is a function of day of year only, and is independent of the location of the observation point. It varies from $23°27'$ on June 21 to $-23°27'$ on December 22. The hour angle is zero at solar noon, and increases by $15°$ for every hour before or after solar noon.

Upon inserting Eq. (2.16) into Eq. (2.15) and denoting the angular velocity of the earth ω by $dh/dt(= 2\pi\,\text{rad}/\text{day})$, Eq. (2.15) yields the form

$$Q = S\left(\frac{d_{\mathrm{m}}}{d}\right)^2 \int_{-H}^{H} (\sin\lambda\sin\delta + \cos\lambda\cos\delta\cos h)\,\frac{dh}{\omega}, \qquad (2.17)$$

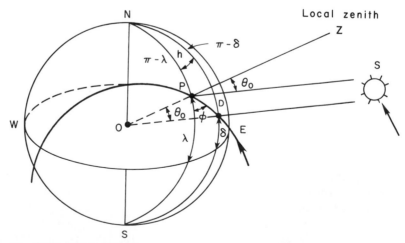

Fig. 2.9 Relationship of the solar zenith angle θ_0 to the latitude λ, the solar inclination angle δ, and the hour angle h. The ϕ here denotes the azimuthal angle of the sun from the south.

where H represents the half-day, i.e., from sunrise or sunset to solar noon. After performing the simple integration we get

$$Q = \frac{S}{\pi}\left(\frac{d_{m}}{d}\right)^{2}(\sin \lambda \sin \delta H + \cos \lambda \cos \delta \sin H). \qquad (2.18)$$

In Eq. (2.18), H in the first term on the right is expressed in units of radians ($180° = \pi$ rad). Note that the factor $(d_{m}/d)^{2}$ never departs by more than 3.5% from unity. It ranges from 1.0344 on January 3 to 0.9674 on July 5.

Equation (2.18) allows us to calculate the distribution of the daily solar energy per unit area over the top of the global atmosphere as functions of latitude and day of year. The results are shown in Fig. 2.10. Since the sun is

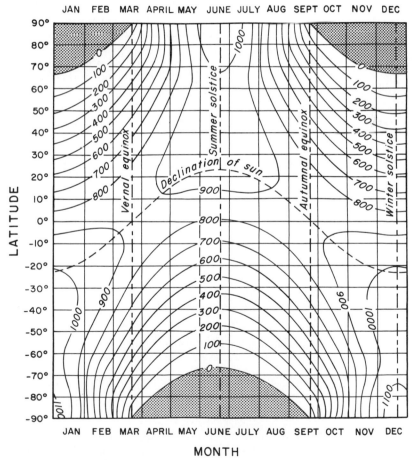

Fig. 2.10 The daily variation of insolation at the top of the atmosphere as a function of latitude and day of year in units of cal cm^{-2} day^{-1} (after List, 1958).

closest to the earth in January (northern hemisphere winter), the distribution of solar energy is slightly asymmetric, and the maximum radiation received in the southern hemisphere is greater than that received in the northern hemisphere. Note that the maximum insolation occurs at summer or winter solstice at either pole owing to the long solar day (24 hours). It should also be noted that after integrating Eq. (2.17) for a period of one year the total annual insolation is the same for the corresponding latitudes in northern and southern hemispheres.

EXERCISES

2.1   By matching the observed solar irradiance curve depicted in Fig. 2.6, show that the sun may be considered as a blackbody in the visible and near infrared wavelengths.

2.2   Given the solar constant 1.94 cal cm^{-2} min^{-1}, the mean earth–sun distance 150×10^6 km, and the sun's radius 0.70×10^6 km, calculate the equilibrium temperature of the sun.

2.3   If the average output of the sun is 6.2×10^7 W m^{-2}, and the radius of the earth is 6.37×10^3 km, what would be the total amount of energy intercepted by the earth in one day?

2.4   Compute the fraction of the emittance that the earth intercepts from the sun.

2.5   Consider a circular cloud whose diameter is 2 km and assume that it is an infinitely thin blackbody with a temperature of 10°C. How much energy does it emit toward the earth? How much energy from this cloud is detected on a square centimeter of the earth's surface when the center of the cloud is 1 km directly over the receiving surface?

2.6   On a clear day, measurements of the direct solar flux density F at the earth's surface in the 1.5- to 1.6-μm-wavelength interval give the following values:

Zenith angle (degree):	40°	50°	60°	70°
F (cal cm^{-2} min^{-1}):	0.020	0.018	0.015	0.011

Find the solar flux density at the top of the atmosphere and the transmissivity of the atmosphere for normal incidence [see Eq. (1.51)] in this wavelength interval.

2.7   Assume that $\bar{r}$ is the mean albedo of the earth (*albedo* is defined as the ratio of the amount of flux reflected to space to the incoming solar flux), and that the earth–atmosphere system is in equilibrium for a long period of time. Show that the equilibrium temperature of the earth–atmosphere system $T = [(1 - \bar{r})S/4\sigma]^{1/4}$

2.8   The following table gives the distances of various planets from the sun and their albedos. Employing the result in Exercise 2.7, compute the equilibrium temperatures of these planets.

Planet	Distance from sun (relative to earth)	Albedo (%)
Mercury	0.39	6
Venus	0.72	78
Earth	1.00	30
Mars	1.52	17
Jupiter	5.20	45

2.9   The height of the earth-synchronous (geostationary) orbiting satellites, such as GOES satellites, is about 35,000 km. Using the solid angle derived from Exercise 1.2, calculate the equilibrium temperature of the satellite in the earth–satellite system, assuming an effective equilibrium temperature of 255°K for the earth and assuming the satellite is a blackbody.

2.10   Show that the change in the earth's equilibrium temperature T_e in terms of the earth–sun distance d is given by $\delta T_e/T_e = -\delta d/2d$. The distance between the earth and the sun varies by about 3.3% with a maximum and minimum on January 3 and July 5, respectively. Compute the seasonal change in the earth's equilibrium temperature.

2.11   Calculate the daily insolation on the top of the atmosphere at (a) the south pole in the winter solstice, and (b) the equator in the vernal equinox. Use the mean earth–sun distance in your calculations and check your values with those shown in Fig. 2.10.

2.12   Compute the solar elevation angle at solar noon at the poles, 60°N(S), 30°N(S), and the equator. Also compute the length of the day (in terms of hours) at the equator and at 45°N in the equinox and solstice.

SUGGESTED REFERENCES

Coulson, K. L. (1975). *Solar and Terrestrial Radiation.* Academic Press, New York. Chapters 3 and 4 give a comprehensive illustration for various kinds of pyrheliometers and pyranometers.

Jastrow, R., and Thompson, M. H. (1974). *Astronomy: Fundamentals and Frontiers,* 2nd ed., Wiley, New York. Chapter 12 provides in-depth discussions on and delightful photos for the structure and composition of the sun.

Sellers, W. D. (1965). *Physical Climatology.* Univ. of Chicago Press, Chicago. Chapter 13 contains discussions on some of the more popular theories of climatic change.

Chapter 3

ABSORPTION AND SCATTERING
OF SOLAR RADIATION IN
THE ATMOSPHERE

3.1 COMPOSITION AND STRUCTURE OF
THE EARTH'S ATMOSPHERE

To describe the interaction of the earth's atmosphere with solar radiation, it is essential that the atmosphere's composition is understood. The atmosphere is composed of a group of nearly *permanent* gases and a group of gases with variable concentration. In addition, the atmosphere also contains various solid and liquid particles such as aerosols, water drops, and ice crystals, which are highly variable in space and time.

Table 3.1 lists the chemical formula and volume ratio for the concentrations of the permanent and variable gases in the earth's atmosphere. It is apparent from this table that nitrogen, oxygen, and argon account for more than 99.99% of the permanent gases. These gases have virtually constant volume ratios up to an altitude of about 60 km in the atmosphere. It should be noted that although carbon dioxide is listed here as a permanent constituent, its concentration varies as a result of the combustion of fossil fuels, absorption and release by the ocean, and photosynthesis. The climatic impact of the increase of carbon dioxide content in the earth's atmosphere will be discussed in Section 4.9. Water vapor concentration varies greatly both in space and time depending upon the atmospheric condition. Its variation is extremely

TABLE 3.1   *The Composition of the Atmosphere[a]*

Permanent constituents		Variable constituents	
Constituent	% by volume	Constituent	% by volume
Nitrogen (N_2)	78.084	Water vapor (H_2O)	0–0.04
Oxygen (O_2)	20.948	Ozone (O_3)	$0-12 \times 10^{-4}$
Argon (Ar)	0.934	Sulfur dioxide (SO_2)[b]	0.001×10^{-4}
Carbon dioxide (CO_2)	0.033	Nitrogen dioxide (NO_2)[b]	0.001×10^{-4}
Neon (Ne)	18.18×10^{-4}	Ammonia (NH_3)[b]	0.004×10^{-4}
Helium (He)	5.24×10^{-4}	Nitric oxide (NO)[b]	0.0005×10^{-4}
Krypton (Kr)	1.14×10^{-4}	Hydrogen sulfide (H_2S)[b]	0.00005×10^{-4}
Xenon (Xe)	0.089×10^{-4}	Nitric acid vapor (HNO_3)	Trace
Hydrogen (H_2)	0.5×10^{-4}		
Methane (CH_4)	1.5×10^{-4}		
Nitrous oxide (N_2O)[b]	0.27×10^{-4}		
Carbon monoxide (CO)[b]	0.19×10^{-4}		

[a] After the U.S. Standard Atmosphere, 1976.
[b] Concentration near the earth's surface.

important in the radiative absorption and emission processes as will be evident in this chapter and the next. Ozone concentration also changes with respect to time and space, and it occurs principally in altitudes from about 15 to about 30 km, where it is both produced and destroyed by photochemical reactions. Most of the ultraviolet radiation is absorbed by ozone, preventing this harmful radiation from reaching the earth's surface. In the next two sections, we will discuss photochemical processes involving ozone. The remaining gases along with several trace gases not listed in the table, enter into many types of reactions with the other gases and particles, and are found in the atmosphere in variable concentration.

All of the gases listed in the table are responsible for the scattering of sunlight and the consequent polarization characteristics. These subjects will also be discussed in this chapter. Finally, we note that the variable solid and liquid particles suspended in the atmosphere play an important role in absorption and scattering of solar radiation, and in the physics of clouds and precipitation.

The vertical temperature profile for the standard atmosphere is depicted in Fig. 3.1. This profile represents typical conditions in middle latitudes. According to the standard nomenclature defined by the International Union of Geodesy and Geophysics (IUGG) in 1960, the vertical profile is divided into four distinct layers as shown in Fig. 3.1. These are the *troposphere*, *stratosphere*, *mesosphere*, and *thermosphere*. The tops of these layers are respectively called the tropopause, stratopause, mesopause, and thermopause.

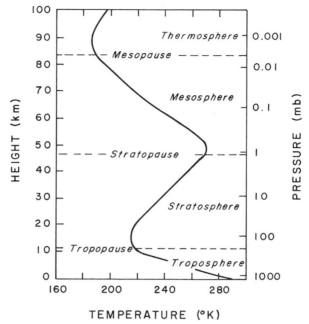

Fig. 3.1 Vertical temperature profile after the U.S. Standard Atmosphere (1976).

The troposphere is characterized by a decrease of temperature with respect to height with a typical lapse rate of 6.5°C/km. The temperature structure in this layer is a consequence of the radiative balance and the convective transport of energy from the surface to the atmosphere. Virtually all the water vapor, cloud, and precipitation are confined in this layer. The stratosphere is characterized by an isothermal layer from the tropopause to about 20 km from where the temperature increases to the stratopause. Ozone occurs chiefly in the stratosphere. In addition, thin layers of aerosol are observed to persist for a long period of time within certain altitude ranges of the stratosphere. Like the troposphere the temperatures in the mesosphere decrease with height from about 50 to about 85 km. Above 85 km and extending upward to an altitude of several hundred kilometers lies the thermosphere where temperatures range from 500°K to as high as 2000°K. The outermost region of the atmosphere above the thermosphere is called the *exosphere*. The term *upper atmosphere* generally is defined as the region of the atmosphere above the troposphere. As will be discussed in Chapter 4, the temperature distribution is the major factor in determining the transfer of thermal infrared radiation in the atmosphere.

3.2   ABSORPTION IN THE ULTRAVIOLET

Before we proceed to discuss the absorption of solar radiation in the ultraviolet and near infrared regions, it would be helpful to introduce the ways in which a molecule can store various energies. Any moving particle has kinetic energy as a result of its motion in space. This is known as *translational energy*. The averaged translational kinetic energy of a single molecule in the X, Y, and Z directions is found to be equal to $KT/2$, where K is the Boltzmann constant and T is the absolute temperature. The molecule which is composed of atoms can rotate, or revolve, about an axis through its center of gravity and, therefore, has *rotational energy*. The atoms of the molecule are bounded by certain forces in which the individual atoms can vibrate about their equilibrium positions relative to one another. The molecule therefore will have *vibrational energy*. These three molecular energy types are based on a rather mechanical model of the molecule that ignores the detailed structure of the molecule in terms of nuclei and electrons. It is possible, however, for the energy of a molecule to change due to a change in the energy state of the electrons of which it is composed. Thus, the molecule has *electronic energy*. The last three energy types are quantized and take discrete values only. As we have pointed out in Section 1.3, absorption and emission of radiation takes place when the atoms or molecules undergo transitions from one energy state to another. In general, these transitions are governed by selection rules. Atoms can exhibit line spectra associated with electronic energy. Molecules, however, can have two additional types of energy which lead to complex band systems.

Solar radiation is mainly absorbed in the atmosphere by $O_2, O_3, N_2, CO_2,$ $H_2O, O,$ and N, although $NO, N_2O, CO,$ and CH_4, which occur in very small quantities, also exhibit absorption spectra. Absorption spectra due to electronic transitions of molecular and atomic oxygen and nitrogen, and ozone occur chiefly in the ultraviolet (UV) region, while those due to the vibrational and rotational transitions of triatomic molecules such as H_2O, O_3, and CO_2 lie in the infrared region. There is very little absorption in the visible region of the solar spectrum. Most of the UV radiation is absorbed in the upper atmosphere by oxygen and nitrogen species. The UV absorption spectrum of molecular oxygen begins at about 2600 Å and continues down to shorter wavelengths. The bands between 2600 and 2000 Å, referred to as the *Herzberg bands*, are very weak and of little importance in the absorption of solar radiation owing to their overlap with the much stronger ozone bands in this spectral region. Nevertheless, the Herzberg bands are considered to be of significance in the formation of ozone. Adjacent to the Herzberg bands are the very strong *Schumann–Runge* band system and continuum which begins at 2000 Å and continues down to about 1250 Å. Also several bands exist between 1250 and 1000 Å. Of particular interest is the Lyman α line at 1216 Å, which is very strong in the solar spectrum. It lies in one of the windows of the O_2 absorption spectrum. The region below 1000 Å is occupied by the very strong O_2 bands, referred to as the *Hopfield bands*.

The absorption spectrum of molecular nitrogen begins at 1450 Å. The regions from 1450 to 1000 Å are called *Lyman–Birge–Hopfield bands*, and consist of narrow and sharp lines. From 1000 to 800 Å, the absorption spectrum of N_2 is occupied by the *Tanaka–Worley bands*. They are very complicated and absorption coefficients are highly variable. Below 800 Å, the absorption spectrum of N_2 is generally made up of the *ionization continuum*. Ionization is a process in which an electron is removed from its orbit. In the ionization process the atom or molecule may absorb more than the minimum energy required to remove the electron. This additional energy is not quantized. As a result the absorption is not selective but is continuous. The ionization contiuum occurs on the high frequency (shorter wavelength) side of the ionization frequency.

Because of the absorption of solar UV radiation, some of the oxygen and nitrogen molecules in the upper atmosphere undergo photochemical *dissociation* and are dissociated into atomic oxygen and nitrogen. Atomic nitrogen exhibits absorption spectrum from about 10 to about 1000 Å. Although atomic nitrogen probably is not abundant enough to be a significant absorber in the upper atmosphere, it may play an important role in the absorption of UV radiation in the thermosphere. Atomic oxygen also shows absorption continuum in the region of 10 to 1000 Å. Owing to the absorption of solar UV radiation, a portion of molecular and atomic oxygen and nitrogen becomes ionized. The ionized layers in the upper atmosphere are formed mainly as a result of these processes.

Of the principal constituents of the upper atmosphere, only O_2 weakly absorbs between 2000 and 3000 Å. This part of the solar spectrum primarily is absorbed by the ozone in the upper stratosphere and mesosphere. The regions which consist of the strongest absorption bands of O_3 are called *Hartley bands*. The bands between 3000 and 3600 Å are called *Huggins bands*, which are not as strong as Hartley bands. O_3 also shows weak absorption bands in the visible and near infrared regions from about 4400 to 11,800 Å, called *Chappuis bands*.

The absorption cross sections of O_2, N_2, O, N, and O_3 have been measured by many workers and they are displayed in Fig. 3.2. The curves illustrated in this figure are intended to indicate the relative significance of various absorbers and should not be taken as a source of quantitative data. (For the original references of these data, see Craig, 1965.) Note that the absorption

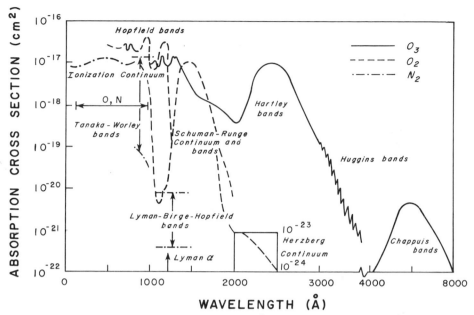

Fig. 3.2 Absorption cross sections σ (in units of cm^2) of ozone, molecular oxygen, and molecular nitrogen in the ultraviolet spectral region. Absorption regions for atomic nitrogen and oxygen are also shown.

cross section σ (in units of cm^2) is related to the absorption coefficient k (in units of $(cm\text{-}atm)^{-1}$) through the Loschmidt's number N_0 (2.687×10^{19} particles cm^{-3} at the standard temperature, $0°C$, and the standard pressure, 1013 mb); i.e., $k = \sigma N_0$. Both σ and k frequently are used in the field of upper atmosphere.

To illustrate the relative absorption effects of O_2, N_2, O, N, and O_3, shown in Fig. 3.3 is the reduction of solar flux density when it penetrates the atmosphere. In this figure, the curve represents the decrease of the flux density of solar radiation at the normal incidence by a factor of e (2.71828); i.e., $\ln(F/F_0) = 1$. For example, the solar flux density F_0 at 1600 Å wavelength is reduced by a factor of 2.71828 when it reaches about 110 km. The figure also shows the variation of the atmospheric transparency below about 3000 Å. From 3000 to 2000 Å the absorption is primarily due to O_3. O_2 is responsible for absorption between about 2000 and 850 Å. Below 850 Å, O_2, O, N_2, and N are responsible for the absorption of solar radiation. The absorption of solar UV radiation represents the prime source for the energetics and dynamics of the upper atmosphere.

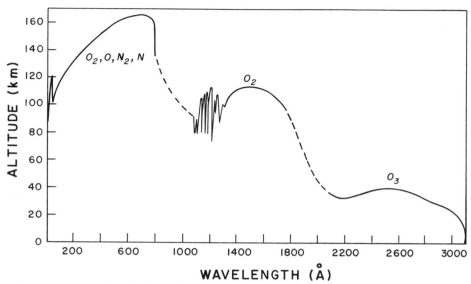

Fig. 3.3 Penetration of solar radiation into the atmosphere. The curve indicates the level at which the solar flux density is reduced by e (after Friedman, 1960).

3.3 PHOTOCHEMICAL PROCESSES AND THE FORMATION OF OZONE LAYERS

Owing to the absorption spectrum of various molecules and atoms in the solar UV region, a great variety of photochemical processes take place in the upper atmosphere. Those involving various forms of oxygen are important in determining the amount of ozone in the stratosphere. The classical photochemistry of the upper atmosphere concerning the ozone problem was first postulated by Chapman (1930) in which five basic reactions were proposed:
Ozone is basically formed by the three-body collision

$$O + O_2 + M \xrightarrow{K_{12}} O_3 + M, \tag{3.1}$$

where M is any third atom or molecule, and K_{12} is the rate coefficient involving O and O_2. Atomic oxygen is produced when the oxygen molecule is dissociated by a quantum of solar energy:

$$O_2 + h\bar{\nu}(\lambda < 2423 \text{ Å}) \xrightarrow{J_2} O + O, \tag{3.2}$$

where J_2 is the dissociating quanta per molecule absorbed by O_2. Ozone is destroyed both by photodissociation

$$O_3 + h\bar{\nu}(\lambda < 11,000 \text{ Å}) \xrightarrow{J_3} O + O_2 \tag{3.3}$$

and also by collision with oxygen atoms

$$O_3 + O \xrightarrow{K_{13}} 2O_2, \tag{3.4}$$

where J_3 is the dissociating quanta per molecule absorbed by O_3, and K_{13} denotes the rate coefficient involving O_3 and O. Meanwhile, oxygen atoms generated by reactions (3.2) and (3.3) may undergo three-body collision,

$$O + O + M \xrightarrow{K_{11}} O_2 + M, \tag{3.5}$$

with K_{11} denoting the rate coefficient involving O and O. Normally, reaction (3.5) may be neglected below 50 to 60 km.

The preceding five reactions take place simultaneously. The number of ozone molecules formed exactly equals the number destroyed in unit volume and time, and the process reaches an equilibrium state. To evaluate the equilibrium amount of ozone, let [O], [O$_2$], [O$_3$], and [M] be the number densities, respectively, for O, O$_2$, O$_3$, and air molecules. Then the photochemical processes given by Eqs. (3.1)–(3.5) may be expressed in terms of the rate of change of the number density of O, O$_2$, and O$_3$ in the following forms:

$$\frac{\partial[O]}{\partial t} = -K_{12}[O][O_2][M] + 2[O_2]J_2 - K_{13}[O][O_3]$$
$$+ [O_3]J_3 - 2K_{11}[O][O][M], \tag{3.6}$$

$$\frac{\partial[O_2]}{\partial t} = -K_{12}[O][O_2][M] - [O_2]J_2 + 2K_{13}[O][O_3]$$
$$+ [O_3]J_3 + K_{11}[O][O][M], \tag{3.7}$$

$$\frac{\partial[O_3]}{\partial t} = K_{12}[O][O_2][M] - K_{13}[O][O_3] - [O_3]J_3, \tag{3.8}$$

with

$$J_2 = \int_0^{0.2423 \ \mu m} k_\lambda(O_2)F_\lambda(\infty)\mathcal{T}_\lambda(O_2)\,d\lambda, \tag{3.9}$$

$$J_3 = \int_0^{1.1 \ \mu m} k_\lambda(O_3)F_\lambda(\infty)\mathcal{T}_\lambda(O_3)\,d\lambda, \tag{3.10}$$

where $F_\lambda(\infty)$ denotes the monochromatic solar flux at the outer edge of the atmosphere in quanta (cm^{-2} sec^{-1} cm^{-1}), k_λ is the absorption cross section (cm^2) per molecule, and $\mathcal{T}_\lambda$ is the nondimensional transmissivity given by Eq. (1.51) for the atmosphere above the volume under consideration in the direction toward the sun. Thus the transmissivity depends upon the solar zenith angle.

Under the assumption of photochemical equilibrium, $\partial[O]/\partial t = \partial[O_2]/\partial t = \partial[O_3]/\partial t = 0$. Thus, three homogeneous equations are obtained.

Further, to a good approximation the values of $[O_2]$ and $[M]$ in Eqs. (3.6) and (3.8) may be considered to be constant. It follows that the equilibrium values of $[O]$ and $[O_3]$ can be evaluated from Eqs. (3.6) and (3.8). In order to make numerical computations, it is necessary to specify the atmospheric density and temperature as a function of altitude, the solar zenith angle, the solar flux at the outer edge of the atmosphere, the oxygen and ozone absorption coefficients, and the rate coefficients.

Figure 3.4 depicts the equilibrium ozone concentration from the classical theory. In the same diagram, the observational range in ozone number densities (shaded area) is also shown. It is seen that the classical theory overestimates the ozone number densities at almost all heights. The total ozone in an atmospheric column from theoretical calculations exceeds the observed values by as much as a factor of three or four. Obviously, additional loss mechanisms are required to explain the observed data.

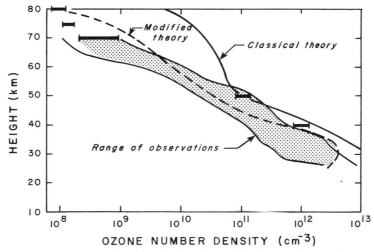

Fig. 3.4 Observational range in ozone number densities (shaded area) and theoretical calculations for equilibrium ozone number densities (after Leovy, 1969; see text for further explanation).

In addition to the photodissociation and collision described previously, the catalytic destruction reactions of ozone have been found to be

$$O_3 + h\bar{\nu} \longrightarrow O + O_2, \qquad\qquad (3.11)$$

$$O + XO \longrightarrow X + O_2, \qquad\qquad (3.12)$$

$$X + O_3 \longrightarrow XO + O_2, \qquad\qquad (3.13)$$

where X may be nitric oxide (NO), chlorine (Cl), hydroxyl radical (OH), or atomic hydrogen (H). The net results of these reactions are $2O_3 + h\tilde{v} = 3O_2$.

The possible sources of NO and OH are produced through the following reactions:

$$O_3 + h\tilde{v}(\lambda < 3100 \text{ Å}) \longrightarrow O(^1D) + O_2, \tag{3.14}$$

$$O(^1D) + M \longrightarrow O + M, \tag{3.15}$$

$$O(^1D) + N_2O \longrightarrow 2NO, \tag{3.16}$$

$$O(^1D) + H_2O \longrightarrow 2OH, \tag{3.17}$$

where $O(^1D)$ denotes the excited atomic oxygen in the ^1D state, which is essential to these reactions. It is clear that the high concentration of ozone from Chapman's theoretical prediction is due to the neglect of these additional loss mechanisms and possibly others. By introducing reactions involving (3.11)–(3.17), the calculated equilibrium ozone concentration labeled as the modified theory is shown in Fig. 3.4 [after Leovy (1969)]. The results from the modified theory appear to match closely with the observed values. However, because of uncertainties on a number of reaction and dissociation rates and the dynamics involved in the ozone formation, the ozone problem is still an area of active research.

Ozone is a natural trace ingredient of the atmosphere that occurs at an average concentration of about 3 parts per million by volume. Its concentration varies with seasons and latitudes. High intensities of the UV radiation shorter than 3200 Å, which are harmful to nearly all forms of life, are largely ($\sim 99\%$) screened out by ozone. Generally it has been agreed that the surface life on the earth did not evolve until after the ozone layer was formed. But the effect of small increases in intensity of ultraviolet radiation due to the reduction of ozone by human activities is still a subject of speculations and scientific debates.

In recent years, it has been speculated that NO and Cl may be increasing due to the industralization of human society. Supersonic transports, aerosol sprays, and nuclear weapons are all probable examples of technological ability carried to excess. The catalytic agents are oxides of nitrogen released into the upper atmosphere by the jet engines of supersonic transports and by nuclear explosions, and the free chlorine derived photolytically from aerosol spray cans ($CFCl_3$) and refrigerant (CF_2Cl_2). Additional oxides of nitrogen also may be produced by the increased use of fixed nitrogen as fertilizer. It seems apparent that the addition of NO and Cl might reduce the ozone concentration in the stratosphere. The yet-to-be-solved problem of the influence of man's expanding industrial and agricultural activities has become a political, economic, and scientific issue which has aroused worldwide interest and concern.

3.4   ABSORPTION IN THE VISIBLE AND INFRARED

The solar spectrum recorded with a low-resolution spectrometer has been shown in Fig. 2.6 in which shaded areas represent the absorption of solar flux by various minor gases in the atmosphere. Molecular oxygen absorbs UV radiation as discussed in the previous section. In addition, it also is found to have two weak bands in the red region of the solar spectrum. The A band of O_2 at 0.7 μm is particularly well known because of the large solar flux contained in this region. The A band also has led to the discovery of the isotopes ^{18}O and ^{17}O.

Absorption bands in the solar near-infrared region chiefly are due to vibrational and rotational transitions. The most important absorber in the near infrared as evident in Fig. 2.6 is water vapor. Carbon dioxide also has weak absorption bands in the solar spectrum, however, the most important band is the one which overlaps with the water vapor 2.7 μm band. There are other minor gases, such as CO, CH_4, N_2O, which also absorb solar infrared radiation. However, the absorption by these gases is insignificant so far as the heat budget of the earth–atmosphere is concerned. In reference to Fig. 2.6 we note that the only significant amount of solar flux lies in wavelengths shorter than about 4 μm.

Water vapor absorbs solar radiation in the vibrational–rotational bands. The absorption bands centered at 0.94, 1.1, 1.38, and 1.87 μm shown in Fig. 2.6 are commonly identified in groups by Greek letters (ρ, σ, τ), ϕ, ψ, and Ω, respectively. These bands arise from ground-state transitions and are called *overtone* and *combination bands*. Although the band at 2.7 μm is most important, the weaker band at 3.2 μm and the overtone and combination bands also contribute significantly in the absorption. The strong 6.3 μm band, to be discussed in Chapter 4, is very important in the thermal infrared region. However, since the band contains very little solar energy, its absorption in the solar spectrum may be ignored.

Carbon dioxide exhibits a number of weak absorption bands in the solar spectrum. The 2.0, 1.6, and 1.4 μm CO_2 bands are so weak that for all practical purposes they can be ignored in solar absorption calculations. The 2.7 μm band of CO_2, which overlaps with the 2.7 μm band of water vapor, is somewhat stronger and should be included in absorption calculations. The 4.3 μm band of CO_2 is more important in the thermal infrared region than the solar region because this band contains very little solar energy.

These absorption bands consist of lines whose intensity varies greatly with the wave number so that the transmissivity cannot be described by the Beer–Bouguer–Lambert law employing an exponential function of the gaseous optical path. Because of the uncertainities on the theoretical knowledge of the line position and intensity, the absorption characteristics of these absorption bands in the solar spectrum have been determined in detail by means of laboratory measurements.

Howard *et al.* (1956) measured the total absorption $\int A_v \, dv$ for water vapor and carbon dioxide bands under simulated atmospheric conditions. For small values of total absorption, the formula

$$A = \int A_v \, dv = c u^{1/2}(P + e)^k, \qquad A < A_c, \tag{3.18}$$

was derived, whereas for large values of total absorption, the empirical equation is given by

$$A = \int A_v \, dv = C + D \log u + K \log(P + e), \qquad A > A_c. \tag{3.19}$$

In these two equations, v represents wave number (cm^{-1}); A the band area in wave number (cm^{-1}); A_v the fractional absorption within the band at v; u the absorbing path $(g \, cm^{-2}$ for H_2O; cm-atm for $CO_2)$; e the partial pressure of absorbing gases (mm Hg; note that 760 mm Hg = 1013 mb, 1 mb = 10^3 dyn cm^{-2}); P the partial pressure of nonabsorbing gases (mm Hg); A_c the critical band area above which the strong band expression becomes applicable; and c, k, C, D, and K are empirically determined constants. Table 3.2 lists these values for H_2O and CO_2 bands.

Although Eqs. (3.18) and (3.19) are derived to calculate the approximate band absorptivity, these two formulas are not continuous when $A = A_c$.

TABLE 3.2   Empirical Constants for H_2O and CO_2 Bands

λ (μm)	c	k	C	D	K	A_c (cm^{-1})	Δv (cm^{-1})	K/D	x_0 $(g \, cm^{-2})$
H_2O band									
0.94	38	0.27	−135	230	125	200	1400	0.54	3.86
1.1	31	0.26	−292	345	180	200	1000	0.52	7.02
1.38	163	0.30	202	460	198	350	1500	0.43	0.36
1.87	152	0.30	127	232	144	257	1100	0.62	0.28
2.7	316	0.32	337	246	150	200	1000	0.62	0.04
3.2	40.2	0.30	−144	295	151	500	540	0.51	3.25
6.3	356	0.30	302	218	157	160	900	0.72	0.41
CO_2 band									
1.4	0.058	0.41	—	—	—	80	600		
1.6	0.063	0.38	—	—	—	80	550		
2.0	0.492	0.39	−536	138	114	80	450		
2.7	3.15	0.43	−137	77	68	50	320		
4.3	—	—	27.5	34	31.5	50	340		
4.8	0.12	0.37	—	—	—	60	180		
5.2	0.024	0.40	—	—	—	30	110		
15.0	3.16	0.44	—	—	—	50	250		

Liou and Sasamori (1975) derived a single formula to approximate the mean absorptivity for both weak and strong absorption in the form

$$A_{\bar{v}} = \frac{A}{\Delta v} = \frac{1}{\Delta v} [C + D \log(x + x_0)], \qquad (3.20)$$

where the same coefficients C and D in Eq. (3.19) are used with a new parameter x_0 to denote each absorption band, and

$$x = uP^{K/D}. \qquad (3.21)$$

For large optical path lengths the absorptivity expressed by Eq. (3.20) approaches that in Eq. (3.19). Moreover, x_0 was chosen in such a way that $A_{\bar{v}}$ in Eq. (3.20) approaches zero as x approaches zero. Thus

$$x_0 = 10^{-C/D}. \qquad (3.22)$$

The numerical values of x_0 for H_2O are also listed in the last column of Table 3.2. Since the partial pressure of water vapor is much smaller than the pressure of dry air, it suffices to use the latter pressure in solar-heating calculations.

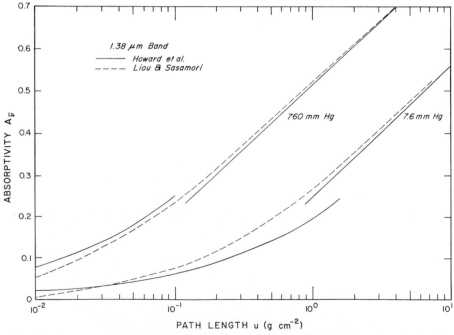

Fig. 3.5 A comparison of the absorptivity of 1.38 μm band calculated from Eqs. (3.18) and (3.19) and from that based on Eq. (3.20) for $P = 760$ and 7.6 mm Hg.

Figure 3.5 illustrates a comparison between the absorptivity calculated from Eq. (3.20) and Eqs. (3.18) and (3.19) for the 1.38 μm water vapor band. It is clear that the new formula gives fairly accurate absorptivity values. The largest deviation takes place at the transition between weak and strong absorption.

As indicated previously, the only important band for CO_2 is the 2.7 μm band which overlaps with the 2.7 μm band of H_2O. Thus, solar heating in the troposphere is mainly generated by water vapor (Note that clouds also play significant roles in generating solar heating.) It should be noted that the 6.3 μm H_2O and 15 μm CO_2 bands are important absorption bands in the thermal infrared spectrum to be discussed in Chapter 4 where band models for evaluations of the absorptivity will be presented in some detail. Also note here that the absorptivity described above may be utilized to calculate the transfer of near infrared solar radiation in cloudy and hazy atmospheres where multiple scattering and absorption take place simultaneously.

3.5   COMPUTATION OF SOLAR HEATING RATES

The importance of the absorption of solar radiation by various gases is the generation of heating in the atmosphere. We will outline the procedures for the evaluation of the solar heating rate. Consider a plane–parallel absorbing and scattering atmosphere illuminated by the solar spectral irradiance $F_{\lambda 0}$ with a solar zenith angle of θ_0. The downward flux density normal to the top of the atmosphere is given by $F_{\lambda 0} \cos \theta_0$. Let the differential thickness within the atmosphere be Δz, and let the spectral downward and upward flux densities centered at wavelength λ be denoted by $F_\lambda^\downarrow$ and $F_\lambda^\uparrow$, respectively. The net flux density (downward) at a given height z is then defined by

$$F_\lambda(z) = F_\lambda^\downarrow(z) - F_\lambda^\uparrow(z). \tag{3.23}$$

Referring to Fig. 3.6, because of absorption, the net flux density decreases from the upper levels to the progressively lower levels. The *loss* of the net flux density, i.e., the net flux density divergence for the differential layer is therefore

$$\Delta F_\lambda(z) = F_\lambda(z) - F_\lambda(z + \Delta z). \tag{3.24}$$

If the spectral absorptivity centered at wavelength λ for the differential layer is denoted by $A_\lambda(\Delta z)$, then Eq. (3.24) can be rewritten as

$$\Delta F_\lambda(z) = -F_\lambda^\downarrow(z + \Delta z) A_\lambda(\Delta z). \tag{3.25}$$

On the basis of the energy conservation principle, the absorbed radiant energy has to be used to heat the layer. Thus, the heating experienced by a layer of air due to radiation transfer may be expressed in terms of the rate of

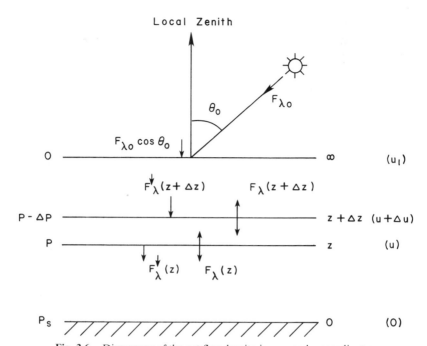

Fig. 3.6 Divergence of the net flux density in z, p, and u coordinates.

temperature changes. It is conventionally given by

$$\Delta F_\lambda(z) = -\rho C_p \Delta z \frac{\partial T}{\partial t}, \tag{3.26}$$

where ρ is the air density in the layer, C_p the specific heat at constant pressure, and t the time. The heating rate for a differential layer Δz is therefore

$$\frac{\partial T}{\partial t} = -\frac{1}{C_p \rho}\frac{\Delta F_\lambda(z)}{\Delta z} = \frac{1}{C_p \rho}\frac{F_\lambda^\downarrow(z + \Delta z)A_\lambda(\Delta z)}{\Delta z}. \tag{3.27}$$

The heating rate may also be expressed in pressure coordinates. By means of the hydrostatic equation

$$dp = -\rho g\, dz \tag{3.28}$$

with g the gravitational acceleration, we have

$$\frac{\partial T}{\partial t} = \frac{g}{C_p}\frac{\Delta F_\lambda(p)}{\Delta p}, \tag{3.29}$$

where g/C_p is the well-known adiabatic lapse rate.

Moreover, it is sometimes convenient to compute the heating rate in terms of the path length of the absorbing gas. The differential path length of a specific gas (say, water vapor) is [see Eq. (1.49)]

$$du = \rho_w \, dz = \frac{\rho_w}{\rho} \rho \, dz = q\rho \, dz = -\frac{q}{g} \, dp, \qquad (3.30)$$

where ρ_w denotes the water vapor density and q represents the specific humidity. Consequently, the radiative heating rate also may be written as

$$\frac{\partial T}{\partial t} = -\frac{q}{C_p} \frac{\Delta F_\lambda(u)}{\Delta u}. \qquad (3.31a)$$

If we divide the solar spectrum into N spectral intervals and carry out the heating rate calculations for each spectral interval i, then the total heating rate due to solar radiation may be written in the form

$$\left(\frac{\partial T}{\partial t}\right)_{tot} = \sum_{i=1}^{N} \left(\frac{\partial T}{\partial t}\right)_i. \qquad (3.31b)$$

One final note may be in order. In Section 1.3.1, we pointed out that the absorption coefficient due to pressure broadening depends strongly on the air pressure because the half width is linearly proportional to it as noted in Eq. (1.37). To account for the pressure dependence of absorption in an inhomogeneous atmosphere, an empirical method has been developed. The method takes into account the effect of atmospheric pressure variations on the absorption process by defining first an effective pressure

$$\bar{P} = \int_0^u P(u) \, du \left/ \int_0^u du \right. . \qquad (3.32)$$

This expression places all the absorbing matter along a pressure gradient at one pressure $\bar{P}$. The empirical adjustment is then carried out by replacing every P in Eqs. (3.18)–(3.21) by $\bar{P}$. In this manner, the variation of the absorption coefficient due to pressure changes in the atmosphere is approximately accounted for.

Figure 3.7 shows the solar heating rate profile up to 30 km using two different atmospheric profiles and three solar zenith angles for a clear atmosphere. Effects of absorption by O_3, H_2O, O_2, and CO_2, multiple scattering by molecules, and the ground reflection are simultaneously taken into consideration in the radiative transfer program covering the entire solar spectrum. In Fig. 3.7a, we see that the tropospheric heating in a tropical atmosphere is much more pronounced than that in a midlatitude winter ·atmosphere due to the higher water vapor concentration. The maximum heating located at a height of about 3 km is seen to be as high as 4°C/day. The heating rate decreases drastically with increasing altitude in phase with

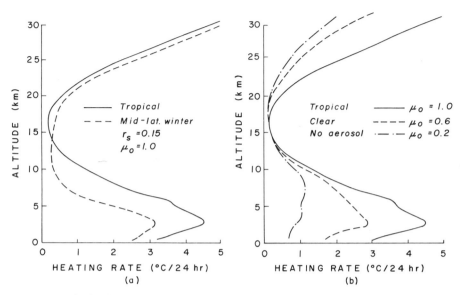

Fig. 3.7 Solar heating rate dependence (a) on the atmosphere and (b) on the solar zenith angle.

the exponential decrease of water vapor concentration and reaches a minimum at about 15 km or so. Above 20 km, the increased solar heating is caused exclusively by the absorption of ozone, which has a maximum concentration at about 25 km. In these calculations, the sun is overhead ($\mu_0 = \cos\theta_0 = 1$) and the surface reflectivity r_s is assumed to be 15%. Effects of the position of the sun on the heating rate are demonstrated in Fig. 3.7b. The solar irradiance available to the atmosphere reduces by a factor of μ_0 as the sun moves away from the zenith. As a result, the heating rate decreases significantly as shown in this figure.

3.6 REPRESENTATION OF POLARIZED LIGHT AND STOKES PARAMETERS

3.6.1 Representation for a Simple Wave

Scattering of sunlight by molecules and particles takes place in the atmosphere. By scattering processes, molecules and particles in the path of electromagnetic waves continuously abstract energy from the incident wave and reradiate that same energy in all directions. Thus, in order to understand the atmospheric scattering of sunlight, it is necessary to describe the representation of electromagnetic waves.

An electromagnetic wave is characterized by electric and magnetic vectors **E** and **H**, which form an orthogonal set with the direction of propagation of the wave. In any medium, **E** and **H** are related, and it is customary to use **E** in scattering discussions. We say light is polarized in a certain direction when the vibration of the electric vector **E** concentrates in that direction. Hence, the direction of polarization is defined as the direction of the electric vector.

The flow of energy and the direction of the wave propagation are represented by the *Poynting vector* depicted in Fig. 3.8 and in Gaussian units it is

$$\mathbf{S} = \frac{c}{4\pi} \mathbf{E} \times \mathbf{H}, \tag{3.33}$$

where $|\mathbf{S}|$ is in the units of flux density. The electric field vector **E** may be decomposed into two components, E_l and E_r, which represent the electric vectors parallel (l) and perpendicular (r) to a plane through the direction of propagation. The plane defined is called the plane of reference and its selection, in principle, is arbitrary. In scattering problems, we choose the plane containing the incident and scattered beams as the common plane of reference for the two beams.

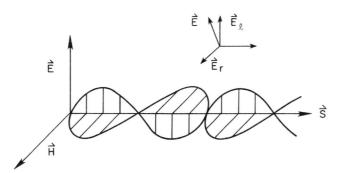

Fig. 3.8   Propagation of an electromagnetic wave. The electric vector can be arbitrarily decomposed into two orthogonal components. (In figures, vectors are indicated by arrows instead of boldface)

Assuming that an electromagnetic wave propagates in the Z direction with a propagation constant $k(2\pi/\lambda)$ and a circular frequency $\omega(kc)$, and that positive amplitudes and phases for the electric field of an electromagnetic wave in the $l(E_l)$ and r(E_r) directions are a_l, a_r and δ_l, δ_r, respectively, then

$$E_l = a_l e^{-i\delta_l} e^{-ikz + i\omega t}, \qquad E_r = a_r e^{-i\delta_r} e^{-ikz + i\omega t}, \tag{3.34}$$

where E_l and E_r are complex, oscillating functions. Let $\zeta = kz - \omega t$ and take the cosine representation for the case when the plane wave is time harmonic, we have

$$E_l = a_l \cos(\zeta + \delta_l), \qquad E_r = a_r \cos(\zeta + \delta_r). \tag{3.35}$$

It follows that

$$\begin{aligned} E_l/a_l &= \cos\zeta\cos\delta_l - \sin\zeta\sin\delta_l, \\ E_r/a_r &= \cos\zeta\cos\delta_r - \sin\zeta\sin\delta_r. \end{aligned} \tag{3.36}$$

We first multiply the first and second equations by $\sin\delta_r$, $\cos\delta_r$ and $\sin\delta_l$, $\cos\delta_l$, respectively, and subtract one from another to obtain

$$\begin{aligned} (E_l/a_l)\sin\delta_r - (E_r/a_r)\sin\delta_l &= \cos\zeta\sin(\delta_r - \delta_l), \\ (E_l/a_l)\cos\delta_r - (E_r/a_r)\cos\delta_l &= \sin\zeta\sin(\delta_r - \delta_l). \end{aligned} \tag{3.37}$$

Upon squaring and adding the above two equations, we obtain

$$(E_l/a_l)^2 + (E_r/a_r)^2 - 2(E_l/a_l)(E_r/a_r)\cos\delta = \sin^2\delta, \tag{3.38}$$

where the phase difference $\delta = \delta_r - \delta_l$.

Equation (3.38) represents the equation of a conic. The associated determinant is

$$\begin{vmatrix} 1/a_l^2 & -\cos\delta/(a_l a_r) \\ -\cos\delta/(a_l a_r) & 1/a_r^2 \end{vmatrix} = \frac{\sin^2\delta}{a_l^2 a_r^2} \geq 0. \tag{3.39}$$

Thus, the conic equation represents an ellipse, and the *elliptically polarized* wave is illustrated in Fig. 3.9a. The ellipse is inscribed into a rectangle whose sides are parallel to the coordinate axes and whose lengths are $2a_l$ and $2a_r$. The ellipse touches the sides at the points $(\pm a_l, \pm a_r\cos\delta)$ and $(\pm a_l\cos\delta, \pm a_r)$.

Two special cases are of particular importance. If $\delta = m\pi$ ($m = 0, \pm 1, \pm 2, \ldots$), then Eq. (3.38) becomes

$$\left(\frac{E_l}{a_l} \pm \frac{E_r}{a_r}\right)^2 = 0, \qquad \text{i.e.,} \qquad \frac{E_l}{a_l} = \mp\frac{E_r}{a_r}. \tag{3.40}$$

This equation describes two lines perpendicular to each other. We call the wave in this case *linearly polarized*. On the other hand, if $\delta = m\pi/2$ ($m = \pm 1, \pm 3, \ldots$) and $a_l = a_r = a$, then we have

$$E_l^2 + E_r^2 = a^2. \tag{3.41}$$

This equation describes a circle, and we call the wave in this case *circularly polarized*. The polarization is called right-handed when $\sin\delta > 0$, whereas it is called left-handed when $\sin\delta < 0$. Right-handed and left-handed refer to the direction of rotation (direction of fingers) when the thumb is pointed in the direction of propagation. Geometrical representations for linear and circular polarization are illustrated in Fig. 3.9b.

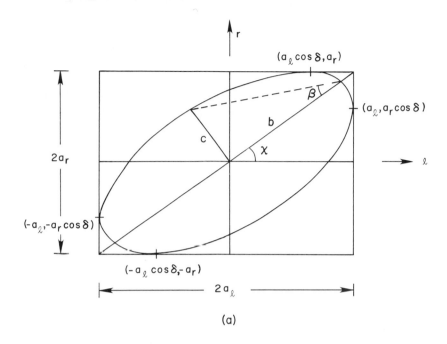

(a)

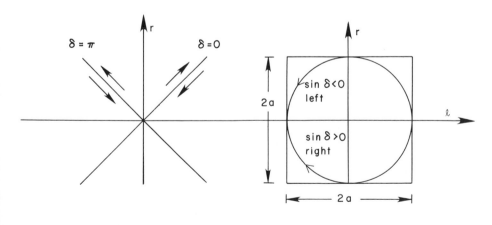

(b)

Fig. 3.9   (a) Geometrical representation of elliptical polarization. (b) Geometrical representations of linear and circular polarization.

To describe the elliptically polarized wave given in Eq. (3.38), three independent parameters a_l, a_r, and δ are needed. However, it is more convenient to use parameters of the same dimension. This can be achieved by a set of four quantities called the *Stokes parameters* first introduced by Stokes in 1852. Since the intensity is proportional to the absolute square of the electric field, we may define, upon neglecting a constant of proportionality, the four parameters

$$
\begin{aligned}
I &= E_l E_l^* + E_r E_r^*, \\
Q &= E_l E_l^* - E_r E_r^*, \\
U &= E_l E_r^* + E_r E_l^*, \\
V &= -i(E_l E_r^* - E_r E_l^*),
\end{aligned}
\tag{3.42}
$$

where an asterisk denotes the complex conjugate value and $i = \sqrt{-1}$. I, Q, U, and V, respectively, give the intensity, the degree of polarization, the plane of polarization, and the ellipticity of the electromagnetic wave at each point and in any given direction. They are real quantities that satisfy

$$
I^2 = Q^2 + U^2 + V^2.
\tag{3.43}
$$

Upon substituting Eq. (3.34) into Eq. (3.42), we have

$$
\begin{aligned}
I &= a_l^2 + a_r^2, \\
Q &= a_l^2 - a_r^2, \\
U &= 2a_l a_r \cos\delta, \\
V &= 2a_l a_r \sin\delta.
\end{aligned}
\tag{3.44}
$$

It is possible to describe the ellipse in Fig. 3.9 in terms of the length of the major (b) and minor (c) axes, and the orientation angle χ, which is the angle between the direction of the major axis, and the l direction. The ellipticity of the ellipse then can be expressed by $\tan\beta = \pm c/b$ with the plus sign for right-handed polarization and the minus sign for left-handed polarization. The four Stokes parameters may be derived in terms of I, χ, and β by direct, but lengthy analyses as

$$
\begin{aligned}
I &= I_l + I_r, \\
Q &= I_l - I_r = I \cos 2\beta \cos 2\chi, \\
U &= I \cos 2\beta \sin 2\chi, \\
V &= I \sin 2\beta.
\end{aligned}
\tag{3.45}
$$

It is seen that I and V are independent of the orientation angle χ. Equation (3.45) may be represented in Cartesian coordinates on a sphere called the *Poincaré sphere* shown in Fig. 3.10. The radius of the sphere is given by I,

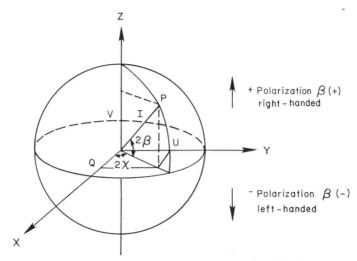

Fig. 3.10   Polarization representation on a Poincaré sphere.

and the zenithal and azimuthal angles are given by $\pi/2 - 2\beta$ and 2χ, respectively. Thus, Q, U, and V denote the lengths in X, Y, and Z directions, respectively. On this sphere, the northern and southern hemispheres represent right-handed and left-handed elliptic polarizations, respectively. The north and south poles denote right-handed and left-handed circular polarizations, respectively, and points on the equatorial plane represent linear polarization.

3.6.2   Representation for a Light Beam

In representing the wave vibration using Eq. (3.34) we have assumed a constant amplitude and phase. However, the actual light consists of many simple waves in very rapid succession. Within a very short duration (on the order of say, one second) more than millions of simple waves are collected by a detector. Consequently, measurable intensities are associated with the superposition of many millions of simple waves with independent phases. Let the operator $\langle \ \rangle$ denote the time average for a time interval (t_1, t_2), then the Stokes parameters of the entire beam of light for this time interval may be expressed by

$$\begin{aligned}
I &= \langle a_l^2 \rangle + \langle a_r^2 \rangle = I_l + I_r, \\
Q &= \langle a_l^2 \rangle - \langle a_r^2 \rangle = I_l - I_r, \\
U &= \langle 2a_l a_r \cos \delta \rangle, \\
V &= \langle 2a_l a_r \sin \delta \rangle.
\end{aligned} \tag{3.46}$$

Based on Eq. (3.46), it is straightforward to prove that (see Exercise 3.5)

$$I^2 \geq Q^2 + U^2 + V^2. \tag{3.47}$$

The degree of polarization of a stream of light can now be defined as

$$P = (Q^2 + U^2 + V^2)^{1/2}/I. \tag{3.48}$$

From the measurement point of view, it is desirable and convenient to represent the Stokes parameters in terms of detectable variables. Referring to Fig. 3.11, we introduce a retardation ε in the r direction with respect to the l direction, and consider the component of the electric field vector in the direction making an angle ψ with the positive l direction. Thus, for a simple wave at time t, we have the representation for the electric field in the form

$$\begin{aligned} E(t; \psi, \varepsilon) &= E_l \cos \psi + E_r e^{-i\varepsilon} \sin \psi \\ &= a_l \cos \psi \, e^{-i\zeta} + a_r e^{-i(\delta + \varepsilon) - i\zeta} \sin \psi \end{aligned} \tag{3.49}$$

The average intensity measured at a time interval (t_1, t_2) is then given by

$$\begin{aligned} I(\psi, \varepsilon) &= \langle E(t; \psi, \varepsilon) E^*(t; \psi, \varepsilon) \rangle \\ &= \langle a_l^2 \rangle \cos^2 \psi + \langle a_r^2 \rangle \sin^2 \psi + \tfrac{1}{2} \langle 2 a_l a_r \cos \delta \rangle \\ &\quad \times \sin 2\psi \cos \varepsilon - \tfrac{1}{2} \langle 2 a_l a_r \sin \delta \rangle \sin 2\psi \sin \varepsilon. \end{aligned} \tag{3.50}$$

Upon making use of Eq. (3.46) and noting that $I_l \cos^2 \psi + I_r \sin^2 \psi = (I + Q \cos 2\psi)/2$, we obtain

$$I(\psi, \varepsilon) = \tfrac{1}{2}[I + Q \cos 2\psi + (U \cos \varepsilon - V \sin \varepsilon) \sin 2\psi]. \tag{3.51}$$

By virtue of Eq. (3.51), we find that the Stokes parameters may be expressed

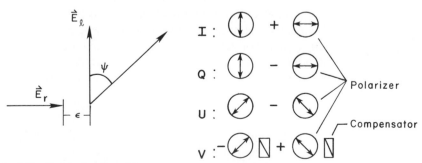

Fig. 3.11   Representation of the electric field in terms of the retardation ε and the polarization angle ψ.

by the retardation and polarization angles as

$$I = I(0°, 0) + I(90°, 0),$$
$$Q = I(0°, 0) - I(90°, 0),$$
$$U = I(45°, 0) - I(135°, 0)$$
$$V = -[I(45°, \pi/2) - I(135°, \pi/2)].$$
(3.52)

Thus, the Stokes parameters of a light beam can be measured by a combination of a number of polarizers and a compensator (e.g., a quarter-wave plate) as illustrated in Fig. 3.11.

On the basis of Eq. (3.51), *natural light* may be defined. Natural light is the light whose intensity remains unchanged and is unaffected by the retardation of one of the orthogonal components relative to the other when resolved in any direction in the transverse plane. That is to say, for natural light we must require $I(\psi, \varepsilon) = I/2$. The intensity is then independent of ψ and ε. Thus, the necessary and sufficient condition that light be natural is $Q = U = V = 0$. Under this condition, the percentage of the degree of polarization defined in Eq. (3.48) for natural light is zero. As a consequence, natural light is also referred to as *unpolarized light*; light emitted from the sun is unpolarized. However, the unpolarized sunlight after interacting with molecules and particles through scattering events generally becomes partially polarized. This will be discussed in the next section. Natural light characterized by $Q = U = V = 0$ can be shown to be equivalent to a mixture of any two independent oppositely polarized streams of half the intensity.

In the atmosphere, light is generally partially polarized and its Stokes parameters (I, Q, U, V) may be decomposed into two independent groups characterized by natural light and elliptically polarized light as

$$\begin{bmatrix} I \\ Q \\ U \\ V \end{bmatrix} = \begin{bmatrix} I - (Q^2 + U^2 + V^2)^{1/2} \\ 0 \\ 0 \\ 0 \end{bmatrix} + \begin{bmatrix} (Q^2 + U^2 + V^2)^{1/2} \\ Q \\ U \\ V \end{bmatrix}. \quad (3.53)$$

Moreover, from Eq. (3.45), the plane of polarization can be determined by $\tan 2\chi = U/Q$, and the ellipticity by $\sin 2\beta = V/(Q^2 + U^2 + V^2)^{1/2}$.

3.7 RAYLEIGH SCATTERING

The simplest and in some ways the most important example of a physical law of light scattering with various applications is that discovered by Rayleigh in 1871. His findings led to the explanation of the blue of the sky. In this section we formulate the scattering of unpolarized sunlight by air molecules and describe its important application to the atmosphere.

3.7.1 Theoretical Development

Consider a small homogeneous, isotropic spherical particle whose radius is much smaller than the wavelength of the incident radiation. The incident radiation produces a homogeneous electric field $\mathbf{E}_0$, called the applied field. Since the particle is very small the applied field generates a dipole configuration on it. The electric field of the particle, caused by the electric dipole, modifies the applied field inside and near the particle. Let $\mathbf{E}$ be the combined field, i.e., the applied field plus the particle's own field. Further, let $\mathbf{p}_0$ be the induced dipole moment, then we apply the electrostatic formula to give

$$\mathbf{p}_0 = \alpha \mathbf{E}_0. \tag{3.54}$$

This equation defines the polarizability α of a small particle. The dimensions of $\mathbf{E}_0$ and $\mathbf{p}_0$ are in units of charge per area and charge times length, respectively, and α has the dimension of volume.

The applied field $\mathbf{E}_0$ generates oscillation of an electric dipole in a fixed direction. The oscillating dipole, in turn, produces a plane polarized electromagnetic wave, the scattered wave. To evaluate the scattered electric field in regions which are far away from the dipole, we let r denote the distance between the dipole and the observational point, γ the angle between the scattered dipole moment $\mathbf{p}$ and the direction of observation, and c the velocity of light. According to the classical electromagnetic solution given by Hertz in 1889, the scattered electric field is proportional to the acceleration of the scattered dipole moment and $\sin\gamma$, but is inversely proportional to the distance r. In Gaussian units (cgs), the electric field in the far field is given by

$$\mathbf{E} = \frac{1}{c^2}\frac{1}{r}\frac{\partial^2 \mathbf{p}}{\partial t^2}\sin\gamma. \tag{3.55}$$

In an oscillating periodic field, the scattered dipole moment may be written in terms of the induced dipole moment as

$$\mathbf{p} = \mathbf{p}_0 e^{-ik(r-ct)}. \tag{3.56}$$

Note that k is the wave number, and $kc = \omega$ is the circular frequency. By combining Eqs. (3.54) and (3.56), Eq. (3.55) yields

$$\mathbf{E} = -\mathbf{E}_0 \frac{e^{-ik(r-ct)}}{r} k^2 \alpha \sin\gamma. \tag{3.57}$$

Now we consider the scattering of unpolarized sunlight by air molecules. Let the plane defined by the directions of incident and scattered waves be the reference plane (plane of scattering). Since any electric vector may be arbitrarily decomposed into orthogonal components, we may choose these two components perpendicular (E_r) and parallel (E_l) to the plane of scattering. From the previous section, we note that the unpolarized sunlight is characterized by the same electric field in **r** and **l** directions and by a random phase relation between these two components. Thus, we may consider separately the scattering of the two electric field components E_{0r} and E_{0l} by molecules assumed to be homogeneous, isotropic spherical particles. And according to Eq. (3.57), we have

$$E_r = -E_{0r} \frac{e^{-ik(r-ct)}}{r} k^2 \alpha \sin \gamma_1, \tag{3.58a}$$

$$E_l = -E_{0l} \frac{e^{-ik(r-ct)}}{r} k^2 \alpha \sin \gamma_2. \tag{3.58b}$$

Referring to Fig. 3.12, we see that $\gamma_1 = \pi/2$ and $\gamma_2 = \pi/2 - \Theta$, where Θ is defined as the scattering angle, which is an angle between the incident and scattered waves. Note that γ_1 is always equal to $90°$ because the scattered dipole moment (or the scattered electric field) in the **r** direction is normal to the scattering plane defined previously.

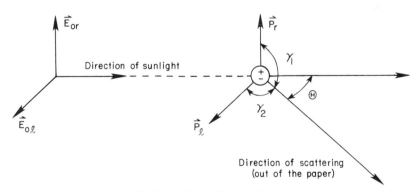

Fig. 3.12   Scattering by a dipole.

The corresponding intensities (per solid angle $\Delta\Omega$) of the incident and scattered radiation in Gaussian units may be written as

$$I_0 = \frac{1}{\Delta\Omega} \frac{c}{4\pi} |E_0|^2, \qquad I = \frac{1}{\Delta\Omega} \frac{c}{4\pi} |E|^2. \tag{3.59}$$

Thus, Eqs. (3.58) and (3.59) can be expressed in the form of intensities as

$$I_r = I_{0r} k^4 \alpha^2 / r^2, \tag{3.60a}$$

$$I_l = I_{0l} k^4 \alpha^2 \cos^2 \Theta / r^2, \tag{3.60b}$$

where I_r and I_l are polarized intensity components perpendicular and parallel to the plane containing the incident and scattered waves, i.e., the plane of scattering. The total scattered intensity of the unpolarized sunlight incident on a molecule in the direction of Θ is then

$$I = I_r + I_l = (I_{0r} + I_{0l} \cos^2 \Theta) k^4 \alpha^2 / r^2. \tag{3.61}$$

But for unpolarized sunlight, $I_{0r} = I_{0l} = I_0/2$, and by noting that $k = 2\pi/\lambda$, we get

$$I = \frac{I_0}{r^2} \alpha^2 \left(\frac{2\pi}{\lambda}\right)^4 \frac{1 + \cos^2 \Theta}{2}. \tag{3.62}$$

This is the original formula derived by Rayleigh, and we call the scattering of sunlight by molecules *Rayleigh scattering*. By this formula, the intensity scattered by a molecule for unpolarized sunlight is proportional to the incident intensity I_0 and is inversely proportional to the square of the distance between the molecule and the point of observation. In addition to these two factors, it also depends on the polarizability, the wavelength of the incident wave, and the scattering angle. The dependence of these three parameters on the scattering of sunlight by molecules introduces a number of significant physical features.

3.7.2   Phase Function, Scattering Cross Section, and Polarizability

On the basis of Eqs. (3.60) and (3.62), the intensity scattered by a molecule depends on the polarization characteristics of the incident light. For vertically (r) polarized incident light, the scattered intensity is independent of the direction in the scattering plane. In this case then, the scattering is isotropic. On the other hand, for horizontally (l) polarized incident light, the scattered intensity is a function of $\cos^2 \Theta$. When the incident light is unpolarized, such as sunlight, the scattered intensity depends on $(1 + \cos^2 \Theta)$. The angular scattering patterns for the three types of incident polarization are illustrated in Fig. 3.13. From this diagram, we see that scattering of unpolarized sunlight by molecules has maxima in the forward $(0°)$ and backward $(180°)$ directions, whereas it shows minima in the side directions $(90°$ and $270°)$. Moreover, it should be pointed out that light scattered by particles or molecules is not limited in the plane of incidence, but is in all directions. Because of the spherical symmetry assumed for molecules, scattering patterns in planes other than the plane of incidence are the same as that depicted in Fig. 3.13. Thus, the three-dimensional scattering pattern for the unpolarized incident light resembles the shape of a doughnut.

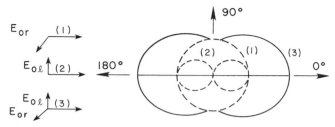

Fig. 3.13 Polar diagram of the scattered intensity for Rayleigh molecules; (1) polarized incident light with electric vector perpendicular to the plane of drawing, (2) polarized incident light with electric vector in the plane of drawing, and (3) unpolarized incident light.

To describe the angular distribution of the scattered energy in conjunction with multiple scattering and radiative transfer analyses and applications for planetary atmospheres, we find it necessary and sufficient to define a non-dimensional parameter called the *phase function* (or sometimes referred to as indicatrix) $P(\cos \Theta)$, such that

$$\int_0^{2\pi} \int_0^{\pi} \frac{P(\cos \Theta)}{4\pi} \sin \Theta \, d\Theta \, d\phi = 1. \qquad (3.63)$$

By this definition, the phase function is said to be normalized to unity. Upon performing simple integrations, the phase function of Rayleigh scattering for incident unpolarized sunlight is given by

$$P(\cos \Theta) = \tfrac{3}{4}(1 + \cos^2 \Theta). \qquad (3.64)$$

Employing the definition of the phase function, Eq. (3.62) may be rewritten in the form

$$I(\Theta) = \frac{I_0}{r^2} \alpha^2 \frac{128\pi^5}{3\lambda^4} \frac{P(\Theta)}{4\pi}. \qquad (3.65a)$$

It follows that the angular distribution of the scattered intensity is directly associated with the phase function.

The scattered flux (or power, in units of energy per time) f can be evaluated by integrating the scattered flux density ($I \, \Delta\Omega$) over the appropriate area a distance r away from the scatterer. Thus,

$$f = \int_{\Omega} (I \, \Delta\Omega) r^2 \, d\Omega \qquad (3.66)$$

where $r^2 \, d\Omega$ represents the area according to the definition of the solid angle. Inserting the expression for the scattered intensity and differential solid angle defined in Eqs. (3.65) and (1.5), respectively, into Eq. (3.66) and carrying out integrations over the solid angle of a sphere, we obtain the equivalent isotropically scattered flux in the form

$$f = F_0\alpha^2 128\pi^5/(3\lambda^4),\tag{3.67}$$

where the incident flux density F_0 is equal to $I_0\,\Delta\Omega$. At this point we may define the scattering cross section σ_s per one molecule as

$$\sigma_s = f/F_0 = \alpha^2 128\pi^5/(3\lambda^4).\tag{3.68}$$

The scattering cross section (in units of area) represents the amount of incident energy which is removed from the original direction due to a single scattering event such that the energy is redistributed isotropically on the area of a sphere whose center is the scatterer and whose radius is r.

In terms of the scattering cross section, the scattered intensity may be expressed by

$$I(\Theta) = I_0 \frac{\sigma_s}{r^2} \frac{P(\Theta)}{4\pi}.\tag{3.65b}$$

This is the general expression for the scattered intensity, which is valid not only for molecules but also for particles whose sizes are larger than the incident wavelength to be discussed in Chapter 5.

The *polarizability* α, which occurred in the preceding equations, can be derived from the principle of the dispersion of electromagnetic waves, and it is given by

$$\alpha = \frac{3}{4\pi N_s}\left(\frac{m^2 - 1}{m^2 + 2}\right),\tag{3.69}$$

where N_s is the total number of molecules per unit volume, and m is the nondimensional refractive index of molecules. This equation is called Lorentz–Lorenz formula and its derivation is given in Appendix D. The refractive index is an optical parameter associated with the velocity change of electromagnetic waves in a medium with respect to vacuum. Its definition and physical meanings also are given in Appendix D. Normally, the refractive indices of atmospheric particles and molecules are composed of a real part m_r and an imaginary part m_i corresponding, respectively, to the scattering and absorption properties of particles and molecules. In the solar visible spectrum, the imaginary parts of the refractive indices for air molecules are insignificantly small so that absorption of solar radiation by air molecules may be neglected in the scattering discussion. The real parts of the refractive indices for air molecules in the solar spectrum are very close to 1, but they significantly depend on the wavelength (or frequency) of the incident radiation as illustrated in Appendix D. Because of this dependence, white light may be *dispersed* by molecules, which function like prisms, into component colors. The real part of the refractive index derived in Appendix D (Eq. (D.17)) may be approximately fitted by

$$(m - 1) \times 10^8 = 6432.8 + \frac{2{,}949{,}810}{146 - \lambda^{-2}} + \frac{25{,}540}{41 - \lambda^{-2}}, \tag{3.70}$$

where λ is in units of μm. Since m_r is close to 1, for all practical purposes, Eq. (3.69) may be approximated by

$$\alpha \approx \frac{1}{4\pi N_s} (m_r^2 - 1). \tag{3.71}$$

Thus, the *scattering cross section* defined in Eq. (3.68) becomes

$$\sigma_s = \frac{8\pi^3 (m_r^2 - 1)^2}{3\lambda^4 N_s^2} f(\delta). \tag{3.72}$$

A correction factor $f(\delta)$ is added in Eq. (3.68) to take into consideration the anisotropic property of molecules, where $f(\delta) = (6 + 3\delta)/(6 - 7\delta)$ with the anisotropic factor δ of 0.035. By anisotropy, we mean that the refractive index of molecules varies along X, Y, and Z directions, and thus is a vector, not a scalar. Hence, the polarizability α is a tensor.

The optical depth [see Eq. (1.62)] of the entire molecular atomosphere at a given wavelength may be calculated from the scattering cross section in the form

$$\tau(\lambda) = \sigma_s(\lambda) \int_0^{z_t} N(Z) \, dz, \tag{3.73}$$

where $N(Z)$ denotes the number density of molecules as a function of height, and z_t is the top of the atmosphere. The optical depth is a physical parameter indicating the attenuation power of molecules with respect to a specific wavelength of the incident light.

3.7.3   Blue Sky and Sky Polarization

Returning to Eq. (3.65a), we see that the scattered intensity depends on the wavelength of incident light, and the index of refraction of air molecules contained in the polarizability term. According to the analyses given in Appendix D and Eq. (3.70), the index of refraction also depends slightly on the wavelength. However, the dependence of the refractive index on the wavelength is relatively insignificant in calculating the scattered intensity as compared with the explicit wavelength term. Thus, the intensity scattered by air molecules in a specific direction may be symbolically expressed in the form

$$I_\lambda \sim 1/\lambda^4. \tag{3.74}$$

The inverse dependence of the scattered intensity on the wavelength to the fourth power is a direct consequence of the theory of Rayleigh scattering, and it is the foundation for the explanation of the blue of the sky.

In accord with the observed solar energy spectrum depicted in Fig. 2.6, it is seen that a large portion of the solar energy is contained in the visible spectrum from blue to red regions. Blue light ($\lambda \approx 0.425$ μm) has a shorter wavelength than red light ($\lambda \approx 0.650$ μm). Consequently, according to Eq. (3.74) blue light scatters about 5.5 more than red light. It is apparent that the λ^{-4} law causes more of the blue light to be scattered than the red, the green, and the yellow, and so the sky, when viewed away from the sun's disk, appears blue. Moreover, since the molecular density decreases drastically with height, it is anticipated that the sky should gradually darken to become completely black in outer space in directions away from the sun. And the sun itself should appear whiter and brighter with increasing height. As the sun approaches the horizon (at sunset or sunrise), sunlight travels through more air molecules, and therefore more and more blue light and light with shorter wavelengths are scattered out of the beam of light, and the luminous sun shows a deeper red color than at the zenith. Since the violet light (~ 0.405 μm) has a shorter wavelength than the blue, why then doesn't the sky appear violet? This is because the energy contained in the violet spectrum is much smaller than that contained in the blue spectrum, and also because the human eye has a much lower response to the violet color.

Larger particles in the atmosphere such as aerosols, cloud droplets, and ice crystals also scatter sunlight and produce many fascinating optical phenomena. However, their single scattering properties are less wavelength-selective and depend largely upon the particle size. As a result of this, clouds in the atmosphere generally appear white instead of blue. In a cloudy atmosphere, the sky appears blue diluted with white scattered light, resulting in a less pure blue sky than would have been expected from pure Rayleigh scattering. Scattering by a spherical particle of arbitrary size has been treated exactly by Mie in 1908 by means of solving the electromagnetic wave equation derived from the fundamental Maxwell equations. To distinguish it from Rayleigh scattering, we call scattering by large particles *Mie scattering*, which will be discussed comprehensively in Chapter 5.

Another important phenomenon resulting from Rayleigh scattering theory is the sky polarization. For many atmospheric sensing applications utilizing polarization, a parameter called the *degree of linear polarization* is commonly used, and it is defined by neglecting U and V components in

Eq. (3.48) in the form

$$LP = -Q/I. \tag{3.75a}$$

Thus, from Eqs. (3.60a) and (3.60b), the degree of linear polarization in the case of Rayleigh scattering simply is given by

$$LP(\Theta) = -\frac{I_l - I_r}{I_l + I_r} = -\frac{\cos^2 \Theta - 1}{\cos^2 \Theta + 1} = \frac{\sin^2 \Theta}{\cos^2 \Theta + 1}. \tag{3.75b}$$

In Fig. 3.14, we plot the angular distribution of the degree of linear polarization generated by molecules for unpolarized light. The polarization pattern reveals that in the forward and backward directions the scattered light remains completely unpolarized, whereas at 90° scattering angle the scattered light becomes completely polarized. In other directions, the scattered light is partially polarized with the percentage of polarization ranging from 0 to 100%.

The theory of Rayleigh scattering developed in Section 3.7.1 is based on the assumption that molecules are homogeneous and isotropic spheres.

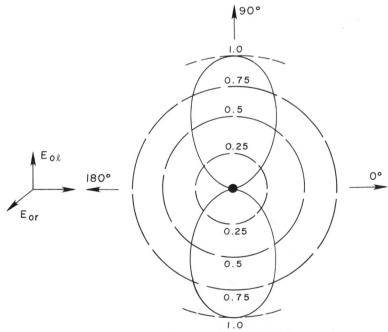

Fig. 3.14 Angular distribution of the degree of linear polarization for a Rayleigh particle in the case of unpolarized incident light. The pattern has axial symmetry around the direction of propagation of the incident light.

However, molecules are in general anisotropic in which polarizability defined in Eq. (3.69) varies in three axes, and hence, is a tensor instead of a scalar. The anisotropic effect of molecules reduces the degree of linear polarization defined in Eq. (3.75a) by only a small percentage. At 90° scattering angle, the degree of linear polarization for dry air is about 0.94. Further, the theory of Rayleigh scattering developed previously considers only single (or primary) scattering, i.e., scattering occurs only once. But in the earth's atmosphere, which contains a large number of molecules and aerosol particles, light may undergo infinite numbers of scattering events. In addition, the earth's surface also reflects light that reaches it. Multiple scattering processes involving the atmosphere and the surface become very complicated and require more advanced treatments based on the radiative transfer theory, which will be discussed in Chapter 6.

The theory of Rayleigh scattering predicts *neutral points*, i.e., points of zero polarization, only at the exact forward and backward directions. However, owing to multiple scattering of molecules and particulates, and reflection of the surface, there normally exists a number of neutral points in cloudless atmospheres. Observations of neutral points and partially polarized sky light go back to 1809 by Arago. He discovered the existence of a neutral point at a position in the sky at about 25° above the antisolar direction (direction exactly opposite to that of the sun). The other two neutral points, which normally occur in the sunlit sky 25° above and 20° below the sun, were discovered by Babinet in 1840 and by Brewster in 1842, respectively. These three neutral points were named to honor these three discoverers, and their relative positions are shown in Fig. 3.15. The neutral points in the sky vary by about 5° or so, depending on the turbidity (an indication of the amount of aerosol loadings in the atmosphere), the sun's elevation angle, and the

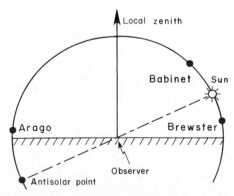

Fig. 3.15 The position of neutral points in the meridian of the sun. Note that the Arago and Brewster points generally are not above the horizons at the same time.

reflection characteristics of the surface at which observations are made. Since the latter two parameters can easily be measured in a planned experiment, variations in neutral points may give an indication of the turbidity of the atmosphere. Use of the polarization technique for the inference of cloud and aerosol properties will be further discussed in Section 7.2.2.2.

EXERCISES

3.1   Under the photoequilibrium condition, derive expressions for the concentrations of ozone and atomic oxygen.

3.2   The principal photochemical reactions involving oxygen in the thermosphere are found to be

$$O_2 + h\tilde{v}\ (\lambda < 1751\ \text{Å}) \xrightarrow{J_2} O + O,$$
$$O + O + M \xrightarrow{K_{11}} O_2 + M,$$
$$O + O \xrightarrow{K'_{11}} O_2 + h\tilde{v}.$$

Express these photochemical processes in terms of the rate of change of the number density of O and O_2. Assuming that the number density of O_2 is constant, derive the number density of O under the photochemical equilibrium condition.

3.3*   Given the following vertical profile of the specific humidity and downward flux densities for various water vapor bands at 600 mb, and assuming that the scattering effect may be neglected, compute the solar heating rate due to water vapor when the sun is overhead.

Pressure (mb)	Specific humidity (%)	Spectral bands (μm)	Fractional solar flux densities
1000	0.82	0.94	0.1346
950	0.49	1.10	0.0892
900	0.43	1.38	0.1021
850	0.42	1.87	0.0622
800	0.41	2.7	0.0300
750	0.30	3.2	0.0218
700	0.20		
650	0.09		
600	0.04		

3.4   What would be the Stokes parameters for unpolarized, linearly polarized, and circularly polarized light?

* Simple computer programming is required.

3.5   Any time-average quantity may be represented by the summation of individual components, e.g., $\langle x \rangle = \sum_{n=1}^{N} t_n x_n$. Utilizing this principle, show that Eq. (3.47) is true based on the relationships given in Eq. (3.46). In doing this exercise, let $N = 2$ for simplicity.

3.6   (a)   By rotating the electric field vector through an angle χ, i.e.,

$$\begin{bmatrix} E'_l \\ E'_r \end{bmatrix} = \begin{bmatrix} \cos \chi & \sin \chi \\ -\sin \chi & \cos \chi \end{bmatrix} \begin{bmatrix} E_l \\ E_r \end{bmatrix}.$$

show that the Stokes parameters in the prime system are given by

$$\begin{bmatrix} I' \\ Q' \\ U' \\ V' \end{bmatrix} = \begin{bmatrix} 1 & 0 & 0 & 0 \\ 0 & \cos 2\chi & \sin 2\chi & 0 \\ 0 & -\sin 2\chi & \cos 2\chi & 0 \\ 0 & 0 & 0 & 1 \end{bmatrix} \begin{bmatrix} I \\ Q \\ U \\ V \end{bmatrix}.$$

(b)   Show that elliptically polarized light can be decomposed into a circularly polarized part and a linearly polarized part. Then rotate the linearly polarized beam through the angle χ and show that χ which makes the intensity maximum (or minimum) in the direction l' is given by $\tan 2\chi = U/Q$.

(c)   Assuming a light beam with 50% linear polarization in the r direction and another independent light beam also with 50% right-handed circular polarization, (1) what would be the Stokes parameters for the mixture and the resulting total intensity and percentage polarization? (2) what would be the measured intensity if a polarizer having a plane of polarization along the r direction is used? and (3) sketch a diagram to denote the resultant polarization.

(d)   With reference to (c), decompose the partially polarized light beam into natural light and 100% elliptically polarized light and compute the plane of polarization χ and ellipticity angle β for the polarized component.

(e)   The natural light is equivalent to any two independent oppositely polarized beams of half the intensity. By virtue of this principle, evaluate the Stokes parameters for these two polarized beams based on results obtained from (d).

(f)   Upon combining the polarized beams derived from (d) and (e), what would be the Stokes parameters corresponding to two independent polarized beams?

3.7   The number of molecules per cubic centimeter of air at sea level in standard atmospheric conditions is about 2.55×10^{19} cm^{-3}. Calculate the scattering cross section of molecules at 0.3, 0.5, and 0.7 μm wavelengths.

3.8   The number density profile as a function of height is given by the following table:

Height(Km):	0	2	4	6	8	10	12	14	16
$N(\times 10^{18} \text{ cm}^{-3})$:	25.5	20.9	17.0	13.7	10.9	8.60	6.49	4.74	3.46

Calculate the optical depth of a clear atmosphere at wavelengths shown in Exercise 3.7.

3.9   For all practical purposes, we find that the refractive index m_r and the molecular density ρ are related by

$$(m_r - 1)_{gas} = \text{const} \times \rho.$$

At sea level the refractive index of air is about 1.000292 for a wavelength of 0.3 μm. Find the refractive indices at heights given in Exercise 3.8. Note that the density (g cm^{-3}) is related to the number density N (cm^{-3}) by $\rho = (M/N_0)N$, where M is the molecular weight of air (28.97 g mole^{-1}), and N_0 is the Avogadro's number (6.02295 $\times$ 10^{23} mole^{-1}). Because the refractive index varies with the density of the atmosphere, light rays bend according to the atmospheric density profile and produce a number of atmospheric optical phenomena such as looming, sinking, and superior and inferior mirages.

3.10   An unpolarized ruby laser operated at 0.7 μm is projected vertically into a clear sky to investigate the density of the atmosphere. A detector located 10 km from the base of the laser is used to receive the flux density scattered from the laser beam by air molecules. Assuming that the laser output has a uniform distribution of flux density F_0 across the beam (i.e., $I_0 = F_0/1$ sr) and if effects of multiple scattering may be neglected, find the scattered flux density at 6 and 10 km received by a detector whose field of view in a plane is 0.05 rad. Use the scattering cross section and molecular density profile obtained from Exercises 3.7 and 3.8.

3.11   (a)   The radar backscattering coefficient (in units of per length) for a volume of identical cloud droplets is defined as

$$\beta_\pi = N_c \sigma_\pi = N_c \sigma_s P(\pi),$$

where N_c is the droplet number density, σ_π the backscattering cross section, and $P(\pi)$ the phase function at backscatter. Employing the expressions for the scattering cross section and phase function, and noting that $N_c = 1/V$, where the volume of a spherical drop with a radius a is $V = \frac{4}{3}\pi a^3$, show that

$$\beta_\pi = \frac{64\pi^5}{\lambda^4} N_c a^6 \left| \frac{m^2 - 1}{m^2 + 2} \right|^2.$$

The dependence of the backscattering coefficient on the sixth power of the droplet radius is a significant consequence for the study of cloud and precipitation echoes by means of a radar reflectivity technique.

  (b) Assuming that the number density and the radius of cloud droplets are 100 cm^{-3} and 20 μm, respectively, calculate β_π for the following two radar wavelengths with the corresponding refractive indices for water:

$$\lambda(\text{cm}): \quad 10 \quad\quad 3.21$$
$$m: \quad 3.99-1.47i \quad 7.14-2.89i$$

Compute β_π again using only the real part of the refractive indices, and show the differences between two computations.

SUGGESTED REFERENCES

Chandrasekhar, S. (1950). *Radiative Transfer*. Dover, New York. Section 15 in Chapter 1 contains more advance treatments for the representation of polarized light.

McCartney, E. J. (1976). *Optics of the Atmosphere*. Wiley, New York. Chapter 4 gives a fairly clear description of Rayleigh scattering by molecules.

McEwan, M. J., and Phillips, L. F. (1975). *Chemistry of the Atmosphere*. Wiley, New York. Chapters 4 and 7 contain a thorough and up-to-date account of photochemical processes involving ozone.

Robinson, N., Ed. (1966). *Solar Radiation*. Elsevier, New York. Chapter 3 provides some discussions on the absorption and scattering of solar radiation in the atmosphere.

Chapter 4
INFRARED RADIATION TRANSFER IN THE ATMOSPHERE

4.1 THE THERMAL INFRARED SPECTRUM AND ATMOSPHERIC EFFECT

The earth-atmosphere reflects about 31% of the incoming solar radiation at the top of the atmosphere and absorbs the remaining part. Absorption and scattering of solar radiation take place in the atmosphere and these processes have been discussed in the last chapter. A large portion of the incoming solar radiation is absorbed by the earth's surface, which is approximately 70% ocean and 30% land. Over a climatological period of time, say, over a year or longer, there is apparently no significant change in global temperatures of the earth. Consequently, radiant energy emitted from the sun that is absorbed in the earth–atmosphere system has to be reemitted to space so that an equilibrium energy state can be maintained.

Just as the sun emits electromagnetic radiation covering all frequencies, so does the earth. However, the global mean temperature of the earth–atmosphere system is only about 250°K. This temperature is obviously much lower than that of the sun's photosphere. As a consequence, we find from Planck's law and Wien's displacement law discussed in Chapter 1 that the intensity of the Planck function is less and the wavelength for the intensity peak of the earth's radiation field is longer. We call the energy emitted from the earth–atmosphere system *thermal infrared* (or *terrestrial*) radiation. We plot the spectral distribution of radiance emitted by a blackbody source at various temperatures in the terrestrial range in terms of wave number, customarily employed in the studies of infrared radiative transfer. It is depicted in Fig. 4.1. In this figure, a measured atmospheric emission spectrum obtained from the Infrared Interferometer Spectrometer (IRIS) instrument on board the Nimbus IV satellite is also shown (after Kunde *et al.*, 1974). The envelope of the emission spectrum is very close to the spectrum emitted from a blackbody with a temperature of about 290°K, which is about the temperature of the surface. Clearly, certain portions of the infrared radiation are trapped by various gases in the atmosphere.

Among these gases, carbon dioxide, water vapor, and ozone are the most important absorbers. Some minor constituents, such as carbon monoxide, nitrous oxide, methane, and nitric oxide, which are not shown in Fig. 4.1, are relatively insignificant absorbers insofar as the heat budget of the earth–atmosphere is concerned. Carbon dioxide absorbs infrared radiation significantly in the 15 μm band from about 600 to 800 cm^{-1}. This spectral region also corresponds to the maximum intensity of the Planck function in the wave number domain. Water vapor absorbs thermal infrared in the 6.3 μm band from about 1200 to 2000 cm^{-1} and in the rotational band (< 500 cm^{-1}). Except for ozone, which has an absorption band in the 9.6 μm region, the atmosphere is relatively transparent from 800 to 1200 cm^{-1}. This region is referred to as the *atmospheric window*. In addition to the 15 μm band, carbon dioxide also has an absorption band in the shorter wavelength of the 4.3 μm region. The distribution of carbon dioxide is fairly uniform over the global space, although there has been observational evidence indicating a continuous global increase over the past century owing to the increase of the combustion of the fossil fuels. This leads to the question of the earth's climate and possible climatic changes due to the increasing carbon dioxide concentration. Unlike carbon dioxide, however, water vapor and ozone are highly variable both with respect to time and the geographical location. These variations are vital to the radiation budget of the earth–atmosphere system and to long-term climatic changes.

In a clear atmosphere without clouds and aerosols, a large portion (about 50%) of solar energy transmits through the atmosphere and is absorbed by the earth's surface (see Fig. 2.6). Energy emitted from the earth, on the contrary, is absorbed largely by carbon dioxide, water vapor, and ozone in the atmosphere as evident in Fig. 4.1. Trapping of thermal infrared radiation by atmospheric gases is typical of the atmosphere and is therefore called the *atmospheric effect*. The atmospheric effect is sometimes referred to as the *greenhouse effect* because in a similar way glass, which covers a greenhouse transmits short-wave solar radiation, but absorbs long-wave thermal infrared radiation. Fleagle and Businger (1963) pointed out that the high temperatures in a greenhouse are caused primarily by the glass cover which prevents the warm air from rising and removing heat from the

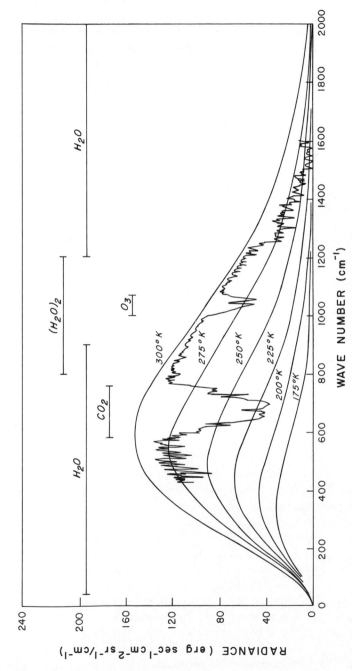

Fig. 4.1   The terrestrial infrared spectra and various absorption bands. Also shown is an acute atmospheric emission spectrum taken by the Nimbus IV IRIS instrument near Guam at 15.1°N and 215.3°W on April 27, 1970.

greenhouse and are not to be attributed to the absorption of thermal infrared radiation.

Solar radiation is also called *short-wave radiation* because solar energy is concentrated in shorter wavelengths with its peak at about 0.5 μm. Thermal infrared radiation from the earth's atmosphere is referred to as *long-wave radiation* because its maximum energy is in the longer wavelength at about 10 μm. The solar and infrared spectra are separated into two spectral ranges above and below about 4 μm, and the overlap between them is relatively insignificant. This distinction makes it possible to treat the two types of radiative transfer and source functions separately and thereby simplify the complexity of the transfer problem. In this chapter, after briefly discussing the general characteristics of absorption spectra of water vapor, carbon dioxide, and ozone, the fundamental theory of infrared radiation is introduced. Absorption-band models, concepts of the broadband emissivity and radiation charts, and computations of infrared cooling rates are further introduced. Lastly, the problem of the carbon dioxide and climate is presented.

4.2   GENERAL CHARACTERISTICS OF INFRARED ABSORPTION SPECTRA OF ATMOSPHERIC GASES

Inspection of high-resolution spectroscopic data reveals that there are thousands of absorption lines within each absorption band noted in the previous section. Figure 4.2 illustrates the fine structure of molecular absorption bands for the 320–380 cm^{-1} region where lines are due to water vapor, and for the 680–740 cm^{-1} region where lines are due to carbon dioxide. The optically active gases of the atmosphere, carbon dioxide, water vapor, and ozone are all triatomic molecules. The band spectroscopic properties of such molecules are discussed comprehensively by Herzberg (1945). Here we shall briefly describe their spectroscopic characteristics as they relate to our later development and discussion.

Spectroscopic evidence indicates that the three atoms of CO_2 form a symmetrical straight-line array having the carbon atom in the middle flanked by oxygen atoms on either side. The length of C–O bond in the fundamental vibration state is 1.1632 Å. Because of linear symmetry it cannot have a static electric dipole moment. Figure 4.3a shows the three normal modes of vibration of such a configuration. The symmetrical motion v_1 should not give rise to an electric dipole moment and therefore should not be optically active. The v_1 vibration mode has been identified in the Raman spectrum near 7.5 μm. In the v_2 vibration mode, the dipole moment is perpendicular to the axis of the molecule. The 15 μm band represents

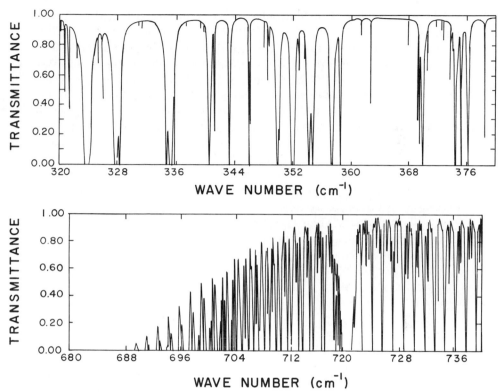

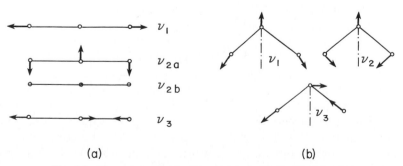

Fig. 4.2   Absorption spectrum of the water vapor rotational band and 15 μm carbon dioxide band at high resolution (after McClatchey and Selby, 1972). (See Section 4.4 for the definition of transmittance.)

Fig. 4.3   Normal modes of a linear (a) and triangle (b) molecules.

this particular vibration. This band is referred to as a *fundamental*, because it is caused by a transition from the ground state to the first excited vibrational state. Another fundamental corresponding to the v_3 vibration mode is the 4.3 μm band, which appears at the short-wave edge of the blackbody curve of atmospheric temperatures.

Because of its straight-line arrangement, the CO_2 molecule has spectroscopic properties close to those of a diatomic molecule. The 15 μm band in particular has all three branches of a typical band of a diatomic molecule, the P, Q, and R branches, corresponding to jumps of the rotational quantum number by -1, 0, and $+1$. The lines of the P and R branches are approximately equidistant. Thus, if the superposition of these two branches with v_2 fundamental could be neglected, the periodic Elsasser model to be discussed in Section 4.5.2 would be an excellent approximation to the transmittance calculations. The Q branch is quite closely clustered near the center of the 15 μm band. For this reason, it does not contribute importantly to radiative transfer.

The water molecule forms an isosceles triangle which is obtuse and has an apical angle of 104.5°. The distance between the oxygen and hydrogen atoms is 0.958 Å. Figure 4.3b shows the three normal modes of vibration for such a structure. The 6.3 μm band has been identified as the v_2 fundamental. The two fundamentals, v_1 and v_3, are found close together in a band near 2.7 μm, i.e., on the short-wave side of the infrared spectral region. This band and the bands representing higher harmonics discussed in Section 3.4 are of meteorological interest in that they give rise to absorption of sunlight by the atmosphere.

The band covering the region from 900 to 40 cm^{-1} shown in Fig. 4.1 represents the purely rotational spectrum of water vapor. The water molecule forms an asymmetrical top with respect to rotation, and the line structure of the spectrum does not have the simplicity of a symmetrical rotator such as found in the CO_2 molecule. Close inspection shows that the absorption lines have no clear-cut regularity. A typical fine structure of the rotational band was shown in Fig. 4.2. Because of the random irregularity, computations of the transmittances for such a spectrum may be modeled by the Goody statistical formula to be discussed in Section 4.5.3. The fine structure of the 6.3 μm band is essentially similar to that of the pure rotational band.

In the region between the two water vapor bands, i.e., between about 8 and 12 μm, the so-called atmospheric window, absorption is continuous and is primarily due to water vapor species. Absorption by carbon dioxide is typically a small part of the total in this region. The overlap of water vapor with ozone in this region is insignificant in the computations of cooling rates since water vapor is important mainly in the lower atmosphere, while cooling due to ozone takes place primarily in the stratosphere and higher. In recent years, there has been evidence that contribution to the continuous absorption in the infrared window regions is chiefly caused by the water dimer $(H_2O)_2$ in addition to the self-broadening contribution in the wings of the water vapor lines. The latter contribution is normally small in the atmosphere. The water dimer is produced by the reaction $2H_2O \rightleftarrows (H_2O)_2 + E$, where E is the binding energy of the dimer molecule. The absorption by the water dimer depends on the water vapor pressure and temperature.

The ozone molecule is of the triatomic nonlinear type (Fig. 4.3b) with a relatively strong rotation spectrum. The apical angle and the distance between atomic oxygen are $116.8°$ and 1.278 Å, respectively. The three fundamental vibrational bands v_1, v_2, and v_3 occur at wavelengths of 9.066, 14.27, and 9.597 μm, respectively. The very strong v_3 and moderately strong v_1 fundamentals combine to make the well-known 9.6 μm band of ozone. The v_2 fundamental is well masked by the 15 μm band of CO_2. The strong band at about 4.7 μm produced by the overtone and combination frequencies of O_3 vibrations is in a weak portion of the Planckian energy distribution for the atmosphere. Note that the absorption bands of O_3 in the UV part of the solar spectrum are due to electronic transitions in the ozone molecule.

4.3 THEORY OF INFRARED TRANSFER IN PLANE–PARALLEL ATMOSPHERES

Consider a nonscattering, plane–parallel atmosphere which is in local thermodynamic equilibrium and assume that thermal infrared radiation from the earth's atmosphere is independent of the azimuthal angle ϕ. The general equation of transfer derived in Eq. (1.63) may then be expressed in the wave number domain as

$$\mu \frac{dI_v(\tau, \mu)}{d\tau} = I_v(\tau, \mu) - B_v(T) \qquad \text{(upward)}, \qquad (4.1)$$

$$-\mu \frac{dI_v(\tau, -\mu)}{d\tau} = I_v(\tau, -\mu) - B_v(T) \qquad \text{(downward)}, \qquad (4.2)$$

where the source function is given by the Planck function. The solution for the upward and downward intensities as given in Eqs. (1.64) and (1.65) are

$$I_v(\tau, \mu) = I_v(\tau_1, \mu)e^{-(\tau_1 - \tau)/\mu} + \int_\tau^{\tau_1} B_v[T(\tau')]e^{-(\tau' - \tau)/\mu} \frac{d\tau'}{\mu}, \qquad (4.3)$$

$$I_v(\tau, -\mu) = I_v(0, -\mu)e^{-\tau/\mu} + \int_0^\tau B_v[T(\tau')]e^{-(\tau - \tau')/\mu} \frac{d\tau'}{\mu}, \qquad (4.4)$$

where the differential normal optical thickness from Eq. (1.62) is given by

$$d\tau = -k_\nu \rho \, dz. \tag{4.5}$$

At the bottom of the atmosphere ($\tau = \tau_1$), the upward radiation simply arises from the emission of earth's surface. To a good approximation, the earth's surface can be considered as a blackbody in the infrared region. Hence, $I_\nu(\tau_1, \mu) = B_\nu(T_s)$, where T_s is the surface temperature. Since there is no downward radiation source at the top of the atmosphere ($\tau = 0$), we shall have $I_\nu(0, -\mu) = 0$.

Thus, the monochromatic upward and downward flux densities defined in Eq. (1.9) are given by

$$F_\nu^\uparrow(\tau) = 2\pi B_\nu(T_s) \int_0^1 e^{-(\tau_1 - \tau)/\mu} \mu \, d\mu$$

$$+ 2 \int_0^1 \int_\tau^{\tau_1} \pi B_\nu[T(\tau')] e^{-(\tau' - \tau)/\mu} \, d\tau' \, d\mu, \tag{4.6}$$

$$F_\nu^\downarrow(\tau) = 2 \int_0^1 \int_0^\tau \pi B_\nu[T(\tau')] e^{-(\tau - \tau')/\mu} \, d\tau' \, d\mu. \tag{4.7}$$

We define the exponential integral as

$$E_n(\tau) \equiv \int_1^\infty \frac{e^{-\tau x}}{x^n} \, dx. \tag{4.8}$$

It is clear that

$$\frac{dE_n(\tau)}{d\tau} = - \int_1^\infty \frac{e^{-\tau x}}{x^{n-1}} \, dx = -E_{n-1}(\tau). \tag{4.9}$$

In Eqs. (4.6) and (4.7), let $x = 1/\mu$, then $d\mu = -dx/x^2$. Further, we note that

$$\int_1^\infty \frac{e^{-(\tau_1 - \tau)x}}{x^3} \, dx = E_3(\tau_1 - \tau) \tag{4.10}$$

and

$$\int_1^\infty \frac{e^{-(\tau' - \tau)x}}{x^2} \, dx = E_2(\tau' - \tau). \tag{4.11}$$

Consequently, the integrations over μ in Eqs. (4.6) and (4.7) may be accomplished in terms of the well-known exponential integral in the forms

$$F_\nu^\uparrow(\tau) = 2\pi B_\nu(T_s) E_3(\tau_1 - \tau) + 2 \int_\tau^{\tau_1} \pi B_\nu[T(\tau')] E_2(\tau' - \tau) \, d\tau', \tag{4.12}$$

$$F_\nu^\downarrow(\tau) = 2 \int_0^\tau \pi B_\nu[T(\tau')] E_2(\tau - \tau') \, d\tau'. \tag{4.13}$$

To evaluate the total upward and downward fluxes at level τ for the entire infrared spectrum, integrations over the wave number are required. Thus,

$$F^{\uparrow}(\tau) = \int_0^{\infty} F_v^{\uparrow}(\tau)\, dv = 2 \int_0^{\infty} \pi B_v(T_s) E_3(\tau_1 - \tau)\, dv$$

$$+ 2 \int_{\tau}^{\tau_1} \int_0^{\infty} \pi B_v[T(\tau')] E_2(\tau' - \tau)\, dv\, d\tau' \quad (4.14)$$

$$F^{\downarrow}(\tau) = \int_0^{\infty} F_v^{\downarrow}(\tau)\, d\tau = 2 \int_0^{\tau} \int_0^{\infty} \pi B_v[T(\tau')] E_2(\tau - \tau')\, dv\, d\tau'. \quad (4.15)$$

At this point the transfer of thermal infrared radiation in clear atmospheres is formally solved. However, there are several practical difficulties of applying Eqs. (4.14) and (4.15) directly to the atmosphere. The chief one is the rapid variation of absorption coefficient with wave number in the vibrational and rotational spectrum of the infrared as illustrated in Fig. 4.2. To obtain the fluxes at a given level τ in the atmosphere, we now have to perform the double integrations over more than thousands of absorption lines. Even with the fast computers currently available, direct line-by-line calculations are still very tedious and not practical. The solution to this catastrophe has been to consider not monochromatic radiation but finite spectral intervals of bands for which the effective *transmission function* (also referred to as *transmittance* or *transmissivity*) can be derived by theory or experiment. These transmission functions are generally much more complicated than the simple exponential attenuation that is valid for monochromatic radiation described in Eq. (1.51). In the following section, we shall discuss the physical meanings of the transmission function, or simply the transmittance, in the theory of infrared radiative transfer.

4.4   CONCEPT OF TRANSMISSION FUNCTION (TRANSMITTANCE)

Let us consider a spectral interval of width Δv which is small enough so that a mean value of the Planck function $B_{\bar{v}}(T)$ may be utilized, but large enough so that it consists of several absorption lines. Then the transmission function may be defined by

$$\mathcal{T}_{\bar{v}}(\tau) = \frac{1}{\Delta v} \int_{\Delta v} e^{-\tau}\, dv, \quad (4.16)$$

where the monochromatic optical depth defined in Eq. (1.62) may be written as

$$\tau = \int_u^{u_1} k_v\, du. \quad (4.17)$$

Note that τ is a function of the wave number and path length u. The total normal path length and optical depth are

$$u_1 = \int_0^\infty \rho \, dz \quad \text{and} \quad \tau_1 = \int_0^{u_1} k_\nu \, du, \tag{4.18}$$

respectively. In order to evaluate Eq. (4.16), we must know how the absorption coefficient k_ν varies with ν within the spectral interval.

We now return to Eq. (4.12). Instead of performing the integration for the entire spectral region, we employ a finite spectral interval $\Delta\nu$ in the wave number integration. Thus, we have

$$F_{\bar\nu}^\uparrow(\tau) = \int_{\Delta\nu} F_\nu^\uparrow(\tau) \frac{d\nu}{\Delta\nu} = 2\pi B_{\bar\nu}(T_s) \int_{\Delta\nu} E_3(\tau_1 - \tau) \frac{d\nu}{\Delta\nu}$$

$$+ 2 \int_\tau^{\tau_1} \pi B_{\bar\nu}(T) \int_{\Delta\nu} E_2(\tau' - \tau) \, d\nu \, d\tau'. \tag{4.19}$$

Analogous to the definition of the transmission function for intensity, we may define the *slab* or *diffuse* transmission function for flux density in the form

$$\mathscr{T}_{\bar\nu}^{\rm f}(\tau) = 2 \int_0^1 \mathscr{T}_{\bar\nu}(\tau/\mu)\mu \, d\mu = 2 \int_{\Delta\nu} E_3(\tau) \frac{d\nu}{\Delta\nu}. \tag{4.20}$$

It is apparent from Eq. (4.9) that

$$\frac{d\mathscr{T}_{\bar\nu}^{\rm f}(\tau)}{d\tau} = -2 \int_{\Delta\nu} E_2(\tau) \frac{d\nu}{\Delta\nu}. \tag{4.21}$$

Inserting the expressions in Eqs. (4.20) and (4.21) into Eq. (4.19), we have

$$F_{\bar\nu}^\uparrow(\tau) = \pi B_{\bar\nu}(T_s)\mathscr{T}_{\bar\nu}^{\rm f}(\tau_1 - \tau) - \int_\tau^{\tau_1} \pi B_{\bar\nu}(\tau') \frac{d\mathscr{T}_{\bar\nu}^{\rm f}(\tau' - \tau)}{d\tau'} \, d\tau'. \tag{4.22}$$

Hence, the spectral upward flux density is now expressed in terms of the spectral averaged Planck irradiance and the slab transmission function. It is sometimes convenient to use the path length u instead of the optical depth τ. By virtue of Eqs. (4.17) and (4.18), Eq. (4.22) may be rewritten to yield

$$F_{\bar\nu}^\uparrow(u) = \pi B_{\bar\nu}(T_s)\mathscr{T}_{\bar\nu}^{\rm f}(u) + \int_0^u \pi B_{\bar\nu}(u') \frac{d\mathscr{T}_{\bar\nu}^{\rm f}(u - u')}{du'} \, du'. \tag{4.23}$$

In a similar manner, the spectral downward flux density may be derived to give

$$F_{\bar\nu}^\downarrow(u) = \int_{u_1}^u \pi B_{\bar\nu}[T(u')] \frac{d\mathscr{T}_{\bar\nu}^{\rm f}(u' - u)}{du'} \, du'. \tag{4.24}$$

Here, we note that we are simply making the coordinate transformation between τ and u. In doing so, the effect of k_ν is not considered. The dependence of the slab transmission function on the absorption coefficient, and hence the temperature will be discussed fully in Section 4.8.2.

If appropriate flux transmission functions can be obtained to represent moderately wide spectral intervals prior to the path length integration, then a great simplification in the transfer calculations can be achieved. Upon changing τ to u and letting $x = u/\mu$, Eq. (4.20) can be rewritten to give

$$\mathcal{T}_{\bar{\nu}}^{f}(u) = 2 \int_{0}^{1} \mathcal{T}_{\bar{\nu}}(u/\mu)\mu \, d\mu = 2u^2 \int_{u}^{\infty} \mathcal{T}_{\bar{\nu}}(x) \, dx/x^3. \qquad (4.25)$$

If in Eq. (4.16), $\mathcal{T}_{\bar{\nu}}$ as a function of the path length u may be derived either from the theoretical method or empirical means, then Eq. (4.25) can readily be computed numerically. The shape of $\mathcal{T}_{\bar{\nu}}^{f}$ is extremely similar to $\mathcal{T}_{\bar{\nu}}^{f}$ in many atmospheric conditions. Thus, for most practical applications, it suffices to set

$$\mathcal{T}_{\bar{\nu}}^{t}(u) = \mathcal{T}_{\bar{\nu}}(1.66u), \qquad (4.26)$$

where the constant 1.66 is called the *diffusivity factor*.

With the assistance of Eq. (4.26), calculations may be carried out for the upward and downward fluxes for any absorption band at a given level whose path length is u. The main question of concern, of course, is how to determine the transmission function $\mathcal{T}_{\bar{\nu}}$ for absorption bands in the infrared spectrum. One approach has been to measure $\mathcal{T}_{\bar{\nu}}$ for a given absorption band in the laboratory employing a number of path lengths under various atmospheric conditions. Then by means of Eq. (4.26) values of $\mathcal{T}_{\bar{\nu}}$ may be incorporated into Eqs. (4.23) and (4.24) to obtain the flux densities in the atmosphere. Another means has been to utilize theoretical band models which are classical, and we shall discuss them in some detail.

4.5 BAND MODELS FOR TRANSMISSION FUNCTIONS (TRANSMITTANCES)

4.5.1 A Single Spectra Line

The shape of spectral lines can depend on a variety of factors, but in the lower part of the atmosphere the infrared lines generally have a Lorentz shape expressed by Eq. (1.34). Assuming that the absorption coefficient k_ν is independent of path length, i.e., the atmosphere is considered to be homogeneous, the transmission function may be written as

$$\mathcal{T}_{\bar{\nu}}(u) = \frac{1}{\Delta\nu} \int_{\Delta\nu} e^{-k_\nu u} \, d\nu. \qquad (4.27)$$

Substitution of k_v in Eq. (1.34) into Eq. (4.27) yields

$$\mathcal{T}_{\bar{v}}(u) = \frac{1}{\Delta v} \int_{\Delta v} dv \exp\left[-\frac{S\alpha u/\pi}{(v - v_0)^2 + \alpha^2}\right]. \tag{4.28}$$

We introduce new variables x and y in the forms

$$x = Su/2\pi\alpha, \qquad \tan y/2 = (v - v_0)/\alpha. \tag{4.29}$$

If the interval of the spectral line is taken wide enough, the limits of the number integration may be extended from $-\infty$ to $+\infty$. Thus, by virtue of Eqs. (4.28) and (4.29) the absorptivity may be written in the form

$$A_{\bar{v}} = 1 - \mathcal{T}_{\bar{v}} = \frac{\alpha}{\Delta v} \int_{-\pi}^{\pi} \{1 - \exp[-x(1 + \cos y)]\}\, d(\tan y/2). \tag{4.30}$$

We now perform integration by parts and carry out further trigonometric manipulations to give

$$A_{\bar{v}} = \frac{\alpha x e^{-x}}{\Delta v} \int_{-\pi}^{\pi} (e^{-x \cos y} - \cos y e^{-x \cos y})\, dy. \tag{4.31}$$

We note that the integral representation of the Bessel function is given by

$$J_n(x) = \frac{i^{-n}}{\pi} \int_0^{\pi} e^{ix \cos \theta} \cos n\theta\, d\theta. \tag{4.32}$$

Moreover, the modified Bessel function of the first kind of order n is

$$I_n(x) = i^{-n} J_n(ix). \tag{4.33}$$

Thus, in terms of the modified Bessel functions, Eq. (4.31) may be written as

$$A_{\bar{v}} = (2\pi\alpha/\Delta v)L(x) = (2\alpha\pi/\Delta v)xe^{-x}[I_0(x) + I_1(x)], \tag{4.34}$$

where $L(x)$ is known as the Ladenberg and Reiche function. The two significant limiting cases are associated with weak and strong absorption. When $x \to 0$, $J_0(ix) \approx 1$, and $iJ_1(ix) \approx -x/2$. Thus, $L(x) \approx x$ and we have

$$A_{\bar{v}} \approx (S/\Delta v)u. \tag{4.35}$$

The absorptivity in this case is directly proportional to the path length, and it is called the region of *linear absorption*. On the other hand, when $x \to \infty$, $J_n(ix) \approx i^n e^x/\sqrt{2\pi x}$. Thus, $L(x) \approx \sqrt{2x/\pi}$ and we have

$$A_{\bar{v}} \approx (2\sqrt{S\alpha}/\Delta v)\sqrt{u}. \tag{4.36}$$

The absorptivity in this case is proportional to the square root of the path length, and it is called the region of *square root absorption*.

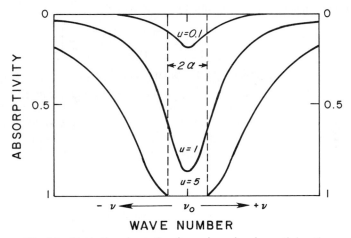

Fig. 4.4 Single line absorption for various absorber path lengths.

The linear and square root approximations for the absorptivity may also be derived directly from Eqs. (4.27) and (4.28). For weak line absorption, $k_\nu u \ll 1$, wave number integration can be carried out analytically. For strong line absorption, $\alpha \ll \nu - \nu_0$, again, integration may also be performed. The absorptivity of a single line for various absorbers in terms of u is shown in Fig. 4.4. It is evident that for strong absorption, the central part of the line has been absorbed completely, and additional absorption can take place only in the wing portions of the absorption line.

4.5.2   Regular (Elsasser) Band Model

Inspection of realistic infrared bands suggests that a single spectral line denoted in Eq. (1.34) may repeat itself periodically (or regularly) as shown in Fig. 4.5. Thus, the absorption coefficient at a wave number displacement

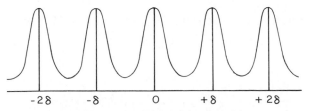

Fig. 4.5   Regular (Elsasser) band model.

v from the center of one particular line is then

$$k_v = \sum_{i=-\infty}^{\infty} \frac{S\alpha/\pi}{(v - i\delta)^2 + \alpha^2}, \tag{4.37}$$

where δ is the line spacing. From the Mittag–Leffler theorem (Whittacker and Watson, 1940), it can be proven that this infinite sum can be expressed in terms of periodic and hyperbolic functions as

$$k_v = \frac{S}{\delta} \frac{\sinh \beta}{\cosh \beta - \cos \gamma}, \tag{4.38}$$

where

$$\beta = 2\pi\alpha/\delta, \qquad \gamma = 2\pi v/\delta. \tag{4.39}$$

Upon variable transformation, the transmission function may be expressed by

$$\mathcal{T}_{\bar{v}} = \frac{1}{\delta} \int_{-\infty}^{\infty} e^{-k_v u} \, dv = \frac{1}{2\pi} \int_{-\pi}^{\pi} e^{-k_v(\gamma)u} \, d\gamma. \tag{4.40}$$

Also we have

$$\frac{d\mathcal{T}_{\bar{v}}}{du} = -\frac{1}{2\pi} \int_{-\pi}^{\pi} e^{-k_v u} k_v \, d\gamma. \tag{4.41}$$

Let

$$\cos \rho = \frac{1 - \cosh \beta \cos \gamma}{\cosh \beta - \cos \gamma}; \tag{4.42}$$

then

$$d\rho = -\frac{\sinh \beta}{\cosh \beta - \cos \gamma} \, d\gamma = -k_v \frac{\delta}{S} \, d\gamma. \tag{4.43}$$

Substituting Eqs. (4.38) and (4.43) into Eq. (4.41), we get

$$\frac{d\mathcal{T}_{\bar{v}}}{du} = -\frac{S}{2\pi\delta} \int_{-\pi}^{\pi} \exp\left(-\frac{Su}{\delta} \frac{\cosh \beta - \cos \rho}{\sinh \beta}\right) d\rho. \tag{4.44}$$

We then define a new variable

$$y = Su/(\delta \sinh \beta) \tag{4.45}$$

to obtain

$$\frac{d\mathcal{T}_{\bar{v}}}{dy} = -\frac{\sinh \beta}{2\pi} \int_{-\pi}^{\pi} \exp(-y \cosh \beta - y \cos \rho) \, d\rho$$

$$= -\sinh \beta e^{-y \cosh \beta} J_0(iy). \tag{4.46}$$

Since $\mathcal{T}_{\bar{\nu}} = 0$ when u (or y) $\to \infty$, it follows that

$$\mathcal{T}_{\bar{\nu}} = \int_0^{\mathcal{T}_{\bar{\nu}}} d\mathcal{T}_{\bar{\nu}} = \sinh \beta \int_y^\infty e^{-y \cosh \beta} J_0(iy)\, dy$$

$$= \int_z^\infty e^{-z \coth \beta} J_0(iz/\sinh \beta)\, dz, \qquad (4.47)$$

where $z = y \sinh \beta$. This is the Elsasser transmission function which can be numerically evaluated. Further approximations and simplifications also can be made on this model. Since $\alpha \ll \delta$ and $\beta \to 0$, we find

$$\coth \beta = \frac{1}{\beta} + \frac{\beta}{3} - \frac{\beta^3}{45} + \cdots \approx \frac{1}{\beta} + \frac{\beta}{3},$$

$$\operatorname{csch} \beta = \frac{1}{\beta} - \frac{\beta}{6} + \frac{7\beta^3}{360} - \cdots \approx \frac{1}{\beta} - \frac{\beta}{6}, \qquad (4.48)$$

and

$$J_0(iz\operatorname{csch}\beta) \approx e^{z\operatorname{csch}\beta}/\sqrt{2\pi z\operatorname{csch}\beta}$$

$$\approx \exp\left[z\left(\frac{1}{\beta} - \frac{\beta}{6}\right)\right]\Big/\sqrt{2\pi z/\beta}. \qquad (4.49)$$

With the approximations given in Eqs. (4.48) and (4.49), Eq. (4.47) becomes

$$\mathcal{T}_{\bar{\nu}} = \frac{1}{\sqrt{2\pi}} \int_z^\infty \sqrt{\beta/z}\, e^{-z\beta/2}\, dz. \qquad (4.50)$$

Finally, we set $x^2 = z\beta/2$; the absorptivity then is given by

$$A_{\bar{\nu}} = 1 - \frac{2}{\sqrt{\pi}} \int_x^\infty e^{-x^2}\, dx. \qquad (4.51)$$

By noting that $(2/\sqrt{\pi}) \int_0^\infty e^{-x^2}\, dx = 1$, we have

$$A_{\bar{\nu}} = \frac{2}{\sqrt{\pi}} \int_0^x e^{-x^2}\, dx = \operatorname{erf}(x) = \operatorname{erf}\left(\frac{\sqrt{\pi S\alpha u}}{\delta}\right). \qquad (4.52)$$

Values of $\operatorname{erf}(x)$ may be obtained from standard mathematical tables. This band model for periodic lines was introduced by Elsasser (1938) who observed fairly regularly spaced absorption lines in the 15 μm CO_2 band. It is sometimes referred to as the Elsasser band. For small values of x, we see that $A_{\bar{\nu}} = 2x/\sqrt{\pi} = 2\sqrt{S\alpha u}/\delta$. This is the region of square root absorption introduced in Eq. (4.36).

4.5.3   Statistical (Goody) Band Model

By inspection of the water vapor rotational band Goody (1952) discovered that the only common feature over a 25 cm^{-1} range is the apparent random line positions. Hence, one should inquire into the absorption of a band with certain random properties.

Let Δv be a spectral interval consisting of n lines of mean distance δ so that $\Delta v = n\delta$. Let $P(S_i)$ be the probability that the ith line has an intensity S_i, and let P be normalized such that

$$\int_0^\infty P(S)\,dS = 1. \tag{4.53}$$

It is assumed that any line has an equal probability of being anywhere in the interval Δv. The mean transmission function is found by averaging over all positions and all intensity of lines. Hence,

$$
\mathcal{T}_{\bar{v}} = \frac{1}{(\Delta v)^n} \int_{\Delta v} dv_1 \cdots \int_{\Delta v} dv_n
$$

$$
\times \int_0^\infty P(S_1)e^{-k_1 u}\,dS_1 \cdots \int_0^\infty P(S_n)e^{-k_n u}\,dS_n, \tag{4.54}
$$

where k_n denotes the absorption coefficient for the nth line. Since all the integrals are alike, we have

$$
\mathcal{T}_{\bar{v}} = \left[\frac{1}{\Delta v} \int dv \int_0^\infty P(S)e^{-ku}\,dS \right]^n = \left[1 - \frac{1}{\Delta v} \int dv \int_0^\infty P(S)(1 - e^{-ku})\,dS \right]^n. \tag{4.55}
$$

Since $\Delta v = n\delta$, when n becomes large, Eq. (4.55) approaches an exponential function, i.e., $(1 - x/n)^n \to e^{-x}$. Thus,

$$
\mathcal{T}_{\bar{v}} \cong \exp\left\{ -\frac{1}{\delta} \int_0^\infty P(S)\left[\int(1 - e^{-ku})\,dv \right]dS \right\}. \tag{4.56}
$$

Let the lines be of different line intensities and consider a simple Poisson distribution for the probability of their intensities, i.e.,

$$
P(S) = \bar{S}^{-1}e^{-S/\bar{S}}, \tag{4.57}
$$

where $\bar{S}$ represents the mean line intensity. By introducing the Lorentz shape for k into Eq. (4.56), and carrying out the line intensity and wave number integration in the domain $(-\infty, \infty)$, the final result of the transmission function is given by

$$
\mathcal{T}_{\bar{v}} = \exp\left[-\frac{\bar{S}u}{\delta}\left(1 + \frac{\bar{S}u}{\pi\alpha} \right)^{-1/2} \right]. \tag{4.58}
$$

Note that the transmission function derived from the random model can be expressed as a function of *two* parameters only, namely, $\bar{S}/\delta$ and $\pi\alpha/\delta$ apart from the path length u. For a given absorption band, these two parameters may be derived by fitting the random model from the laboratory or quantum-mechanical data. It is clear that the transmission function is now reduced to exponential attenuation. Because of its computational simplicity and relatively high accuracy in the transmittance calculations, the random model has been widely used in atmospheric cooling rate computations and satellite-sensing applications.

To derive $\bar{S}/\delta$ and $\pi\alpha/\delta$ from the quantum-mechanical data, we first define the equivalent width for n absorption lines as

$$W = \frac{1}{n}\sum_{i=1}^{n}W_i = \int_0^\infty P(S)\left[\int(1 - e^{-ku})\,dv\right]dS = \bar{S}u\left(1 + \frac{\bar{S}u}{\pi\alpha}\right)^{-1/2} \quad (4.59)$$

In view of the definition of the absorptivity [see Eqs. (4.27) and (4.30)], the equivalent width is essentially the spectral absorptivity. We then consider the random model in the limit of the strong and weak absorption regions. For weak absorption $\bar{S}u/\pi\alpha \ll 1$ and by virtue of Eq. (4.35), we must have

$$\frac{1}{n\delta}\sum W_i(\text{weak}) = \frac{\bar{S}u}{\delta} = \frac{1}{\Delta v}\sum S_i u. \quad (4.60a)$$

Note that $n\delta = \Delta v$, and S_i is the line intensity for the ith individual line. For strong absorption $\bar{S}u/\pi\alpha \gg 1$ and by virtue of Eq. (4.36), we must also have

$$\frac{1}{n\delta}\sum W_i(\text{strong}) = \frac{1}{\delta}\sqrt{\pi\bar{S}\alpha u} = \frac{2}{\Delta v}\sum\sqrt{S_i\alpha_i u}. \quad (4.60b)$$

Thus, from Eqs. (4.60a) and (4.60b), we find

$$\frac{\bar{S}}{\delta} = \frac{\sum S_i}{\Delta v}, \qquad \frac{\bar{S}\pi\alpha}{\delta^2} = \left(\frac{2\sum\sqrt{S_i\alpha_i}}{\Delta v}\right)^2. \quad (4.61)$$

On the other hand, the transmission function as a function of the gaseous path length may be measured in the laboratory. Moreover, the random model [Eq. (4.58)] can be expressed in the form

$$\left(\frac{u}{\ln \mathscr{T}_{\bar{v}}}\right)^2 = \frac{\delta^2}{\bar{S}^2} + \frac{\delta^2}{\bar{S}\pi\alpha}u. \quad (4.62)$$

It follows that a statistical regression analysis in $(u/\ln \mathscr{T}_{\bar{v}})^2$ and u for the laboratory data will give the slope $\delta^2/\bar{S}\pi\alpha$ and the intercept $\delta^2/\bar{S}^2$. Consequently $\bar{S}/\delta$ and $\pi\alpha/\delta$ can be determined.

TABLE 4.1   *Random Model Band Parameters in the Infrared Region*

Band	Interval (cm^{-1})	$\bar{S}/\delta$ $(cm^2\,g^{-1})$	$\pi\alpha/\delta$
H_2O rotational	40–160	7210.30	0.182
	160–280	6024.80	0.094
	280–380	1614.10	0.081
	380–500	139.03	0.080
	500–600	21.64	0.068
	600–720	2.919	0.060
	720–800	0.386	0.059
	800–900	0.0715	0.067
CO_2 15 μm	582–752	718.7	0.448
O_3 9.6 μm	1000.0–1006.5	6.99×10^2	5.0
	1006.5–1013.0	1.40×10^2	5.0
	1013.0–1019.5	2.79×10^3	5.0
	1019.5–1026.0	4.66×10^3	5.5
	1026.0–1032.5	5.11×10^3	5.8
	1032.5–1039.0	3.72×10^3	8.0
	1039.0–1045.5	2.57×10^3	6.1
	1045.5–1052.0	6.05×10^3	8.4
	1052.0–1058.5	7.69×10^3	8.3
	1058.5–1065.0	2.79×10^3	6.7
H_2O 6.3 μm	1200–1350	12.65	0.089
	1350–1450	134.4	0.230
	1450–1550	632.9	0.320
	1550–1650	331.2	0.296
	1650–1750	434.1	0.452
	1750–1850	136.0	0.359
	1850–1950	35.65	0.165
	1950–2050	9.015	0.104
	2050–2200	1.529	0.116

    Utilizing the laboratory data for the H_2O 6.3 μm vibrational–rotational band and the quantum-mechanical data for the H_2O rotational and CO_2 15 μm bands, Rodgers and Walshaw (1966) derived the random model parameters for these bands. These parameters are summarized in Table 4.1. In this table, random model parameters for the O_3 9.6 μm band computed by Goldman and Kyle (1968) from the quantum-mechanical data are also listed. In the ozone band, the band parameters have been calculated for the interval 1000–1060 cm^{-1} at a temperature of 233°K using an average interval of 6.5 cm^{-1}.

4.6 CURTIS—GODSON APPROXIMATION
FOR INHOMOGENEOUS ATMOSPHERES

In the analysis of transmission functions and band models in the previous section, assumption was made that the absorption coefficient k_v is independent of the path length. However, as noted earlier in Section 1.3.1, the collision broadening characterized by the Lorentz profile depends, in general, on the pressure and temperature. Therefore, in order to apply the formulations of band models presented in the previous sections to an inhomogeneous atmosphere whose pressure and temperature vary with height, certain physical adjustments are needed. Among the various methods, there is a simplified procedure known as the Curtis–Godson (C–G) approximation for the application of the homogeneous transmission to an in-inhomogeneous path length. It has been illustrated that the C–G approximation is fairly accurate for the infrared transfer calculations involving water vapor and carbon dioxide atmospheres.

We first note that the finite width of a Lorentz line is produced by the collision of the radiating molecule with other molecules. As has been discussed in Section 1.3.1 the number of collisions is proportional to $PT^{-1/2}$ based on the kinetic theory, and the expression for the half width has been given in Eq. (1.37). It has been found that the pressure effect shown in Eq. (1.37) is extremely significant in atmospheric infrared transfer calculations, whereas the temperature effect is less important. This is owing to the fact that in the earth's atmosphere the pressure has a much larger variation than the temperature. The line intensity S is also a function of the path length through the temperature dependence, and it can be expressed by

$$S = S_0 \left(\frac{T_0}{T}\right)^m \exp\left[-\frac{E}{K}\left(\frac{1}{T} - \frac{1}{T_0}\right)\right], \tag{4.63}$$

where E is the energy of the lower state, K the Boltzmann constant, m the numerical factor related to the absorber ($m = 1, \frac{3}{2}$, and $\frac{5}{2}$ for CO_2, H_2O, and O_3, respectively), T_0 the standard atmospheric temperature, and S_0 the line intensity at standard atmospheric conditions.

Consider now a path length in the atmosphere containing integrated mass u such that

$$\int_0^u k_v \, du \neq k_v u,$$

where we write

$$k_v = \frac{S}{\pi} \frac{\alpha}{(v - v_0)^2 + \alpha^2} = k_v(P, T).$$

The temperature and pressure dependence of the absorption coefficient is associated with variations of the half width and line intensity in the atmosphere. We define two new parameters

$$\bar{S} = \int_0^u S(T)\,du/u, \qquad\qquad\qquad (4.64a)$$

$$\bar{\alpha} = \int_0^u S(T)\alpha(P, T)\,du \bigg/ \int_0^u S(T)\,du, \qquad\qquad (4.64b)$$

which denote the mean line intensity and half width, respectively, over an inhomogeneous path $(0, u)$.

Equations (4.64a) and (4.64b) constitute the so-called Curtis–Godson approximation for transfer of infrared radiation through an inhomogeneous path in the atmosphere. It states that the transmission of the inhomogeneous atmospheric path is approximately equal to the transmission of a homogeneous path (constant temperature and pressure) whose integrated absorber amount is u with a mean half width $\bar{\alpha}$ and a mean line intensity $\bar{S}$. The C–G approximation eliminates the integration over the exact line shapes which would otherwise be required as evident in Section 4.4. It has been found to be a reasonably good approximation for the strong and weak lines, and also gives reasonably good results in the intermediate cases for the 15 μm CO_2 band and H_2O rotational band. However, the C–G approximation does not give accurate results for the 9.6 μm O_3 band owing to the fact that ozone increases to the stratosphere with decreasing pressure. This is contrary to the water vapor and carbon dioxide cases in which the optical mass decreases with decreasing pressure. We note that if the temperature effect is neglected, Eq. (4.64b) is essentially equivalent to Eq. (3.32) discussed in Section 3.4.

4.7   COMPUTATION OF INFRARED COOLING RATES

In Section 4.4, flux formulations were made for a spectral interval Δv. To obtain the flux density covering the entire infrared spectrum, summation over the band flux densities is required. Assuming that the infrared spectrum consists of N spectral intervals $\Delta v_i (i = 1, \ldots, N)$, then the total upward infrared flux density based on Eq. (4.23) is given by

$$F^\uparrow(u) = \sum_{i=1}^N F^\uparrow_{\bar{v}_i}(u)$$

$$= \sum_{i=1}^N \left\{ \pi B_{\bar{v}_i}(T_s)\mathcal{T}^f_{\bar{v}_i}(u) + \int_0^u \pi B_{\bar{v}_i}[T(u')]\,d\mathcal{T}^f_{\bar{v}_i}(u - u') \right\}. \quad (4.65a)$$

Likewise, the total downward infrared flux density is

$$F^{\downarrow}(u) = \sum_{i=1}^{N} F^{\downarrow}_{\bar{\nu}_i}(u) = \sum_{i=1}^{N} \int_{u_1}^{u} \pi B_{\bar{\nu}_i}[T(u')]\, d\mathcal{T}^{f}_{\bar{\nu}_i}(u' - u). \qquad (4.65b)$$

In Section 3.5, we have introduced the formulation of solar heating rate in which the net transfer of radiation is downward. On the other hand, however, thermal infrared radiation may be thought of as initiating from the earth's surface (upward). Consequently, we may define the net flux density at a given height as

$$F(z) = F^{\uparrow}(z) - F^{\downarrow}(z). \qquad (4.66a)$$

Let the two plane-parallel levels in the atmosphere be denoted by z and $z + \Delta z$ (Fig. 4.6), then the net loss of radiant energy per unit area per unit time suffered by the layer Δz is

$$\Delta F = F(z + \Delta z) - F(z). \qquad (4.66b)$$

The radiative cooling or warming experienced by a layer of air whose thickness is Δz may be evaluated from the principle of conservation of energy as pointed out in Section 3.5. If the net flux density at the top of the layer is smaller than that at the bottom, the difference must be used to heat the layer, and vice versa. On the basis of the discussions in Section 3.5, the heating or cooling rate may be expressed by

$$\left(\frac{\partial T}{\partial t}\right)_{\text{IR}} = -\frac{1}{c_p \rho}\frac{\Delta F}{\Delta z} = \frac{g}{c_p}\frac{\Delta F}{\Delta p} = -\frac{g}{c_p}\frac{\Delta F}{\Delta u}. \qquad (4.67)$$

Infrared cooling rates as a function of height in a typical clear tropical atmosphere are illustrated in Fig. 4.7 based on band-by-band calculations

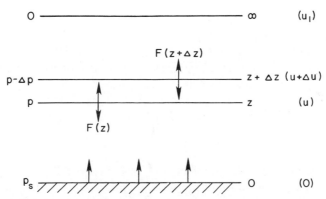

Fig. 4.6 Divergence of net flux density in u, z, and p coordinates.

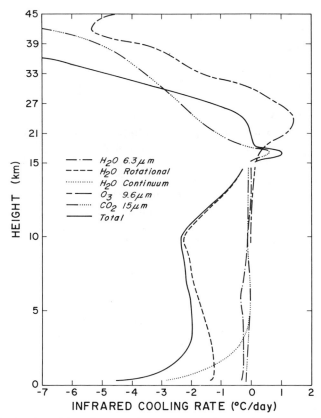

Fig. 4.7 Total and partial cooling rates in a clear tropical atmosphere (after Roewe and Liou, 1978).

(Roewe and Liou, 1978). The random model parameters for O_3, CO_2, and H_2O bands listed in Table 4.1 were used in the cooling rate calculations in which the Curtis–Godson approximation for inhomogeneous atmospheres was also utilized. In addition, the recent data presented by Roberts *et al.* (1976) for the water continuum were also incorporated in the calculations. According to their analyses, the continuum absorption coefficient at a standard temperature of 296°K in the 8–12 μm window region is given by

$$k(v, 296) = a + be^{-\beta v}, \tag{4.68a}$$

where $a = 4.2 \ \mathrm{cm^2 \, g^{-1}}$, $b = 5588 \ \mathrm{cm^2 \, g^{-1}}$, and $\beta = 7.87 \times 10^{-3}$ cm. The expression accounts for the continuum absorption contributions due to both water dimer and water vapor. Furthermore, the temperature dependence of the absorption coefficient has been found to be quite significant, and it can be taken into account by the empirical formula

$$k(v, T) = k(v, 296) \exp\left[T_0\left(\frac{1}{T} - \frac{1}{296} \right) \right] \tag{4.68b}$$

with a best empirical value of $1800°\,K$ for T_0.

Figure 4.7 shows the band-by-band as well as total atmospheric cooling rates. In the lower 2 km the most important band influencing cooling is the water vapor continuum. This is due to the rapid increase in the temperature and partial pressure of water vapor as the surface is approached. However, above 5 km the continuum contributes little to the total cooling rate. The contribution of the H_2O 6.3 μm rotational–vibrational band to the total cooling rate is always small compared to other bands. The reason is that the Planckian curves for temperatures representative of the earth's atmosphere contain only a small amount of energy at these wavelengths compared with bands closer to the peak of the Planckian curve. While the water vapor continuum is seen to dominate the lower tropospheric cooling, the cooling in the middle and upper troposphers is primarily caused by absorption in the water vapor rotational band. The rather strong heating effect of ozone between 18 and 27 km results from a large increase in the ozone concentration at these levels, which are strongly warmed by radiation from the ground. Above 30 km the cooling rate begins to increase rapidly in the CO_2 15μm and O_3 9.6 μm bands, and cooling to space becomes increasingly significant and effective. In order to correctly compute the cooling rates in the upper atmosphere, transition from the Lorentz line shape to the Doppler line profile should be taken into consideration.

4.8   INFRARED FLUX IN TERMS OF STEFAN—BOLTZMANN LAW AND RADIATION CHART

4.8.1   Concept of Broadband Flux Emissivity

In reference to Eqs. (4.14) and (4.15) [see also Eqs. (4.23) and (4.24)], the total upward and downward infrared flux densities may be expressed by

$$F^{\uparrow}(u) = \int_0^{\infty} \pi B_v(T_s) \mathcal{T}_v^{f}(u)\,dv + \int_0^{\infty} \int_0^{u} \pi B_v[T(u')] \frac{d\mathcal{T}_v^{f}(u - u')}{du'}\,du'\,dv, \tag{4.69}$$

$$F^{\downarrow}(u) = \int_0^{\infty} \int_{u_1}^{u} \pi B_v[T(u')] \frac{d\mathcal{T}_v^{f}(u' - u)}{du'}\,du'\,dv, \tag{4.70}$$

where $\mathcal{T}_\nu^f$ denotes the monochromatic slab transmission function. From the Stefan–Boltzmann law introduced in Section 1.2.2, we have

$$\int_0^\infty \pi B_\nu(T)\,d\nu = \sigma T^4. \tag{4.71}$$

Utilizing this relation, Eqs. (4.69) and (4.70) may be rewritten in the forms

$$F^\uparrow(u) = \sigma T_s^4 t^f(u, T_s) + \int_0^u \sigma T^4(u') \frac{dt^f(u - u', T)}{du'}\,du', \tag{4.72}$$

$$F^\downarrow(u) = \int_{u_1}^u \sigma T^4(u') \frac{dt^f(u' - u, T)}{du'}\,du', \tag{4.73}$$

where we define the isothermal *broadband flux transmissivity*, which is a function of temperature and path length in the form

$$t^f(u, T) = \int_0^\infty \pi B_\nu(T)\mathcal{T}_\nu^f(u)\,d\nu/(\sigma T^4). \tag{4.74}$$

In defining this parameter, we assume that the plane-parallel atmosphere may be divided into many infinitesimal layers such that each layer may be thought of as an isothermal layer where the temperature is a constant. Moreover, the isothermal *broadband flux emissivity* is defined by

$$\varepsilon^f(u, T) = 1 - t^f(u, T) = \int_0^\infty \pi B_\nu(T)[1 - \mathcal{T}_\nu^f(u)]\,d\nu/(\sigma T^4). \tag{4.75a}$$

The slab transmission functions $\mathcal{T}_{\bar\nu}^f$ derived from either theory or experiment normally are available for small spectral intervals but not for monochromatic wave numbers as discussed in the previous sections. Thus, in practice, the broadband emissivity is given by

$$\varepsilon^f(u, T) = \sum_{i=1}^N \pi B_{\bar\nu_i}(T)[1 - \mathcal{T}_{\bar\nu_i}^f(u)]\,\Delta\nu_i/(\sigma T^4), \tag{4.75b}$$

where the infrared spectrum is divided into N subspectral intervals $\Delta\nu_i$ $(i - 1, \ldots, N)$. Flux emissivity values for H_2O, CO_2, and O_3 have been determined empirically by Elsasser and Culbertson (1960) and presented correctly by Staley and Jurica (1970).

Figures 4.8a and b depict the broadband flux emissivities for water vapor and carbon dioxide as functions of the path length for a number of temperatures. Note that the path length units for carbon dioxide are in cm atm. For large carbon dioxide path lengths, it is seen that the temperature dependence of the flux emissivity is quite significant. To correct for the water vapor and carbon dioxide overlap, let the path lengths for water vapor and carbon dioxide be u_w and u_c, respectively. Hence, the monochromatic flux transmission function taking into account both water vapor and carbon dioxide

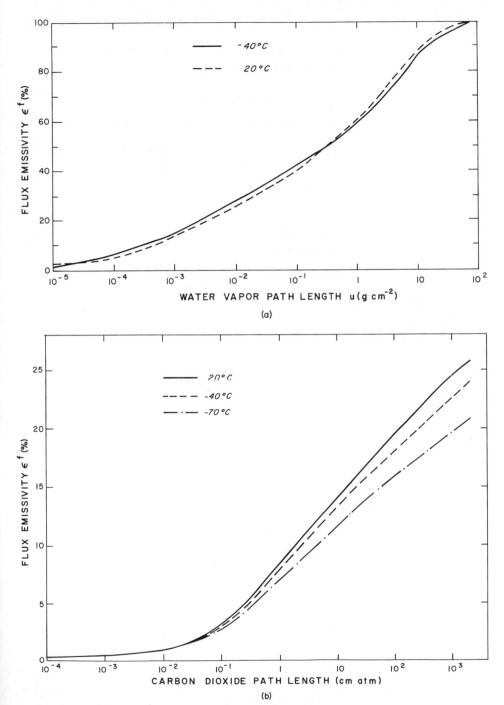

Fig. 4.8 (a) Broadband flux emissivity as a function of path length for water vapor. (b) Broadband flux emissivity as a function of path length for carbon dioxide.

absorption may be written as

$$\mathscr{T}_\nu^f(u_w, u_c) = \mathscr{T}_\nu^f(u_w)\mathscr{T}_\nu^f(u_c).$$

Thus, Eq. (4.75a) may be rewritten in the form

$$\varepsilon^f(u_w, u_c, T) = \varepsilon^f(u_w, T) + \varepsilon^f(u_c, T) - \Delta\varepsilon^f(u_w, u_c, T), \qquad (4.75c)$$

where the last term represents a correction due to overlap of the wings of water vapor and carbon dioxide radiation, and is given by

$$\Delta\varepsilon^f(u_w, u_c, T) = \int_0^\infty \pi B_\nu(T)[1 - \mathscr{T}_\nu^f(u_w)][1 - \mathscr{T}_\nu^f(u_c)]\, d\nu/(\sigma T^4).$$

H_2O–CO_2 overlap correction quantities $\Delta\varepsilon^f$ were also provided by Staley and Jurica for a number of temperatures.

Having obtained the empirical broadband flux emissivity, total downward and upward infrared flux densities can be computed from Eqs. (4.72) and (4.73), respectively, for given atmospheric temperature and gaseous profiles. Subsequently, infrared cooling rates due to various gases can be evaluated. However, since the slope of the flux emissivity $d\varepsilon^f(u)/du$ in Eqs. (4.72) and (4.73) is generally difficult to obtain accurately from curves shown in Fig. 4.8, it is desirable to perform integration by parts for these two equations to remove the differentiation of the broadband flux transmissivity with respect to the path length. Exercise 4.8 involves the computation of infrared flux densities and cooling rates utilizing the broadband emissivity values.

4.8.2 Radiation Chart

For the purpose of developing the concept of the radiation chart, we define the monochromatic slab transmission function for flux density in the form [see Eq. (4.20)]

$$\mathscr{T}_\nu^f(\tau) = 2E_3(\tau), \qquad \frac{d\mathscr{T}_\nu^f(\tau)}{d\tau} = -2E_2(\tau). \qquad (4.76)$$

Thus, the monochromatic upward and downward flux densities shown in Eqs. (4.12) and (4.13) may be rewritten as

$$F_\nu^\uparrow(\tau) = \pi B_\nu(T_s)\mathscr{T}_\nu^f(\tau_1 - \tau) - \int_\tau^{\tau_1} \pi B_\nu(\tau')\frac{d\mathscr{T}_\nu^f(\tau' - \tau)}{d\tau'}\, d\tau', \qquad (4.77)$$

$$F_\nu^\downarrow(\tau) = \int_0^\tau \pi B_\nu(\tau')\frac{d\mathscr{T}_\nu^f(\tau - \tau')}{d\tau'}\, d\tau'. \qquad (4.78)$$

Integration by parts yields the form

$$F_\nu^\uparrow(\tau) = \{\pi B_\nu(T_s) - \pi B_\nu[T(\tau_1)]\}\mathcal{T}_\nu^f(\tau_1 - \tau) + \pi B_\nu[T(\tau)]$$

$$+ \int_\tau^{\tau_1} \mathcal{T}_\nu^f(\tau' - \tau)\frac{d\pi B_\nu(\tau')}{d\tau'}\,d\tau', \tag{4.79}$$

$$F_\nu^\downarrow(\tau) = \pi B_\nu[T(\tau)] - \pi B_\nu[T(0)]\,\mathcal{T}_\nu^f(\tau) - \int_0^\tau \mathcal{T}_\nu^f(\tau - \tau')\frac{d\pi B_\nu(\tau')}{d\tau'}\,d\tau'. \tag{4.80}$$

In reference to Fig. 4.9, we change the optical depth coordinate to the temperature and path length coordinate and assume that the surface temperature is the same as air temperature immediately above the surface, i.e., $T_s = T(\tau_1)$. By noting that $T(0) = T_t$, Eqs. (4.79) and (4.80) may be rewritten to give

$$F^\uparrow(u) = \int_0^\infty F_\nu^\uparrow(u)\,d\nu = \int_0^\infty \pi B_\nu(T)\,d\nu$$
$$+ \int_T^{T_s}\left[\int_0^\infty \mathcal{T}_\nu^f(u - u', T')\frac{d\pi B_\nu(T')}{dT'}\,d\nu\right]dT', \tag{4.81}$$

$$F^\downarrow(u) = \int_0^\infty F_\nu^\downarrow(u)\,d\nu = \int_0^\infty \pi B_\nu(T)\,d\nu - \int_0^\infty \pi B_\nu(T_t)\mathcal{T}_\nu^f(u_1 - u, T_t)\,d\nu$$
$$- \int_{T_t}^T\left[\int_0^\infty \mathcal{T}_\nu^f(u' - u, T')\frac{d\pi B_\nu(T')}{dT'}\,d\nu\right]dT'. \tag{4.82}$$

To construct the radiation chart, we define

$$Q(u, T) = \int_0^\infty \mathcal{T}_\nu^f(u, T)\frac{d\pi B_\nu(T)}{dT}\,d\nu. \tag{4.83}$$

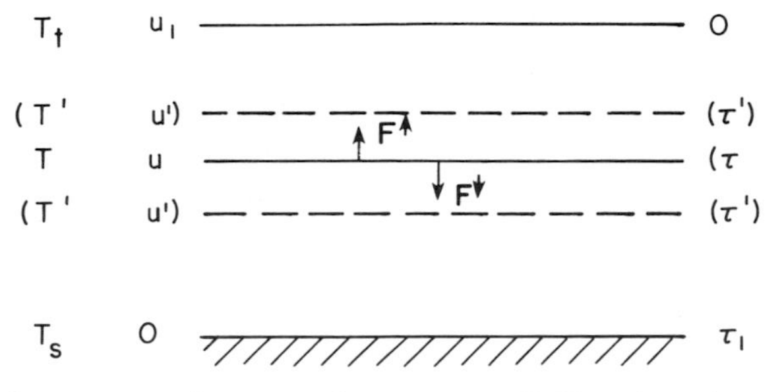

Fig. 4.9 Coordinate systems for τ, u, and temperature T. Note that τ increases downward whereas u increases upward.

Further, we note $[\mathcal{T}_\nu^f(0, T) = 1]$

$$\int_0^\infty \pi B_\nu(T)\,d\nu = \int_0^\infty d\nu \int_0^T \mathcal{T}_\nu^f(0, T')\frac{d\pi B_\nu(T')}{dT'}\,dT'$$

$$= \int_0^T Q(0, T')\,dT' = \sigma T^4 \qquad (4.84)$$

and

$$\int_0^\infty \pi B_\nu(T_t)\mathcal{T}_\nu^f(u_1 - u, T_t)\,d\nu$$

$$= \int_0^\infty d\nu \int_0^{T_t} \frac{d}{dT'}\left[\pi B_\nu(T')\mathcal{T}_\nu^f(u_1 - u, T')\right]dT'$$

$$= \int_0^\infty d\nu \int_0^{T_t}\left[\mathcal{T}_\nu^f(u_1 - u, T')\frac{d\pi B_\nu(T')}{dT'}\right.$$

$$\left. + \pi B_\nu(T')\frac{d\mathcal{T}_\nu^f(u_1 - u, T')}{dT'}\right]dT'. \qquad (4.85)$$

If the temperature dependence of the slab transmittance can be neglected, i.e., $d\mathcal{T}_\nu^f/dT \approx 0$, we should have

$$\int_0^\infty \pi B_\nu(T_t)\mathcal{T}_\nu^f(u_1 - u, T_t)\,d\nu \cong \int_0^{T_t} Q(u_1 - u, T')\,dT'. \qquad (4.86)$$

By virtue of Eqs. (4.84) and (4.86), we obtain

$$F^\downarrow(u) = \int_0^T Q(0, T')\,dT' + \int_T^{T_t} Q(u' - u, T')\,dT' + \int_{T_t}^0 Q(u_1 - u, T')\,dT', \quad (4.87)$$

$$F^\uparrow(u) = \int_0^T Q(0, T')\,dT' + \int_T^{T_s} Q(u - u', T')\,dT'. \qquad (4.88)$$

It is evident that the upward and downward flux densities are now given by a closed integration in Q–T domain and thus, they may be estimated by means of the graphical method on a diagram. Equations (4.87) and (4.88) originally were derived by Elsasser (1942). We now describe how the diagram, the radiation chart, is constructed.

We first define the abscissa and ordinate of the radiation chart, respectively, as

$$x = aT^2, \qquad y = Q/(2aT), \qquad (4.89)$$

where a is an arbitrary constant, so that

$$y\,dx = \frac{Q}{2aT}\,2aT\,dT = Q\,dT. \qquad (4.90)$$

Hence, an area on the Q–T diagram is equal to the flux densities as defined by Eqs. (4.87) and (4.88). We next find the boundaries of the diagram. From

Eq. (4.83), the quantity Q is a function of two variables u and T and has a maximum value when $\mathscr{T}_v^f = 1$; i.e., $u = 0$. Thus,

$$Q_{\max} = \int_0^\infty \frac{d}{dT} \pi B_v(T)\,dv = \frac{d}{dT} \int_0^\infty \pi B_v(T)\,dv = \frac{d}{dT}(\sigma T^4) = 4\sigma T^3. \quad (4.91)$$

It follows that $y_{\max} = (2\sigma/a^2)x$. Since y is linearly proportional to x in this case, $u = 0$ is represented by a straight line on the diagram and forms the upper edge of the radiation chart. Moreover, when $u = \infty$, $\mathscr{T}_v^f = 0$; this implies $Q = 0$; i.e., $y = 0$. We may interpret that at this line there is an infinite path length which absorbs all the radiation falling upon it, the so-called blackbody. The radiation chart is schematically shown in Fig. 4.10. The abscissa is $x = aT^2$ which increases from the right to the left, and the ordinate is $y = Q/(2aT)$. The vertical lines are isotherms T, while the slanting curves represent lines of constant path length u (isopleths).

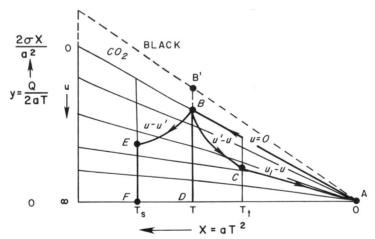

Fig. 4.10   Schematic diagram of the Elsasser's radiation chart.

To evaluate the downward and upward flux densities due to water vapor defined by Eqs. (4.87) and (4.88) in the radiation chart, we proceed as follows:

$$\left.\begin{array}{l} 0 \to T \text{ (along } u = 0\text{), } AB \\ T \to T_t \text{ (along } (u' - u)\text{, } u' = u_1, u\text{), } BC \\ T_t \to 0 \text{ (along } (u_1 - u)\text{), } CA \end{array}\right\} \quad F^\downarrow(u) = \text{area } (ABCA),$$

$$\left.\begin{array}{l} 0 \to T \text{ (along } u = 0\text{), } AB \\ T \to T_s \text{ (along } (u - u')\text{, } u' = u, 0\text{), } BE \\ T_s = \text{black surface, } EF \end{array}\right\} \quad F^\uparrow(u) = \text{area } (ABEFA).$$

We note that in the last evaluation of the upward flux density, the temperature integration reaches the surface. Since the surface follows the blackbody radiation and may be thought of as composed of an infinitely thick isothermal layer of temperature T_s, point E has to reach the point where $u = \infty$ along the isotherm T_s. We further note that the flux coming from an isothermal layer of infinite thickness is given by the area of the triangle along $u = 0$ to the right of the isotherm corresponding to the temperature of the layer.

To compute the upward and downward flux densities due to carbon dioxide, an empirical method is utilized. It is assumed that CO_2 absorbs so strongly in the 15 μm band that a thin atmosphere may be considered as a blackbody in this spectral interval. Thus, $F^{\uparrow\downarrow}(CO_2)$ at any reference level must originate in the thin layer of temperature T immediately adjacent to the level, and it is approximately given by $0.185\sigma T^4$. It is now straightforward to evaluate the net flux density and the infrared cooling rate on the radiation chart. Since the upward and downward flux densities are the same for CO_2 emission, the cooling rate calculation employing the Elsasser's radiation chart is due to water vapor only. Various graphical methods to estimate the infrared fluxes based on the principle already discussed also have been presented by Möller (1943) and Yamamoto (1952).

In reference to Eq. (4.86), we see that

$$\int_0^\infty \pi B_\nu(T) \mathscr{T}_\nu^f(u, T)\, d\nu = \int_0^T \tilde{Q}(u, T')\, dT' \qquad (4.92)$$

represents the transmission of an isothermal layer of path length u. Thus, the emission of this isothermal layer may be written as

$$\int_0^\infty \pi B_\nu(T)[1 - \mathscr{T}_\nu^f(u, T)]\, d\nu = \int_0^T \tilde{R}(u, T')\, dT', \qquad (4.93)$$

where a new parameter $\tilde{R}(u, T)$ is defined in terms of the flux transmission function. Moreover, from the broadband flux transmissivity and emissivity defined in Eqs. (4.74) and (4.75), it is apparent that

$$\int_0^T \tilde{Q}(u, T')\, dT' = t^f(u, T)\sigma T^4, \quad \int_0^T \tilde{R}(u, T')\, dT' = \varepsilon^f(u, T)\sigma T^4. \quad (4.94)$$

It should be noted that in the construction of Elsasser's radiation chart, effect of the temperature dependence on the slab transmittance has been neglected so that Eq. (4.86) may be applied. Zdunkowski *et. al.* (1966), Sasamori (1968), and Charlock and Herman (1976) have pointed out this shortcoming and have discussed the validity of using the numerical tables presented by Elsasser and Culbertson (1960) in the flux density calculations.

4.9   CARBON DIOXIDE AND CLIMATE

In recent years, one of the major concerns in climate studies has been the steady increase in the carbon dioxide content of the atmosphere produced by the rapid burning of fossil fuels and its impact on the atmospheric temperature and climate changes of the earth–atmosphere system. Since the beginning of the industrial revolution more than a century ago, man-made carbon dioxide has been released increasingly and continuously to the atmosphere through the combustion of fossil fuels (primarily coal, petroleum, and natural gas). The combustion of fossil carbon produces CO_2 via the oxidation reaction $C + O_2 \rightarrow CO_2$.

The atmospheric CO_2 concentrations recorded at Mauna Loa, Hawaii, and other locations show a steady increase in the annual average. Figure 4.11 depicts the monthly values of CO_2 concentration at Mauna Loa (19°N). The yearly increase in concentration is also given in the abscissa. There are

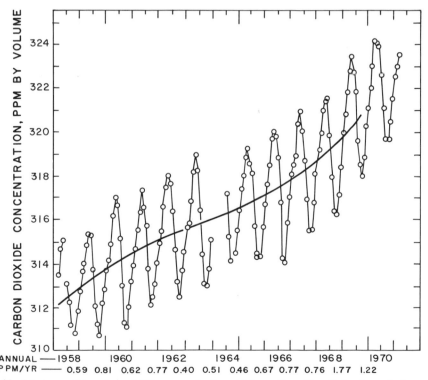

Fig. 4.11   Mean monthly values of CO_2 concentration at Mauna Loa, Hawaii, for the period 1958–1971. The solid line is the best-fit to the data (from Study of Man's Impact on Climate, 1971).

two distinct features in these observations. The first is a seasonal variation in which the CO_2 concentration decreases during the summer growing season of the northern hemisphere. The second is a long-term upward trend presumed to be a consequence of the combustion of fossil fuels. The increase in the annual average amounts to about 4% rise in total CO_2 between about 1958 and 1972. The present-day CO_2 excess relative to the year of 1850 (~ 290 ppm, i.e., parts per million by volume) is estimated to be at about 13%. Moreover, it has been estimated that between 50 and 75% of the fossil CO_2 input to the atmosphere from human activities has stayed in the atmosphere with the remaining part gone into the ocean and the biosphere, mostly the forests. It is also estimated that the present CO_2 concentration with a value of about 330 ppm will reach some 380–390 ppm by the year 2000.

As discussed in Section 3.4, CO_2 is virtually transparent to solar radiation. However, it is a strong absorber in the 15 μm band (~ 12–18 μm) of the thermal infrared spectrum described in various sections of this chapter. Consequently, an increase of the atmospheric CO_2 content could result in trapping the outgoing thermal infrared radiation emitted from the lower atmosphere and could produce the greenhouse effect, raising its temperature. To what degree and extent will the increase of the CO_2 concentration influence the atmospheric and earth's surface temperatures?

There have been a number of investigations on the effect of the changes of atmospheric CO_2 on the atmospheric radiative budget. But a reliable determination of the changes of atmospheric temperature due to the variation of the CO_2 concentration must take into consideration not only the radiative aspect of CO_2 properties, but also the convective nature of the lower atmosphere (see Section 8.5.2). Based on a radiative-convective equilibrium model (see also Section 8.5.2), Manabe and Wetherald (1967) concluded that increases of CO_2 resulted in a warming of the entire lower atmosphere. Using the assumptions of constant relative humidity and fixed cloudiness and the broadband emissivity values for H_2O, CO_2, and O_3, they found that the changes of mean atmospheric temperature due to CO_2 are such that a 10% increase of CO_2 concentration (from 300 to 330 ppm) would lead to a warming of 0.3°K. Doubling of CO_2 concentration from 300 to 600 ppm produces a 2.36°K increase in the equilibrium temperature of the earth's surface. More recently, by means of a three-dimensional general circulation model, Manabe and Wetherald (1975) showed an increase of 2.39°K by doubling the CO_2 concentration, a slightly higher value than their earlier results. In addition, they also found that large cooling occurs in the model stratosphere caused by the increasing emission to space resulting from the increasing CO_2 concentration. Also, the tropospheric warming is most pronounced in high latitudes of the lower troposphere due to the fact that the vertical mixing by convection is suppressed in the stable layer of the tropo-sphere in the polar regions. It should be noted that in none of these models, simple one-dimensional radiative-convective models or more sophisticated general circulation models, has allowance been made for the significant thermal feedback effects as the cloudcover changes. Until such time as the modeling of the cloud interaction effects can be effectively incorporated in the model, it is very likely that the thermal effects of CO_2 changes in the real atmosphere may be different than those predicted by the current models.

EXERCISES

4.1 Show that the spectral infrared cooling rate may be expressed by

$$-\frac{c_p}{q}\left(\frac{\partial T}{\partial t}\right)_{\bar{v}} = \pi B_{\bar{v}}(T_t)\frac{d\mathcal{T}^f_{\bar{v}}(u_1 - u)}{du} - \int_0^{u_1}\frac{\partial \mathcal{T}^f_{\bar{v}}(u - u')}{\partial u}\frac{d\pi B_{\bar{v}}(u')}{du'}\,du',$$

where T_t denotes the temperature at the top of the atmosphere.

4.2 Derive the regions of linear and square root absorption directly from Eqs. (4.27) and (4.28). Note that for weak absorption $k_v u \ll 1$, while for strong absorption $\alpha \ll v - v_0$. (*Hint:* For strong absorption, let $\eta = s\alpha u/[\pi(v - v_0)^2]$.)

4.3 Compute and plot the monochromatic transmission function for a Lorentz line as a function of $(v - v_0)/\alpha$ for $x = su/(2\pi\alpha)$ of 0.1, 1, and 5. Compare your results with those in Fig. 4.4.

4.4 Derive Eq. (4.31) from Eq. (4.30).

4.5 Derive Eq. (4.58) from Eq. (4.56) and show that if the half widths of lines are much smaller than their mutual distance, the transmission function reduces to

$$\mathcal{T}_{\bar{v}}(u) = \exp(-\sqrt{\pi S_0 \alpha u}/\delta).$$

This is the square root approximation for the random model.

4.6 On the basis of Eq. (4.56) we define the equivalent width of the lines as

$$W = \int_0^\infty P(S)\,dS \int_{-\infty}^\infty (1 - e^{-k_v u})\,dv.$$

It has been found that the realistic distributions of line strengths do not follow the Poission function expressed by Eq. (4.57), but rather are given by

$$P(S) = (N_0/S)e^{-S/\bar{S}},$$

where N_0 is a normalization factor. Show that the transmission function is given by

$$\mathscr{T}_{\bar{v}} = e^{-W/\delta},$$

where

$$W = 2\pi\alpha N_0\left[\left(1 + \frac{\bar{S}u}{\pi\alpha}\right)^{1/2} - 1\right].$$

4.7   Using the square root approximation for the random model, show that the precipitable water in a clear atmosphere (no cloud and aerosol) may be derived from

$$PW = (c/m)[\ln(F_{\Delta\lambda}/F_{0,\Delta\lambda})]^2,$$

where c is a constant related to the band and known atmospheric parameters, m denotes the air mass ($\sim$ pressure), and $F_{\Delta\lambda}$ and $F_{0,\Delta\lambda}$ represent the observed solar flux in 0.94 μm band at the ground and at the top of the atmosphere, respectively. This is the principle of the sun photometer for the measurement of precipitable water.

4.8   Given the following vertical distribution of temperature and specific humidity from radiosonde observations:

Pressure (mb)	Temperature (°C)	Specific humidity (%)
1000	12.0	0.82
950	10.3	0.49
900	8.1	0.43
850	4.1	0.42
800	0.1	0.41
750	− 3.7	0.30
700	− 7.5	0.20
650	− 9.6	0.09
600	− 11.8	0.04

compute the upward and downward fluxes using Eqs. (4.72) and (4.73), and the heating rate at 800 mb. Use the empirical values for the emissivity given in Fig. 4.8. For the convenience of calculations, perform integration by parts for Eqs. (4.72) and (4.73), and remove the differentiation of the flux emissivity with respect to the path length.

4.9   Utilizing the schematic diagram of the Elsasser radiation chart depicted in Fig. 4.10, sketch the areas corresponding to (a) the cooling rate for a layer having the top and bottom temperatures of T_1 and T_2 ($T_1 < T_2$); (b) the net flux densities at the surface; and (c) the net flux densities below a black cloud whose base temperature is T_b ($T_b < T_s$).

SUGGESTED REFERENCES

Elsasser, W. M., and Culbertson, M. F. (1960). Atmospheric radiation tables. *Meteorol. Monog.* **4**, No. 23, 1–43. This monograph provides essential information on the transfer of infrared radiation and broadband emissivity values.

Goody, R. M. (1964). *Atmospheric Radiation*, Vol. I: *Theoretical Basis*, Oxford Univ. Press (Clarendon), London and New York. Chapters 3–5 contain authoritative and fundamental discussions on the theory of gaseous absorption and band models.

Kondratyev, K. Ya. (1969). *Radiation in the Atmosphere* (International Geophysics Series, Vol. 12). Academic Press, New York. Gaseous absorption and infrared radiative transfer are discussed in Chapters 3 and 9.

Tiwari, S. N. (1978). *Models for infrared atmospheric radiation. Adv. Geophys.* **20**, 1–80. This review paper gives updated information on various band models and transmittance calculations.

Chapter 5

LIGHT SCATTERING BY PARTICULATES IN THE ATMOSPHERE

The earth's atmosphere contains cloud and aerosol particles whose sizes are much larger than the wavelengths of the incoming visible sunlight. Thus, the dipole mode of the electric field, which leads to the development of the Rayleigh scattering theory, is not applicable. Because of the large particle size, the incident beam of light induces high-order modes of polarization configuration, which require more advanced treatment. This chapter presents the scattering of electromagnetic waves by a homogeneous isotropic sphere from the classical wave equations, the so-called Mie scattering theory (Mie, 1908). Maxwell's equations, which are the fundamentals in theoretical optics, are introduced first. The formal solution of the scattering problem is presented following the derivation of the solution of the vector wave equation in spherical coordinates. Far field solutions are then given, and we show how the scattering matrix, in reference to a plane containing scattered and incident waves, is obtained. Extinction and scattering cross sections for a single sphere and for a polydispersion of spheres are further discussed. The final section is concerned with the asymptotic ray-optics approach to the scattering problem. The approximation is based on the localization principle in which the incident beam of light may be thought of as consisting of separate rays of light pursuing their own path. It includes discussions of Fraunhofer diffraction and geometrical reflection and refraction leading to the explanation of the corona, and rainbow and halo phenomena, respectively. Comparisons be-

122

tween the formal Mie scattering theory and ray optics approximation are further carried out to understand pronounced features that occur in the scattered intensity and polarization. Lastly, this chapter also provides some discussions on the scattering of light by nonspherical ice crystals that are typically observed in cirrus clouds.

5.1 MAXWELL EQUATIONS

The state of excitation which is established in space by the presence of electric charges is said to constitute an electromagnetic field. It is represented by two vectors $\mathbf{E}$ and $\mathbf{B}$, called the electric vector and magnetic induction, respectively. It is necessary to introduce a second set of vectors, the electric current density $\mathbf{j}$, the electric displacement $\mathbf{D}$, and the magnetic vector $\mathbf{H}$, to describe effects of the electromagnetic field on material objects. At every point where the physical properties of the medium are continuous in its neighborhood, the space and time derivatives of these five vectors can be related by Maxwell equations:

$$\nabla \times \mathbf{H} = \frac{1}{c} \frac{\partial \mathbf{D}}{\partial t} + \frac{4\pi}{c} \mathbf{j}, \tag{5.1}$$

$$\nabla \times \mathbf{E} = -\frac{1}{c} \frac{\partial \mathbf{B}}{\partial t}, \tag{5.2}$$

$$\nabla \cdot \mathbf{D} = 4\pi\rho, \tag{5.3}$$

$$\nabla \cdot \mathbf{B} = 0, \tag{5.4}$$

where t denotes time, c the velocity of light, and ρ the density of charge. Equation (5.3) may be regarded as a defining equation for the electric charge density ρ, and Eq. (5.4) implies that no free magnetic poles exist. Here the Gaussian (cgs) system of units is used.

From Eq. (5.1), since $\nabla \cdot \nabla \times \mathbf{H} = 0$, dot product operation leads to

$$\nabla \cdot \mathbf{j} = -\frac{1}{4\pi} \nabla \cdot \frac{\partial \mathbf{D}}{\partial t}. \tag{5.5}$$

Hence, differentiating Eq. (5.3) with respect to t, we obtain

$$\frac{\partial \rho}{\partial t} + \nabla \cdot \mathbf{j} = 0. \tag{5.6}$$

This is the equation of continuity in an electromagnetic field.

To allow a unique determination of the field vectors from a given distribution of current and charges, these equations must be supplemented by relations describing the behavior of substances under the influence of the field.

These relations are given by

$$\mathbf{j} = \sigma \mathbf{E}, \tag{5.7}$$

$$\mathbf{D} = \varepsilon \mathbf{E}, \tag{5.8}$$

$$\mathbf{B} = \mu \mathbf{H}, \tag{5.9}$$

where σ is the specific conductivity, ε the permittivity, and μ the magnetic permeability.

We shall now confine our attention to the field where there are no charges ($\rho = 0$) and currents ($|\mathbf{j}| = 0$), and to the medium which is homogeneous so that ε and μ are constants. Thus, Maxwell equations reduce to

$$\nabla \times \mathbf{H} = \frac{\varepsilon}{c} \frac{\partial \mathbf{E}}{\partial t}, \tag{5.10}$$

$$\nabla \times \mathbf{E} = \frac{-\mu}{c} \frac{\partial \mathbf{H}}{\partial t}, \tag{5.11}$$

$$\nabla \cdot \mathbf{E} = 0, \tag{5.12}$$

$$\nabla \cdot \mathbf{H} = 0. \tag{5.13}$$

Equations (5.10)–(5.13) will be used to derive the electromagnetic wave equation. Note here that Eqs. (5.12) and (5.13) can be obtained immediately from Eqs. (5.10) and (5.11) by carrying out the dot operation.

5.2   THE ELECTROMAGNETIC WAVE EQUATION AND ITS SOLUTION

We consider a plane electromagnetic wave in a periodic field with a circular frequency ω so that we may write

$$\mathbf{E} \rightarrow \mathbf{E}e^{i\omega t}, \tag{5.14}$$

$$\mathbf{H} \rightarrow \mathbf{H}e^{i\omega t}. \tag{5.15}$$

On the basis of these transformations, Eqs. (5.10) and (5.11) become

$$\nabla \times \mathbf{H} = ikm^2\mathbf{E}, \tag{5.16}$$

$$\nabla \times \mathbf{E} = -ik\mathbf{H}, \tag{5.17}$$

where $k = 2\pi/\lambda\,(=\omega/c)$ is the wave number denoting the propagation constant in vacuum, λ is the wavelength in vacuum, $m = \sqrt{\varepsilon}$ is the complex refractive index of the medium at the frequency ω, and the permeability $\mu \approx 1$ for air.

We now perform the curl operation on Eq. (5.17) to obtain

$$\nabla \times \nabla \times \mathbf{E} = -ik\nabla \times \mathbf{H}. \tag{5.18}$$

Moreover, by noting that $\nabla \cdot \nabla \times \mathbf{E} = 0$ and that $\nabla \cdot \mathbf{E} = 0$, we get

$$\nabla^2 \mathbf{E} = -k^2 m^2 \mathbf{E}. \tag{5.19}$$

In a similar way we have from Eqs. (5.16) and (5.13)

$$\nabla^2 \mathbf{H} = -k^2 m^2 \mathbf{H}. \tag{5.20}$$

Equations (5.19) and (5.20) indicate that the electric vector and magnetic induction in a homogeneous medium satisfy the vector wave equation

$$\nabla^2 \mathbf{A} + k^2 m^2 \mathbf{A} = 0, \tag{5.21}$$

where $\mathbf{A}$ may be either $\mathbf{E}$ or $\mathbf{H}$.

Now, if ψ satisfies the scalar wave equation

$$\nabla^2 \psi + k^2 m^2 \psi = 0, \tag{5.22}$$

vectors $\mathbf{M}_\psi$ and $\mathbf{N}_\psi$ in spherical coordinates (r, θ, ϕ) defined by

$$\mathbf{M}_\psi = \nabla \times [\mathbf{a}_r(r\psi)] = \left(\mathbf{a}_r \frac{\partial}{\partial r} + \mathbf{a}_\theta \frac{1}{r}\frac{\partial}{\partial \theta} + \mathbf{a}_\phi \frac{1}{r\sin\theta}\frac{\partial}{\partial \phi} \right) \times [\mathbf{a}_r(r\psi)]$$

$$= \mathbf{a}_\theta \frac{1}{r\sin\theta}\frac{\partial(r\psi)}{\partial \phi} - \mathbf{a}_\phi \frac{1}{r}\frac{\partial(r\psi)}{\partial \theta},$$

$$\tag{5.23}$$

$$mk\mathbf{N}_\psi = \nabla \times \mathbf{M}_\psi$$

$$= \mathbf{a}_r \left[\frac{\partial^2(r\psi)}{\partial r^2} + m^2 k^2(r\psi) \right] + \mathbf{a}_\theta \frac{1}{r}\frac{\partial^2(r\psi)}{\partial r \partial \theta} + \mathbf{a}_\phi \frac{1}{r\sin\theta}\frac{\partial^2(r\psi)}{\partial r \partial \phi} \tag{5.24}$$

satisfy the vector wave equation defined in Eq. (5.21) subject to Eq. (5.22). Vectors $\mathbf{a}_r, \mathbf{a}_\theta$, and $\mathbf{a}_\phi$ are unit vectors in spherical coordinates. To obtain Eq. (5.24), we have used Eq. (5.29) defined below.

Assuming that u and v are two independent solutions of the scalar wave equation defined in Eq. (5.22), then the electric and magnetic field vectors expressed by

$$\mathbf{E} = \mathbf{M}_v + i\mathbf{N}_u, \tag{5.25}$$

$$\mathbf{H} = m(-\mathbf{M}_u + i\mathbf{N}_v) \tag{5.26}$$

satisfy Eqs. (5.16) and (5.17). Employing Eqs. (5.23) and (5.24), $\mathbf{E}$ and $\mathbf{H}$ can be written explicitly as

$$\mathbf{E} = \mathbf{a}_r \frac{i}{mk}\left[\frac{\partial^2(ru)}{\partial r^2} + m^2k^2(ru)\right]$$

$$+ \mathbf{a}_\theta\left[\frac{1}{r\sin\theta}\frac{\partial(rv)}{\partial\phi} + \frac{i}{mkr}\frac{\partial^2(ru)}{\partial r\,\partial\theta}\right]$$

$$+ \mathbf{a}_\phi\left[-\frac{1}{r}\frac{\partial(rv)}{\partial\theta} + \frac{1}{mkr\sin\theta}\frac{\partial^2(ru)}{\partial r\,\partial\phi}\right], \qquad (5.27)$$

$$\mathbf{H} = \mathbf{a}_r \frac{i}{k}\left[\frac{\partial^2(rv)}{\partial r^2} + m^2k^2(rv)\right]$$

$$+ \mathbf{a}_\theta\left[-\frac{m}{r\sin\theta}\frac{\partial(ru)}{\partial\phi} + \frac{i}{kr}\frac{\partial^2(rv)}{\partial r\,\partial\theta}\right]$$

$$+ \mathbf{a}_\phi\left[\frac{m}{r}\frac{\partial(ru)}{\partial\theta} + \frac{i}{kr\sin\theta}\frac{\partial^2(rv)}{\partial r\,\partial\phi}\right]. \qquad (5.28)$$

The scalar wave equation defined in Eq. (5.22) in spherical coordinates is given by

$$\frac{1}{r^2}\frac{\partial}{\partial r}\left(r^2\frac{\partial\psi}{\partial r}\right) + \frac{1}{r^2\sin\theta}\frac{\partial}{\partial\theta}\left(\sin\theta\frac{\partial\psi}{\partial\theta}\right)$$

$$+ \frac{1}{r^2\sin^2\theta}\frac{\partial^2\psi}{\partial\phi^2} + k^2m^2\psi = 0. \qquad (5.29)$$

This equation is separable by letting

$$\psi(r,\theta,\phi) = R(r)\Theta(\theta)\Phi(\phi). \qquad (5.30)$$

Upon substituting Eq. (5.30) into Eq. (5.29) and dividing the entire equation by $\psi(r,\theta,\phi)$, we obtain

$$\frac{1}{r^2}\frac{1}{R}\frac{\partial}{\partial r}\left(r^2\frac{\partial R}{\partial r}\right) + \frac{1}{r^2\sin\theta}\frac{1}{\Theta}\frac{\partial}{\partial\theta}\left(\sin\theta\frac{\partial\Theta}{\partial\theta}\right)$$

$$+ \frac{1}{r^2\sin^2\theta}\frac{1}{\Phi}\frac{\partial^2\Phi}{\partial\phi^2} + k^2m^2 = 0. \qquad (5.31)$$

If Eq. (5.31) is multiplied by $r^2\sin^2\theta$, we get

$$\left[\sin^2\theta\frac{1}{R}\frac{\partial}{\partial r}\left(r^2\frac{\partial R}{\partial r}\right) + \sin\theta\frac{1}{\Theta}\frac{\partial}{\partial\theta}\left(\sin\theta\frac{\partial\Theta}{\partial\theta}\right) + k^2m^2r^2\sin^2\theta\right]$$

$$+ \frac{1}{\Phi}\frac{\partial^2\Phi}{\partial\phi^2} = 0. \qquad (5.32)$$

Since the first three terms in this equation consist of variables r and θ, but not ϕ, the only possibility that Eq. (5.32) may be valid is when

$$\frac{1}{\Phi}\frac{d^2\Phi}{d\phi^2} = \text{const} = -l^2, \tag{5.33}$$

where we set the constant equal to $-l^2$ (l denotes an integer) for mathematical convenience. In view of Eqs. (5.32) and (5.33) it is also clear that

$$\sin^2\theta \frac{1}{R}\frac{\partial}{\partial r}\left(r^2\frac{\partial R}{\partial r}\right) + \sin\theta\frac{1}{\Theta}\frac{\partial}{\partial\theta}\left(\sin\theta\frac{\partial\Theta}{\partial\theta}\right)$$

$$+ k^2m^2r^2\sin^2\theta - l^2 = 0. \tag{5.34}$$

Upon dividing Eq. (5.34) by $\sin^2\theta$, we obtain

$$\frac{1}{R}\frac{\partial}{\partial r}\left(r^2\frac{\partial R}{\partial r}\right) + k^2m^2r^2 + \frac{1}{\sin\theta}\frac{1}{\Theta}\frac{\partial}{\partial\theta}\left(\sin\theta\frac{\partial\Theta}{\partial\theta}\right) - \frac{l^2}{\sin^2\theta} = 0. \tag{5.35}$$

Thus, we must have

$$\frac{1}{R}\frac{d}{dr}\left(r^2\frac{dR}{dr}\right) + k^2m^2r^2 = \text{const} = n(n+1), \tag{5.36}$$

$$\frac{1}{\sin\theta}\frac{1}{\Theta}\frac{d}{d\theta}\left(\sin\theta\frac{d\Theta}{d\theta}\right) - \frac{l^2}{\sin^2\theta} = \text{const} = -n(n+1) \tag{5.37}$$

in order to satisfy Eq. (5.35), where n is an integer. The selection of the constant here is also for mathematical convenience. Rearranging Eqs. (5.33), (5.36), and (5.37), we have

$$\frac{d^2(rR)}{dr^2} + \left[k^2m^2 - \frac{n(n+1)}{r^2}\right](rR) = 0, \tag{5.38}$$

$$\frac{1}{\sin\theta}\frac{d}{d\theta}\left(\sin\theta\frac{d\Theta}{d\theta}\right) + \left[n(n+1) - \frac{l^2}{\sin^2\theta}\right]\Theta = 0, \tag{5.39}$$

$$\frac{d^2\Phi}{d\phi^2} + l^2\Phi = 0. \tag{5.40}$$

The single value solution for Eq. (5.40) is simply

$$\Phi = a_l\cos l\phi + b_l\sin l\phi, \tag{5.41}$$

where a_l and b_l are arbitrary constants. Equation (5.39) is the well-known equation for spherical harmonics. For convenience we introduce a new variable $\mu = \cos\theta$ so that

$$\frac{d}{d\mu}\left[(1-\mu^2)\frac{d\Theta}{d\mu}\right] + \left[n(n+1) - \frac{l^2}{1-\mu^2}\right]\Theta = 0. \tag{5.42}$$

The solutions of Eq. (5.42) can be expressed by the associated Legendre polynomials (spherical harmonics of the first kind) in the form

$$\Theta = p_n^l(\mu) = p_n^l(\cos\theta). \tag{5.43}$$

Finally, in order to solve the remaining equation (5.38), we set

$$kmr = \rho, \qquad R = (1/\sqrt{\rho})Z(\rho) \tag{5.44}$$

to obtain

$$\frac{d^2Z}{d\rho^2} + \frac{1}{\rho}\frac{dZ}{d\rho} + \left[1 - \frac{(n+\frac{1}{2})^2}{\rho^2}\right]Z = 0. \tag{5.45}$$

The solution of this equation can be expressed by the general cylindrical function of order $n + \frac{1}{2}$ and is given by

$$Z = Z_{n+1/2}(\rho). \tag{5.46}$$

Thus, the solution of Eq. (5.38) is then

$$R = \frac{1}{\sqrt{kmr}} Z_{n+1/2}(kmr). \tag{5.47}$$

Upon combining Eqs. (5.41), (5.43), and (5.47), the elementary wave functions at all points on the surface of a sphere, therefore, are given by

$$\psi(r, \theta, \phi) = \frac{1}{\sqrt{kmr}} Z_{n+1/2}(kmr) p_n^l(\cos\theta)(a_l\cos l\phi + b_l\sin l\phi). \tag{5.48}$$

Each cylindrical function denoted in Eq. (5.47) may be expressed as a linear combination of two cylindrical functions of standard type, e.g., the Bessel functions $J_{n+1/2}(\rho)$ and the Neumann functions $N_{n+1/2}(\rho)$. We define

$$\psi_n(\rho) = \sqrt{\pi\rho/2}J_{n+1/2}(\rho), \qquad \chi_n(\rho) = -\sqrt{\pi\rho/2}N_{n+1/2}(\rho). \tag{5.49}$$

The functions ψ_n are regular in every finite domain of the ρ plane including the origin, whereas the functions χ_n have singularities at the origin $\rho = 0$ where they become infinite. Hence, we may use ψ_n, but not χ_n to represent the wave inside the sphere. On utilizing the definitions in Eq. (5.49), Eq. (5.47) can be rewritten in the form

$$rR = c_n\psi_n(kmr) + d_n\chi_n(kmr), \tag{5.50}$$

where c_n and d_n are arbitrary constants. Equation (5.50) now represents the general solution of Eq. (5.38).

It follows that the general solution of the scalar wave equation (5.29) then can be expressed by

$$r\psi(r, \theta, \phi) = \sum_{n=0}^{\infty} \sum_{l=-n}^{n} p_n^l(\cos\theta)[c_n\psi_n(kmr) + d_n\chi_n(kmr)]$$
$$\times (a_l \cos l\phi + b_l \sin l\phi). \tag{5.51}$$

Note that the electric and magnetic field vectors of the electromagnetic waves subsequently can be derived from Eqs. (5.27) and (5.28).

Moreover, when $c_n = 1$, and $d_n = i$, we note that

$$\psi_n(\rho) + i\chi_n(\rho) = \sqrt{\pi\rho/2} H_{n+1/2}^{(2)}(\rho) = \xi_n(\rho), \tag{5.52}$$

where $H_{n+1/2}^{(2)}$ is the half integral order Hankel function of the second kind. It has the property of vanishing at infinity in the complex plane and is suitable for the representation of the scattered wave.

5.3   FORMAL SCATTERING SOLUTION

Having the vector wave equation solved, we may now discuss the scattering of a plane wave by a homogeneous sphere. For simplicity, we assume that outside the medium is vacuum ($m = 1$), that the material of the sphere has an index of refraction m, and that the incident radiation is linearly polarized. We select the origin of a rectangular system of coordinates at the center of the sphere, with the positive Z axis along the direction of propagation of the incident wave. If the amplitude of the incident wave is normalized to unity, the incident electric and magnetic field vectors are

$$\mathbf{E}^i = \mathbf{a}_x e^{-ikz}, \qquad \mathbf{H}^i = \mathbf{a}_y e^{-ikz}, \tag{5.53}$$

where $\mathbf{a}_x$ and $\mathbf{a}_y$ are unit vectors along the X and Y axes, respectively.

The components of any vector, say $\mathbf{a}$, in the Cartesian system may be transformed to the spherical polar coordinates (r, θ, ϕ) defined by

$$x = r \sin\theta \cos\phi, \qquad y = r \sin\theta \sin\phi, \qquad z = r \cos\theta. \tag{5.54}$$

According to the geometrical relationship shown in Fig. 5.1, we find

$$\mathbf{a}_r = \mathbf{a}_x \sin\theta \cos\phi + \mathbf{a}_y \sin\theta \sin\phi + \mathbf{a}_z \cos\theta,$$
$$\mathbf{a}_\theta = \mathbf{a}_x \cos\theta \cos\phi + \mathbf{a}_y \cos\theta \sin\phi - \mathbf{a}_z \sin\theta, \tag{5.55}$$
$$\mathbf{a}_\phi = -\mathbf{a}_x \sin\phi + \mathbf{a}_y \cos\phi,$$

where $\mathbf{a}_x$, $\mathbf{a}_y$, and $\mathbf{a}_z$ are unit vectors along x, y, and z, respectively, and $\mathbf{a}_r$, $\mathbf{a}_\theta$, and $\mathbf{a}_\phi$ are unit vectors in spherical coordinates.

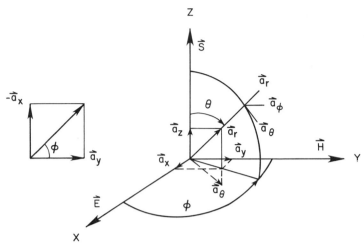

Fig. 5.1   Transformation of rectangular to spherical coordinates. S is the Poynting vector, and **a** is an arbitrary unit vector.

Thus, the electric and magnetic field vectors of the incident wave are

$$E_r^i = e^{-ikr\cos\theta}\sin\theta\cos\phi,$$
$$E_\theta^i = e^{-ikr\cos\theta}\cos\theta\cos\phi, \tag{5.56}$$
$$E_\phi^i = -e^{-ikr\cos\theta}\sin\phi,$$

$$H_r^i = e^{-ikr\cos\theta}\sin\theta\sin\phi,$$
$$H_\theta^i = e^{-ikr\cos\theta}\cos\theta\sin\phi, \tag{5.57}$$
$$H_\phi^i = e^{-ikr\cos\theta}\cos\phi.$$

On the basis of Bauer's Formula (Watson, 1944) the first factor on the right-hand side of this equation may be expressed in the following different-iable series of Legendre polynomials

$$e^{-ikr\cos\theta} = \sum_{n=0}^{\infty} (-i)^n(2n+1)\frac{\psi_n(kr)}{kr}P_n(\cos\theta), \tag{5.58}$$

where ψ_n is defined in Eq. (5.49). Also we have the mathematical identities

$$e^{-ikr\cos\theta}\sin\theta = \frac{1}{ikr}\frac{\partial}{\partial\theta}(e^{-ikr\cos\theta}), \tag{5.59}$$

$$\frac{\partial}{\partial\theta}P_n(\cos\theta) = -P_n^1(\cos\theta); \qquad P_0^1(\cos\theta) = 0. \tag{5.60}$$

Equation (5.60) relates the Legendre polynomial P_n with the associated Legendre polynomial P_n^1.

To determine the potentials u and v, only one of the components in Eq. (5.27) is needed. The first of them is ($m = 1$)

$$E_r^i = e^{-ikr\cos\theta}\sin\theta\cos\phi = \frac{i}{k}\left[\frac{\partial^2(ru^i)}{\partial r^2} + k^2(ru^i)\right]. \qquad (5.61)$$

In view of Eqs. (5.58)–(5.60), we have

$$e^{-ikr\cos\theta}\sin\theta\cos\phi = \frac{1}{(kr)^2}\sum_{n=1}^{\infty}(-i)^{n-1}(2n+1)\psi_n(kr)P_n^1(\cos\theta)\cos\phi. \qquad (5.62)$$

Accordingly, we take a trial solution in Eq. (5.61) in a series of a similar form

$$ru^i = \frac{1}{k}\sum_{n=1}^{\infty}\alpha_n\psi_n(kr)P_n^1(\cos\theta)\cos\phi. \qquad (5.63)$$

Upon substituting Eqs. (5.62) and (5.63) into Eq. (5.61) and comparing coefficients, we obtain

$$\alpha_n\left[k^2\psi_n(kr) + \frac{\partial^2\psi_n(kr)}{\partial r^2}\right] = (-i)^n(2n+1)\frac{\psi_n(kr)}{r^2}. \qquad (5.64)$$

In Eq. (5.50), since $\chi_n(kr)$ become infinite at the origin through which the incident wave must pass, we may let $c_n = 1$, and $d_n = 0$. Hence

$$\psi_n(kr) = rR \qquad (5.65)$$

is a solution of Eq. (5.38) (with $m = 1$)

$$\frac{d^2\psi_n}{dr^2} + \left[k^2 - \frac{\alpha}{r^2}\right]\psi_n = 0 \qquad (5.66)$$

provided that $\alpha = n(n+1)$. Comparing Eq. (5.66) with (5.64), we find

$$\alpha_n = (-i)^n\frac{2n+1}{n(n+1)}. \qquad (5.67)$$

Utilizing the similar procedures, v^i can be derived from Eq. (5.28). Thus, for incident waves outside the sphere, we have

$$ru^i = \frac{1}{k}\sum_{n=1}^{\infty}(-i)^n\frac{2n+1}{n(n+1)}\psi_n(kr)P_n^1(\cos\theta)\cos\phi,$$

$$rv^i = \frac{1}{k}\sum_{n=1}^{\infty}(-i)^n\frac{2n+1}{n(n+1)}\psi_n(kr)P_n^1(\cos\theta)\sin\phi. \qquad (5.68)$$

In order to match u^i and v^i with those of the internal and scattered waves whose potentials already have been derived in Eq. (5.51), the latter must be expressed in a series of similar form but with arbitrary coefficients. For internal waves, because the function $\chi_n(kmr)$ becomes infinite at the origin, only the function $\psi_n(kmr)$ may be used. Thus, for internal waves we have

$$ru^t = \frac{1}{mk} \sum_{n=1}^{\infty} (-i)^n \frac{2n+1}{n(n+1)} c_n \psi_n(kmr) P_n^1(\cos\theta) \cos\phi,$$

$$rv^t = \frac{1}{mk} \sum_{n=1}^{\infty} (-i)^n \frac{2n+1}{n(n+1)} d_n \psi_n(kmr) P_n^1(\cos\theta) \sin\phi.$$

(5.69)

For scattered waves, they must vanish at infinity, and the Hankel functions expressed in Eq. (5.52) will impart precisely this property. Thus, for scattered waves we have

$$ru^s = -\frac{1}{k} \sum_{n=1}^{\infty} (-i)^n \frac{2n+1}{n(n+1)} a_n \xi_n(kr) P_n^1(\cos\theta) \cos\phi,$$

$$rv^s = -\frac{1}{k} \sum_{n=1}^{\infty} (-i)^n \frac{2n+1}{n(n+1)} b_n \xi_n(kr) P_n^1(\cos\theta) \sin\phi.$$

(5.70)

The coefficients a_n, b_n, c_n, and d_n have to be determined from the boundary conditions at the surface of the sphere. The boundary conditions are that the tangential components of $\mathbf{E}$ and $\mathbf{H}$ be continuous across the spherical surface $r = a$. So we have

$$
\begin{array}{ll}
E_\theta^i + E_\theta^s = E_\theta^t, & H_\theta^i + H_\theta^s = H_\theta^t, \\
E_\phi^i + E_\phi^s = E_\phi^t, & H_\phi^i + H_\phi^s = H_\phi^t,
\end{array}
\quad \text{at} \quad r = a.
$$

(5.71)

In view of Eqs. (5.27), (5.29), and (5.68)–(5.70), it is evident that apart from common factors and differentiations with respect to θ and ϕ, which are the same for the wave inside and outside the sphere, both of the field components E_θ and E_ϕ contain the expressions v and $\partial(ru)/m\,\partial r$. It is also clear that components H_θ and H_ϕ contain mu and $\partial(ru)/\partial r$. Equation (5.71) implies that these four expressions have to be continuous at $r = a$. Consequently,

$$\frac{\partial}{\partial r}[r(u^i + u^s)] = \frac{1}{m}\frac{\partial}{\partial r}(ru^t), \qquad u^i + u^s = mu^t,$$

$$\frac{\partial}{\partial r}[r(v^i + v^s)] = \frac{\partial}{\partial r}(rv^t), \qquad v^i + v^s = v^t.$$

(5.72)

From these equations, it is now apparent that

$$
\begin{aligned}
m[\psi_n'(ka) - a_n \xi_n'(ka)] &= c_n \psi_n'(kma), \\
[\psi_n'(ka) - b_n \xi_n'(ka)] &= d_n \psi_n'(kma), \\
[\psi_n(ka) - a_n \xi_n(ka)] &= c_n \psi_n(kma), \\
m[\psi_n(ka) - b_n \xi_n(ka)] &= d_n \psi_n(kma),
\end{aligned}
$$

(5.73)

where the prime denotes differentiation with respect to the argument. Upon eliminating c_n and d_n, we obtain the coefficients for the scattered waves in the forms

$$a_n = \frac{\psi_n'(y)\psi_n(x) - m\psi_n(y)\psi_n'(x)}{\psi_n'(y)\xi_n(x) - m\psi_n(y)\xi_n'(x)},$$

$$b_n = \frac{m\psi_n'(y)\psi_n(x) - \psi_n(y)\psi_n'(x)}{m\psi_n'(y)\xi_n(x) - \psi_n(y)\xi_n'(x)}, \qquad (5.74)$$

where $x = ka$, and $y = mx$. As for c_n and d_n, fractions with the same respective denominators as those of a_n and b_n are found with $m[\psi_n'(x)\xi_n(x) - \psi_n(x)\xi_n'(x)]$ as a common numerator. At this point, solution of the scattering of electromagnetic waves by a sphere whose radius is $r = a$ and whose index of refraction is m is complete. The electric and magnetic field vectors expressed in Eqs. (5.27) and (5.28) at any point inside or outside the sphere are now expressed in terms of the known mathematical functions given by Eqs. (5.68)–(5.70). We have assumed up to this point that the suspending medium is a vacuum for simplicity. Now let the outside medium and the sphere have the refractive indices m_2 (real part) and m_1 (maybe complex), respectively. Replacing the m by m_1/m_2 and the wave number k by $m_2 k$ (vacuum), the results in Eq. (5.74) can be generalized to cases where a sphere is suspended in a medium.

5.4   THE FAR FIELD SOLUTION AND EXTINCTION PARAMETERS

We shall now consider the scattered field at very large distances from the sphere. We note that for practical applications, all light scattering observations are normally carried out in the far-field zone. In the far field, the Hankel functions denoted in Eq. (5.52) reduce to the form

$$\xi_n(kr) \approx i^{n+1} e^{-ikr}, \qquad kr \gg 1. \qquad (5.75)$$

With this simplification Eq. (5.70) becomes

$$ru^s \approx -\frac{ie^{-ikr}\cos\phi}{k} \sum_{n=1}^{\infty} \frac{2n+1}{n(n+1)} a_n P_n^1(\cos\theta),$$

$$rv^s \approx -\frac{ie^{-ikr}\sin\phi}{k} \sum_{n=1}^{\infty} \frac{2n+1}{n(n+1)} b_n P_n^1(\cos\theta). \qquad (5.76)$$

The three components of the electric and magnetic field vectors in Eqs. (5.27) and (5.28) then are given by

$$E_r^s = H_r^s \approx 0,$$

$$E_\theta^s = H_\phi^s \approx \frac{-i}{kr} e^{-ikr} \cos \phi \sum_{n=1}^{\infty} \frac{2n+1}{n(n+1)} \left[a_n \frac{dP_n^1(\cos \theta)}{d\theta} + b_n \frac{P_n^1(\cos \theta)}{\sin \theta} \right],$$

$$-E_\phi^s = H_\theta^s \approx \frac{-i}{kr} e^{-ikr} \sin \phi \sum_{n=1}^{\infty} \frac{2n+1}{n(n+1)} \left[a_n \frac{P_n^1(\cos \theta)}{\sin \theta} + b_n \frac{dP_n^1(\cos \theta)}{d\theta} \right].$$

$$(5.77)$$

We find that the radial components E_r^s and H_r^s may be neglected in the far-field zone. To simplify Eq. (5.77), we define two *scattering functions* of the forms

$$S_1(\theta) = \sum_{n=1}^{\infty} \frac{2n+1}{n(n+1)} [a_n \pi_n(\cos \theta) + b_n \tau_n(\cos \theta)],$$

$$S_2(\theta) = \sum_{n=1}^{\infty} \frac{2n+1}{n(n+1)} [b_n \pi_n(\cos \theta) + a_n \tau_n(\cos \theta)],$$

$$(5.78)$$

where

$$\pi_n(\cos \theta) = \frac{1}{\sin \theta} P_n^1(\cos \theta),$$

$$\tau_n(\cos \theta) = \frac{d}{d\theta} P_n^1(\cos \theta).$$

$$(5.79)$$

Thus, we may write

$$E_\theta^s = -\frac{i}{kr} e^{-ikr} \cos \phi S_2(\theta),$$

$$-E_\phi^s = -\frac{i}{kr} e^{-ikr} \sin \phi S_1(\theta).$$

$$(5.80)$$

These fields represent an outgoing spherical wave with amplitude and state of polarization as functions of the scattering angle θ. It is convenient to define the *perpendicular* and *parallel* components of the electric field as E_r and E_l, respectively. In reference to Fig. 5.2, the scattered perpendicular and parallel electric fields are given by

$$E_r^s = -E_\phi^s, \qquad E_l^s = E_\theta^s. \qquad (5.81)$$

Also, the normalized incident electric vector [see Eq. (5.53)] may be decomposed into perpendicular and parallel components as

$$E_r^i = e^{-ikz} \sin \phi, \qquad E_l^i = e^{-ikz} \cos \phi. \qquad (5.82)$$

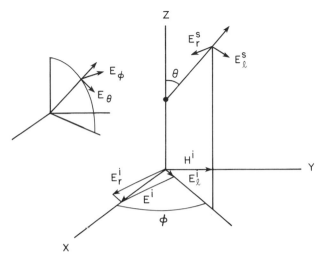

Fig. 5.2 Decomposition of the incident and scattered electric vectors into perpendicular and parallel components.

Equation (5.80) can then be expressed by

$$\begin{bmatrix} E_l^s \\ E_r^s \end{bmatrix} = \frac{e^{-ikr+ikz}}{ikr} \begin{bmatrix} S_2(\theta) & 0 \\ 0 & S_1(\theta) \end{bmatrix} \begin{bmatrix} E_l^i \\ E_r^i \end{bmatrix}. \tag{5.83}$$

Equation (5.83) is the fundamental equation for the study of radiation scattered by spheres including polarization.

The scattered intensity components in the far-field zone can now be written in terms of the incident intensity components in the form

$$I_l^s = I_l^i \frac{i_2}{k^2 r^2}, \qquad I_r^s = I_r^i \frac{i_1}{k^2 r^2}, \tag{5.84}$$

where

$$i_1(\theta) = |S_1(\theta)|^2, \qquad i_2(\theta) = |S_2(\theta)|^2, \tag{5.85}$$

and they are called the *intensity functions* for the perpendicular and parallel components, respectively. Each of these components of the scattered light can be thought of as arising from that component of the incident beam polarized in the same direction. The computational problem involved in Mie scattering is to compute i_1 and i_2 as functions of the scattering angle, the index of refraction m, and the particle size parameter $x = 2\pi a/\lambda$.

In the far-field zone, we would like to evaluate the reduction of the incident energy due to the absorption and scattering of light by a sphere. For this purpose we consider incident light polarized linearly in the perpendicular direction. From Eq. (5.83) the scattered electric field is given by

$$E_r^s = \frac{e^{-ikr+ikz}}{ikr} S_1(\theta)E_r^i. \tag{5.86}$$

Next, we consider a point (x, y, z) in the forward direction, i.e., $\theta \approx 0$. In the far field, since $x(y) \ll z$, we have in the forward direction

$$r = (x^2 + y^2 + z^2)^{1/2} \approx z + \frac{x^2 + y^2}{2z}. \tag{5.87}$$

Upon superimposing the incident and scattered electric fields in the forward direction, we obtain

$$E_r^i + E_r^s \approx E_r^i \left\{ 1 + \frac{S_1(0)}{ikz} e^{-ik(x^2+y^2)/2z} \right\}. \tag{5.88}$$

The far-field combined flux density in the forward direction then is proportional to

$$|E_r^i + E_r^s|^2 \approx |E_r^i|^2 \left\{ 1 + \frac{2}{kz} \text{Re}\left[\frac{S_1(0)}{i} e^{-ik(x^2+y^2)/2z} \right] \right\}, \tag{5.89}$$

where $\text{Re}[\ ]$ represents the real part of the argument (note that $z \gg x$ and y).

Integrating the combined flux density over the cross section area of a sphere whose radius is $r = a$, we obtain the total power of the combined image:

$$\frac{1}{|E_r^i|^2} \iint |E_r^i + E_r^s|^2 dx\, dy = \pi a^2 + \sigma_e, \tag{5.90}$$

where the first term on the right-hand side of Eq. (5.90) represents the cross section area of the sphere. The physical interpretation of the second term σ_e is that the total light received in the forward direction is reduced by the presence of the sphere, and the amount of the reduction is as if an area σ_e of the objective had been covered up. The double integral over $dx\, dy$, by which σ_e is defined, contains two Fresnel integrals, and if the limits are assumed to extend to ∞, we get

$$\int\int_{-\infty}^{\infty} e^{-ik(x^2+y^2)/2z} dx\, dy = \frac{2\pi z}{ik}. \tag{5.91}$$

Thus, the extinction cross section is

$$\sigma_e = (4\pi/k^2)\, \text{Re}[S(0)]. \tag{5.92}$$

We note here that in the forward direction

$$S_1(0) = S_2(0) = S(0) = \tfrac{1}{2} \sum_{n=1}^{\infty} (2n + 1)(a_n + b_n). \tag{5.93}$$

The fact that there is only one $S(0)$ is because of the symmetry of the forward scattering in which the extinction is independent of the state of polarization of the incident light. It should be noted that Eq. (5.92) is valid only when the sphere is isotropic and homogeneous. Furthermore, we define the extinction efficiency for a sphere with a radius of $r = a$ as

$$Q_e = \frac{\sigma_e}{(\pi a^2)} = \frac{2}{x^2} \sum_{n=1}^{\infty} (2n + 1) \operatorname{Re}[a_n + b_n], \qquad (5.94)$$

where $x = ka$ as denoted earlier, and it is called the *size parameter*.

The scattering cross section can be derived by the following procedures. From Eq. (5.80), the flux density of the scattered light in an arbitrary direction is given by

$$F(\theta, \phi) = \frac{F_0}{k^2 r^2} [i_2(\theta) \cos^2 \phi + i_1(\theta) \sin^2 \phi] \qquad (5.95)$$

with $F_0 = 1$ (unit incident amplitude). The total flux (or power) of the scattered light is therefore

$$f = \int_0^{2\pi} \int_0^{\pi} F(\theta, \phi) r^2 \sin \theta \, d\theta \, d\phi, \qquad (5.96)$$

where $\sin \theta \, d\theta \, d\phi$ is the differential solid angle $d\Omega$, and $r^2 \, d\Omega$ denotes the differential area. Hence, the scattering cross section may be defined as

$$\sigma_s = \frac{f}{F_0} = \frac{\pi}{k^2} \int_0^{\pi} [i_1(\theta) + i_2(\theta)] \sin \theta \, d\theta. \qquad (5.97)$$

In a similar way as in the extinction case, we define the scattering efficiency for a sphere

$$Q_s = \frac{\sigma_s}{\pi a^2} = \frac{1}{x^2} \int_0^{\pi} [i_1(\theta) + i_2(\theta)] \sin \theta \, d\theta. \qquad (5.98)$$

We note (see Appendix E) the following orthogonal and recurrence properties of the associated Legendre polynomials:

$$\int_0^{\pi} \left(\frac{dP_n^1}{d\theta} \frac{dP_m^1}{d\theta} + \frac{1}{\sin^2 \theta} P_n^1 P_m^1 \right) \sin \theta \, d\theta = \begin{cases} 0, & \text{if } n \neq m \\ \dfrac{2n(n+1)}{2n+1} \dfrac{(n+1)!}{(n-1)!}, & \text{if } n = m \end{cases} \qquad (5.99)$$

and

$$\int_0^{\pi} \left(\frac{P_n^1}{\sin \theta} \frac{dP_m^1}{d\theta} + \frac{P_m^1}{\sin \theta} \frac{dP_n^1}{d\theta} \right) \sin \theta \, d\theta = [P_n^1(\theta) P_m^1(\theta)]_0^{\pi} = 0. \qquad (5.100)$$

The scattering efficiency can be evaluated with the help of these two equations to yield

$$Q_s = \frac{2}{x^2} \sum_{n=1}^{\infty} (2n + 1)(|a_n|^2 + |b_n|^2). \qquad (5.101)$$

Finally, the absorption cross section and efficiency of a sphere can be calculated from

$$\sigma_a = \sigma_e - \sigma_s, \qquad Q_a = Q_e - Q_s. \qquad (5.102)$$

For an absorbing sphere, it is convenient to define the index of refraction as $m = m_r - im_i$, with m_r and m_i representing the real and imaginary parts of the refractive index, respectively.

Figure 5.3 shows the scattering efficiency factor Q_s as a function of the size parameter x for a real index of refraction of 1.33 with several values of the imaginary part. For $m_i = 0$, i.e., a perfect reflector, there is no absorption so that $Q_s = Q_e$. Q_s in this case shows a series of major maxima and minima and ripples. The major maxima and minima are due to interference of light diffracted and transmitted by the sphere, whereas the ripple arises from edge

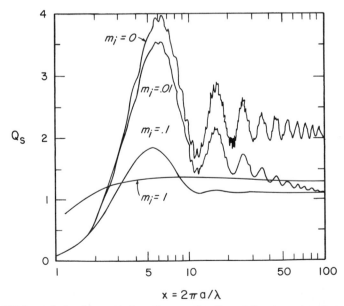

Fig. 5.3 Efficiency factor for scattering, Q_s, as a function of the size parameter $x = 2\pi a/\lambda$. The refractive index in $m_r = 1.33$, with results shown for four values of m_i (after Hansen and Travis, 1974).

rays that are grazing and traveling the sphere, spewing off energy in all directions. Q_s (or Q_e) increases rapidly when the size parameter reaches about five and approaches an asymptotic value of two. This implies that a large particle removes from the incident beam exactly twice the amount of light that it can intercept. Physically, the removal of the incident light beam includes the diffracted component, which passes by the particle, plus the light scattered by reflection and refraction within the particle to be discussed in Section 5.6. Both the ripples and the major maxima and minima damp out as absorption within the particle increases.

For nonabsorbing cases, the scattering efficiency factor may be expanded in terms of the expansion of the scattering coefficients a_n and b_n. It is given by (Penndorf, 1962)

$$Q_s = \frac{8x^4}{3}\left(\frac{m_r^2 - 1}{m_r^2 + 2}\right)^2\left[1 + \tfrac{6}{5}x^2\left(\frac{m_r^2 - 1}{m_r^2 + 2}\right) + x^4\left\{\frac{3}{175}\frac{m_r^6 + 41m_r^4 - 284m_r^2 + 284}{(m_r^2 + 2)^2}\right.\right.$$

$$\left.\left. + \frac{1}{900}\left(\frac{m_r^2 + 2}{2m_r^2 + 3}\right)^2 [15 + (2m_r^2 + 3)^2]\right\} + \cdots\right]. \tag{5.103}$$

The leading term is the dipole mode contribution, i.e., the Rayleigh scattering. This term is the same as that shown in Eq. (3.65) if πa^2 is divided by the scattering cross section in that equation; note that $N_s = 1/V$ where $V = \tfrac{4}{3}\pi a^3$ represents the volume.

5.5   THE SCATTERING PHASE MATRIX

On the basis of the Stokes parameters defined in Eq. (3.42), we may now express the incident and scattered electric vectors given by Eq. (5.83) in terms of the intensity components. Letting the subscript 0 denote the incident component, it is straightforward to show that

$$\begin{bmatrix} I \\ Q \\ U \\ V \end{bmatrix} = \mathbf{M}\begin{bmatrix} I_0 \\ Q_0 \\ U_0 \\ V_0 \end{bmatrix}, \tag{5.104}$$

where

$$\mathbf{M} = \begin{bmatrix} M_{11} & M_{12} & 0 & 0 \\ M_{12} & M_{11} & 0 & 0 \\ 0 & 0 & M_{33} & -M_{34} \\ 0 & 0 & M_{34} & M_{33} \end{bmatrix} \tag{5.105}$$

and

$$M_{11} = \frac{1}{2k^2r^2} \left[S_1(\theta)S_1^*(\theta) + S_2(\theta)S_2^*(\theta) \right],$$

$$M_{12} = \frac{1}{2k^2r^2} \left[S_2(\theta)S_2^*(\theta) - S_1(\theta)S_1^*(\theta) \right],$$

$$M_{33} = \frac{1}{2k^2r^2} \left[S_2(\theta)S_1^*(\theta) + S_1(\theta)S_2^*(\theta) \right],$$
(5.106)

$$-M_{34} = \frac{1}{2k^2r^2} \left[S_1(\theta)S_2^*(\theta) - S_2(\theta)S_1^*(\theta) \right].$$

M here is called the transformation matrix of a single sphere. For incident unpolarized light ($Q_0 = U_0 = V_0 = 0$), Eq. (5.104) reduces to Eq. (5.84).

In conjunction with the transformation matrix, we can define a parameter called phase matrix in such a way that

$$\mathbf{M}(\theta) = C\mathbf{P}(\theta)$$
(5.107)

and that

$$\int_0^{2\pi} \int_0^{\pi} \frac{P_{11}(\theta)}{4\pi} \sin\theta \, d\theta \, d\phi = 1.$$
(5.108)

On the basis of Eqs. (5.107) and (5.108), it is evident that

$$C = \tfrac{1}{2} \int_0^{\pi} M_{11}(\theta) \sin\theta \, d\theta = \frac{1}{4k^2r^2} \int_0^{\pi} \left[i_1(\theta) + i_2(\theta) \right] \sin\theta \, d\theta.$$
(5.109)

According to the definition of the scattering cross section in Eq. (5.97), the constant of proportionality C is

$$C = \sigma_s / (4\pi r^2).$$
(5.110)

Thus,

$$\frac{P_{11}}{4\pi} = \frac{1}{2k^2\sigma_s} (i_1 + i_2) = \frac{1}{2} \left(\frac{P_1}{4\pi} + \frac{P_2}{4\pi} \right),$$
(5.111a)

$$\frac{P_{12}}{4\pi} = \frac{1}{2k^2\sigma_s} (i_2 - i_1) = \frac{1}{2} \left(\frac{P_2}{4\pi} - \frac{P_1}{4\pi} \right),$$
(5.111b)

$$\frac{P_{33}}{4\pi} = \frac{1}{2k^2\sigma_s} (i_3 + i_4),$$
(5.111c)

$$-\frac{P_{34}}{4\pi} = \frac{1}{2k^2\sigma_s} (i_4 - i_3),$$
(5.111d)

where

$$i_1 = S_1 S_1^* = |S_1|^2, \tag{5.112a}$$

$$i_2 = S_2 S_2^* = |S_2|^2, \tag{5.112b}$$

$$i_3 = S_2 S_1^*, \tag{5.112c}$$

$$i_4 = S_1 S_2^*. \tag{5.112d}$$

The scattering phase matrix for a single homogeneous sphere is then

$$\mathbf{P} = \begin{bmatrix} P_{11} & P_{12} & 0 & 0 \\ P_{12} & P_{11} & 0 & 0 \\ 0 & 0 & P_{33} & -P_{34} \\ 0 & 0 & P_{34} & P_{33} \end{bmatrix}. \tag{5.113}$$

In general, if no assumption is made about the shape and position of the scatterer, the scattering phase matrix consists of 16 independent elements. For a single sphere, it is clear that the independent elements reduce to only four. Graphs of P_1 and P_2, as functions of the scattering angle for a real part of the refractive index of 1.5 and a size parameter of 60, are shown in Fig. 5.4a. The phase functions of a Mie particle are characterized by the strong forward scattering. Also, the large back scattering is noticeable. The scattering patterns consist of rapid fluctuation due to interference effects, which depend upon the size parameter. Clearly, the scattering behavior of a Mie particle differs greatly from that of a Rayleigh molecule as described in Section 3.7. Since a spherical particle is symmetrical with respect to the incident light, the scattering pattern is also symmetrical in the intervals $(0°, 180°)$ and $(180°, 360°)$. Thus, we may present the Mie scattering phase function in a polar diagram similar to the one depicted in Figs. 3.13 and 1.4. Figure 5.4b illustrates graphs of P_{33} and P_{34} as functions of the scattering angle. P_{33} has the same behavior as those of P_1 and P_2, but P_{34} shows negative values resulting from the differences of the cross components of S_1 and S_2.

All the developments discussed in the previous sections are concerned with the scattering of electromagnetic waves by a single homogeneous sphere. We shall now extend these developments to a sample of cloud or aerosol particles so that practical equations for the calculations of extinction parameters and phase functions may be derived. We *assume* that particles are sufficiently far from each other and that the distance between them is much greater than the incident wavelength. Thus, it is possible to study the scattering by one particle without reference to the other ones. Consequently, intensities scattered by various particles may be added without regard to the phase of the scattered waves. This particular scattering phenomenon is called *independent*

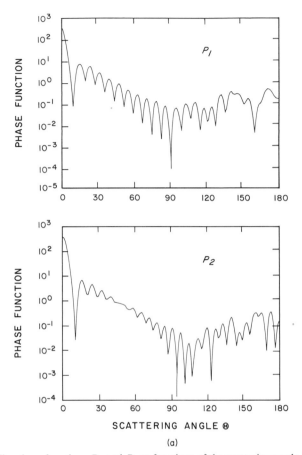

Fig. 5.4   (a) The phase functions P_1 and P_2 as functions of the scattering angle for a refractive index $m_r = 1.5$ and a size parameter $x = 60$. (b) The phase functions P_{33} and P_{34} as functions of the scattering angle for a refractive index $m_r = 1.5$ and a size parameter $x = 60$.

scattering. It is in the context of the independent scattering concept that the following discussions are based.

We consider a sample of cloud particles whose size spectrum can be described by $dn(a)/da$ (in units, say, $\mathrm{cm}^{-3}\,\mu\mathrm{m}^{-1}$). Assume that the size range of particles is from a_1 to a_2; then the total number of particles is given by

$$N = \int_{a_1}^{a_2} \frac{dn(a)}{da}\, da. \qquad (5.114)$$

With the particle size distribution prescribed, we can define the extinction and scattering parameters for a sample of particles. The extinction and

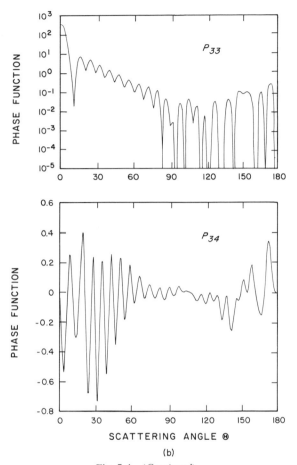

Fig. 5.4   (*Continued*)

scattering coefficients (in units of per length) are defined respectively, as

$$\beta_{\mathrm{e}} = \int_{a_1}^{a_2} \sigma_{\mathrm{e}}(a) \frac{dn(a)}{da} \, da, \qquad (5.115)$$

$$\beta_{\mathrm{s}} = \int_{a_1}^{a_2} \sigma_{\mathrm{s}}(a) \frac{dn(a)}{da} \, da. \qquad (5.116)$$

Lastly, we define the single scattering albedo for a sample of particles as

$$\tilde{\omega} = \beta_{\mathrm{s}}/\beta_{\mathrm{e}}. \qquad (5.117)$$

It is apparent that the single-scattering albedo represents the percentage of light beam which will undergo scattering in a single scattering event. The remaining part of this section defines the phase matrix for a sample of particles.

Since the phase matrix is a nondimensional physical parameter denoting the scattered intensity and polarization state for a sample of particles in the particle range (a_1, a_2), it is independent of the particle size distribution $dn(a)/da$. Hence, we rearrange Eq. (5.111a) and perform particle size integration to obtain

$$\frac{P_{11}}{4\pi} \int_{a_1}^{a_2} \sigma_s \frac{dn(a)}{da} \, da = \frac{1}{2k^2} \int_{a_1}^{a_2} [i_1(a) + i_2(a)] \frac{dn(a)}{da} \, da. \qquad (5.118)$$

From Eq. (5.115) we find

$$\frac{P_{11}}{4\pi} = \frac{1}{2k^2 \beta_s} \int_{a_1}^{a_2} [i_1(a) + i_2(a)] \frac{dn(a)}{da} \, da. \qquad (5.119)$$

Similarly, we have

$$\frac{P_{22}}{4\pi} = \frac{1}{2k^2 \beta_s} \int_{a_1}^{a_2} [i_2(a) - i_1(a)] \frac{dn(a)}{da} \, da, \qquad (5.120)$$

$$\frac{P_{33}}{4\pi} = \frac{1}{2k^2 \beta_s} \int_{a_1}^{a_2} [i_3(a) + i_4(a)] \frac{dn(a)}{da} \, da, \qquad (5.121)$$

$$\frac{P_{34}}{4\pi} = \frac{1}{2k^2 \beta_s} \int_{a_1}^{a_2} [i_4(a) - i_3(a)] \frac{dn(a)}{da} \, da. \qquad (5.122)$$

Note here that $i_j (j = 1, 2, 3, 4)$ are functions of the particle radius a, the index of refraction m, the incident wavelength λ, and the scattering angle θ.

5.6   RAY OPTICS

The laws of geometrical optics may be used to compute the angular distribution of light, which is scattered when a plane electromagnetic wave is incident on a particle much larger than the wavelength of the incident light. Such a computation is an approximation based on the assumption that the light may be thought of as consisting of separate localized rays which travel along straight-line paths; it is an asymptotic approach which becomes increasingly accurate in the limit as the size-to-wavelength ratio approaches infinity. Processes involving geometrical optics include rays externally reflected by the particle and rays refracted into the particle; the latter rays may

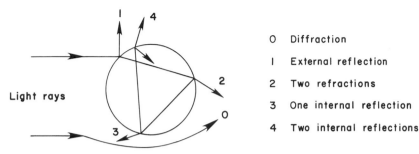

Fig. 5.5   Representations of light rays scattered by a sphere according to ray optics.

be absorbed in the particle, or they may emerge from it after possibly suffering several internal reflections. Hence the total energy scattered and absorbed by the particle is equal to that impinging on the cross section of the particle presented to the incident beam.

Particles much larger than the incident wavelength also scatter light by means of diffraction, which removes energy from the light wave passing by the particle. The diffraction is concentrated in a narrow lobe around the forward direction, and like geometrical reflection and refraction, it contains an amount of energy equal to that incident on the cross section of the particle. In the far field, the diffracted component of the scattered light may be approximated by Fraunhofer diffraction theory. The diffraction pattern depends only upon the shape of the cross section of the particle.

We use the term *ray optics* to describe both geometrical reflection and refraction plus Fraunhofer diffraction. Figure 5.5 illustrates the geometrical configuration for different contributions to light scattered by a large sphere. In the following subsections, we shall discuss the theoretical foundations for the treatment of geometrical optics and diffraction.

5.6.1   Diffraction: Corona

We will now present the theoretical development for diffraction on the basis of Babinet's principle, which states that diffraction pattern in the far field, i.e., Fraunhofer diffraction, from a circular aperture is the same as that from an opaque disk or sphere of the same radius. Let the Z axis be in the direction of propagation of the incident light, and let the wave disturbance be sought at a distance point P from the geometrical aperture A. In reference to Fig. 5.6, the distance from P to point $0'(x, y)$ on the aperture area and the origin 0 are denoted as r and r_0, respectively. Thus, the phase difference of the disturbance at P for waves passing through points 0 and 0' is given by (see

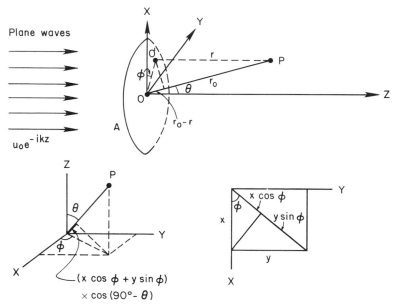

Fig. 5.6 Diffraction by a circular aperture with a geometrical area A. The geometrical relationship between the phase difference and the coordinate systems is also shown.

Fig. 5.6)

$$\delta = k(r - r_0) = k(x\cos\phi + y\sin\phi)\sin\theta, \qquad (5.123)$$

where $k = 2\pi/\lambda$, and λ is the wavelength.

In the far field the light wave disturbance at P can be derived from the Fraunhofer diffraction theory and is given by

$$u_p = -\frac{iu_0}{r\lambda} \iint_A e^{-ikr}\, dx\, dy. \qquad (5.124)$$

Here u_0 represents the disturbance in the original wave at point 0 on the plane wave front whose wavelength is λ. Upon utilizing Eq. (5.123), it follows that

$$u_p = -\frac{iu_0}{r\lambda} e^{-ikr_0} \iint_A e^{-ik(x\cos\phi + y\sin\phi)\sin\theta}\, dx\, dy. \qquad (5.125)$$

For a circular aperture, we may change rectangular coordinates to polar coordinates (ρ, ψ) to give $x = \rho\cos\psi$ and $y = \rho\sin\psi$. Thus,

$$u_p = -\frac{iu_0}{r\lambda} e^{-ikr_0} \int_0^a \int_0^{2\pi} e^{-ik\rho\cos(\psi - \phi)\sin\theta}\, \rho\, d\rho\, d\psi. \qquad (5.126)$$

We first note that the zero-order Bessel function is defined by

$$J_0(y) = \frac{1}{2\pi} \int_0^{2\pi} e^{-iy \cos \alpha} \, d\alpha. \tag{5.127}$$

This gives

$$u_p = -\frac{iu_0}{r\lambda} e^{-ikr_0} 2\pi \int_0^a J_0(k\rho \sin \theta)\rho \, d\rho. \tag{5.128}$$

In addition there exists a well-known recurrence relation involving Bessel functions

$$\frac{d}{dy}[yJ_1(y)] = yJ_0(y) \tag{5.129}$$

giving

$$\int_0^y y'J_0(y')\,dy' = yJ_1(y). \tag{5.130}$$

From Eqs. (5.128) and (5.130), it follows that

$$u_p = -\frac{iu_0}{r\lambda} e^{-ikr_0} A \frac{2J_1(x \sin \theta)}{x \sin \theta}, \tag{5.131}$$

where the geometrical shadow area $A = \pi a^2$ and the size parameter $x = ka$. Hence, the scattered intensity in terms of the incident intensity $I_0 = |u_0|^2$ is given by

$$I_p = |u_p|^2 = I_0 \frac{i_p}{k^2 r^2}, \tag{5.132}$$

where the angular intensity function for diffraction analogous to the Mie scattering theory for a single sphere is

$$i_p = \frac{x^4}{4} \left[\frac{2J_1(x \sin \theta)}{x \sin \theta} \right]^2. \tag{5.133}$$

It is clear that diffraction depends only on the particle size parameter and is independent of the index of refraction.

Figure 5.7 shows a plot of $D^2 = [2J_1(y)/y]^2$ versus y. It has a principal maximum of 1 at $y = 0$ (i.e., $\theta = 0$), and with increasing y it oscillates with gradually diminishing amplitude. Note that θ is the scattering angle denoting the angle between the incident and scattered waves. When $J_1(y) = 0$, then $D^2 = 0$; this gives the minima of the diffraction pattern. The positions of the maxima are given by values of y that satisfy

$$\frac{d}{dy}\{J_1(y)/y\} = 0. \tag{5.134}$$

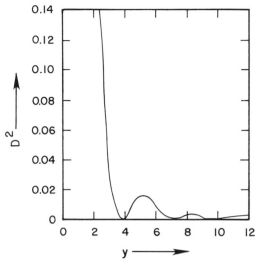

Fig. 5.7   The relative diffraction pattern as a function of $y = x \sin \theta$.

Table 5.1 lists these maxima and minima. The minima or dark rings can be approximated by

$$y = ka \sin \theta = (n + 0.22)\pi \tag{5.135}$$

or $(n = 1, 2, \ldots)$

$$\sin \theta = (n + 0.22)\lambda/(2a). \tag{5.136}$$

The first maximum at $y = 0$ usually is obscured by the finite size of the source. Thus, the first observable maximum diffraction ring is when $y = 5.136$.

The diffraction theory developed above for a single sphere can be employed to explain the optical phenomenon known as the corona. The corona is associated with the illumination frequently observed near the sun, the moon,

TABLE 5.1   *The First Few Maxima and Minima of the Diffraction Pattern*

y	D^2	Max or min
0	1	Max
3.832	0	Min
5.136	0.0175	Max
7.016	0	Min
8.417	0.0042	Max
10.174	0	Min
11.620	0.0016	Max

or other luminous objects when they are seen through a mist or thin cloud. It is usually in the form of circles, or near circles, concentric with the luminous body and situated within a few layers around it. The corona is usually very bright and of a white or bluish-white color with a reddish or brownish tinge. The colors are diluted with a great deal of white light. As many as four corona rings have been recorded, though only the first ring has been frequently observed around the sun and the moon. The condition during which the corona may be produced by thin clouds is when such thin clouds are composed of particles of almost equal size, which is said to be *monodisperse*. Applying the resulting diffraction theory one may evaluate the angular positions of the corona if the wavelength of the visible sunlight and the mean particle size are known. Based on Eq. (5.136), we see clearly that red color, having a longer wavelength, is to be observed in the outer ring of the corona with blue and green colors inside the ring. Also it is evident that the angular width depends on the particle diameter. These theoretical analyses are in agreement with observations.

5.6.2 Geometrical Reflection and Refraction: Rainbows and Halos

When a plane wave falls on to a boundary between two homogeneous media of different optical properties, it is split into two waves; a transmitted wave proceeding into the second medium, and a reflected wave propagating back into the first medium. From the part of the wave that hits the plane, we can isolate a narrow beam much smaller as compared to the surface. Such a beam is called a *ray* as it is used in geometrical optics. Let v_1 and v_2 be the velocities of propagation in the two media ($v_1 > v_2$), and let θ_i and θ_t be the angles corresponding to the incident and refracted waves. Referring to Fig. 5.8, we find

$$\sin \theta_i / \sin \theta_t = v_1/v_2 = m, \tag{5.137}$$

where m is the index of refraction for the second medium with respect to the first medium. This is Snell's law, which relates the incident and refracted angles through the index of refraction.

Let $\mathbf{E}^i$ be the electric vector of the incident field. As shown in Fig. 5.8, the components of the incident electric field vector perpendicular (r) and parallel (*l*) to the plane containing the incident and refracted fields mapped in rectangular coordinates are

$$
\begin{aligned}
E_x^i &= -E_l^i \cos \theta_i, \\
E_y^i &= E_r^i, \\
E_z^i &= E_l^i \sin \theta_i.
\end{aligned}
\tag{5.138}
$$

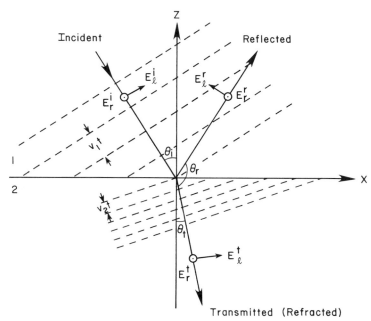

Fig. 5.8   Illustration of the reflection and refraction of a plane wave. The choice of the positve directions for the parallel components (*l*) of the electric vectors is indicated in the diagram. The perpendicular components are at right angles into the plane of reference.

From the Maxwell equation, the relation between the electric and magnetic vectors can be shown to be $\mathbf{H} = \sqrt{\varepsilon}\,\mathbf{a} \times \mathbf{E}$, or $\mathbf{E} = -\sqrt{1/\varepsilon}\,\mathbf{a} \times \mathbf{H}$, where $\mathbf{a}$ is an unit vector in the direction of propagation. Thus, the components of the magnetic vector are ($\mu \approx 1, m = \sqrt{\varepsilon}$)

$$H^i_x = -E^i_r \cos\theta_i m_1,$$
$$H^i_y = -E^i_l m_1, \qquad\qquad (5.139)$$
$$H^i_z = E^i_r \sin\theta_i m_1,$$

where m_1 is the refractive index of the first medium with respect to vacuum.

Similarly, if $\mathbf{E}^t$ and $\mathbf{E}^r$ denote the transmitted (refracted) and reflected electric vectors, respectively, we find the relations

$$
\begin{array}{ll}
E^t_x = -E^t_l \cos\theta_t, & H^t_x = -E^t_r \cos\theta_t m_2, \\
E^t_y = E^t_r, & H^t_y = -E^t_l m_2, \qquad (5.140) \\
E^t_z = E^t_l \sin\theta_t, & H^t_z = E^t_r \sin\theta_t m_2,
\end{array}
$$

$$
\begin{array}{ll}
E^r_x = -E^r_l \cos\theta_r, & H^r_x = -E^r_r \cos\theta_r m_1, \\
E^r_y = E^r_r, & H^r_y = -E^r_l m_1, \qquad (5.141) \\
E^r_z = E^r_l \sin\theta_r, & H^r_z = E^r_r \sin\theta_r m_1,
\end{array}
$$

where m_2 is the refractive index of the second medium with respect to vacuum, and we note that $\theta_r = 180° - \theta_i$.

The boundary conditions require that the tangential components of **E** and **H** be continuous. Hence we must have

$$E_x^i + E_x^r = E_x^t, \qquad H_x^i + H_x^r = H_x^t$$
$$E_y^i + E_y^r = E_y^t, \qquad H_y^i + H_y^r = H_y^t. \tag{5.142}$$

Upon substituting into Eq. (5.142) for all the components, we obtain four relations:

$$\cos \theta_i (E_l^i - E_l^r) = \cos \theta_t E_l^t, \tag{5.143a}$$

$$E_r^i + E_r^r = E_r^t, \tag{5.143b}$$

$$m_1 \cos \theta_i (E_r^i - E_r^r) = m_2 \cos \theta_t E_r^t, \tag{5.143c}$$

$$m_1 (E_l^i + E_l^r) = m_2 E_l^t. \tag{5.143d}$$

By virtue of these equations, the solutions of the components of the reflected and transmitted waves in terms of the incident wave are

$$E_l^r = R_2 E_l^i, \qquad E_l^t = T_2 E_l^i, \qquad E_r^r = R_1 E_r^i, \qquad E_r^t = T_1 E_r^i, \tag{5.144}$$

where the amplitude coefficients

$$R_1 = \frac{\cos \theta_i - m \cos \theta_t}{\cos \theta_i + m \cos \theta_t}, \qquad R_2 = \frac{m \cos \theta_i - \cos \theta_t}{m \cos \theta_i + \cos \theta_t},$$

$$T_1 = \frac{2 \cos \theta_i}{\cos \theta_i + m \cos \theta_t}, \qquad T_2 = \frac{2 \cos \theta_i}{m \cos \theta_i + \cos \theta_t}, \tag{5.145}$$

with $m = m_2/m_1$, the refractive index of the second medium with respect to the first medium. Equations (5.145) are called *Fresnel formulas*, which were first derived by Fresnel in 1823. When absorption is involved, the amplitude coefficients become much more complicated, and they can be derived by means of straightforward mathematical analyses (see Exercise 5.8).

In regard to the energy, from the Poynting vector equation denoted in Eq. (3.33) and the relation between **E** and **H** stated previously, we find that the flux density $|\mathbf{S}| = (c/4\pi)\sqrt{\varepsilon}|\mathbf{E}|^2$ ($\mu = 1$). Thus, the amount of energy incident, reflected, and transmitted on a unit area of the boundary per unit time is

$$F^i = |\mathbf{S}^i| \cos \theta_i = (c/4\pi) m_1 |E^i|^2 \cos \theta_i,$$
$$F^r = |\mathbf{S}^r| \cos \theta_i = (c/4\pi) m_1 |E^r|^2 \cos \theta_i, \tag{5.146}$$
$$F^t = |\mathbf{S}^t| \cos \theta_t = (c/4\pi) m_2 |E^t|^2 \cos \theta_t.$$

Therefore, the reflected and transmitted portions of the energy in two polarization components, with respect to the incident energy, are proportional to

$R_{1,2}^2$ and $T_{1,2}^2 m \cos \theta_t / \cos \theta_i$. It can be easily verified that $R_{1,2}^2 + T_{1,2}^2 m \cos \theta_t / \cos \theta_i = 1$, which is in agreement with the energy conservation principle. Consequently, the transmitted (or refracted) parts of the energy simply can be written as $(1 - R_{1,2}^2)$.

Consider now a large sphere, and let $\wp = 0$ for the external reflection, $\wp = 1$ for two refractions, and $\wp \geq 2$ for internal reflections. We define the amplitude coefficients

$$\begin{aligned} \varepsilon_1 &= R_1 & \text{for} \quad \wp = 0 \\ \varepsilon_1 &= (1 - R_1^2)^{1/2}(-R_1)^{\wp - 1}(1 - R_1^2)^{1/2} & \text{for} \quad \wp \geq 1, \end{aligned} \tag{5.147}$$

where $-R_1$ denotes the amplitude coefficient for an internal reflection. These definitions also apply to index $l(2)$ for other polarization.

Next we discuss the effect of the curvature on the reflected and refracted intensity. We consider a finite pencil of light characterized by $d\theta_i$ and $d\phi$ with ϕ being the azimuthal angle. Let I_0 denote the incident intensity of the light pencil plane-polarized in one of the two main directions. Thus, the flux of energy contained in this pencil is $I_0 a^2 \cos \theta_i \sin \theta_i \, d\theta_i \, d\phi$, where a denotes the radius of the sphere. This flux of energy is devided by successive reflection and refraction. The emergent pencil spreads into a solid angle $\sin \theta \, d\theta \, d\phi$ at a large distance r from the sphere. As a result, the scattered intensity is given by

$$I_r = \frac{\varepsilon_1^2 I_0 a^2 \cos \theta_i \sin \theta_i \, d\theta_i \, d\phi}{r^2 \sin \theta \, d\theta \, d\phi}. \tag{5.148}$$

The pencil of light emergent from the sphere is characterized by a small range $d\theta$ around the scattering angle θ. In reference to Fig. 5.9, the total deviation from the original direction is

$$\theta' = 2(\theta_i - \theta_t) + 2(\wp - 1)(\pi/2 - \theta_t). \tag{5.149a}$$

The scattering angle defined in the interval $[0, \pi]$ may be expressed by

$$\theta' = 2\pi n - q\theta, \tag{5.149b}$$

where n is an integer and $q = +1$ or -1. Hence,

$$\frac{d\theta}{d\theta_i} = \left| \frac{d\theta'}{d\theta_i} \right| = 2 - 2\wp \frac{\cos \theta_i}{m \cos \theta_t}. \tag{5.150}$$

We define the divergence factor due to the curvature effect in the form

$$D = \frac{\cos \theta_i \sin \theta_i}{\sin \theta \, d\theta/d\theta_i}. \tag{5.151}$$

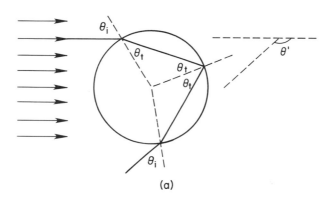

(a)

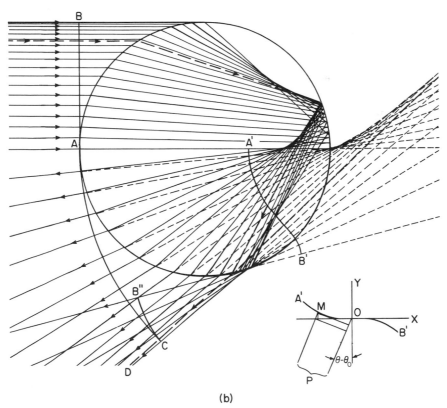

(b)

Fig. 5.9 (a) Geometrical reflection and refraction by a sphere and the definition of the deviation angle. (b) Refraction and reflection of light rays, which have a plane wave surface, in a spherical water drop (after Humphreys, 1954).

Thus from Eq. (5.148), which is also valid for index l, we obtain

$$I_{r,l} = \frac{I_0}{k^2 r^2} x^2 \varepsilon_{1,2}^2 D.$$ (5.152)

In comparison with Mie scattering theory we find

$$i_{1,2} = x^2 \varepsilon_{1,2}^2 D.$$ (5.153)

5.6.2.1 *Rainbows* The rainbow is probably the best-known phenomenon of atmospheric optics. It has inspired art and mythology in all peoples and has been a challenge to mathematical physicists. We see a rainbow in the sky usually on summer afternoons after a rainshower is over. Rainbows are produced by the geometrical reflections of the sun's rays within the raindrops. The sun's rays undergo minimum deviation within the drop and generate the maximum intensity at a specific angle that is much stronger than those at neighboring angles.

To evaluate the angles at which rainbows are formed, we return to the equation denoting the total deviation from the original direction. The minimum deviation of a bundle of rays may be found by differentiating Eq. (5.149) with respect to the incident angle and setting the result equal to zero. Thus,

$$\frac{d\theta'}{d\theta_i} = 0 = 2\left(1 - \not{p}\,\frac{d\theta_t}{d\theta_i}\right).$$ (5.154a)

Furthermore, differentiating Snell's law leads to

$$\frac{d\theta_t}{d\theta_i} = \frac{\cos\theta_i}{m\cos\theta_t}.$$ (5.155)

It follows then that

$$\not{p}\cos\theta_i = m\cos\theta_t.$$ (5.154b)

On eliminating the refracted angle θ_t from Eq. (5.154b) and Snell's equation, we obtain the incident angle at which the minimum deviation takes place as

$$\cos^2\theta_i = (m^2 - 1)/(\not{p}^2 - 1), \qquad \not{p} > 2.$$ (5.156)

Once the refractive index has been given, we may evaluate the incident angle corresponding to the minimum deviation for a given $\not{p}$. The refracted angle and the scattering angle also may be calculated subsequently from Snell's law and Eqs. (5.149a) and (5.149b), respectively.

Table 5.2 gives the incident and scattering angles for rainbows with various indices of refraction. Minimum deviation also can occur in one case other than the rainbows, namely in the glory at $\theta = 180°$. The condition for having

TABLE 5.2 *Incident (θ_i) and Scattering (θ) Angles for the Glory and Rainbows*

m	Glory for $\mu = 2$		Rainbow for $\mu = 2$		Rainbow for $\mu = 3$		Rainbow for $\mu = 4$		Rainbow for $\mu = 5$	
	θ_i	θ	θ_i	θ	θ_i	θ	θ_i	θ	θ_i	θ
1.10	—	—	75°	84°	81°	139°	83°	170°	85°	119°
1.33	—	—	60	137	72	130	77	43	80	42
1.45	87°	180°	53	152	68	102	74	4	78	92
1.50	83	180	50	157	67	93	73	9	77	109
1.54	79	180	47	161	66	86	72	19	76	121
1.75	60	180	34	173	59	58	68	60	73	175
2.00	33	180	0	180	52	35	63	94	69	140

a glory for the $\mu = 2$ rays is that the refractive index has to be between $\sqrt{2}$ and 2. Since the refractive index for water drops is about 1.33, we see that no glory would be produced by geometrical ray tracing. The fact that a glory is observed in nature from clouds points out one of the greatest discrepancies between ray optics and Mie theory. It has been suggested that the glory phenomenon is produced by the back scattering from edge rays apparently connected with surface waves generated on the sphere. Primary ($\mu = 2$) and secondary ($\mu = 3$) rainbows frequently are observed in the atmosphere. Owing to the variation of the index of refraction of the raindrop with respect to the incident visible wavelengths, various color sequences are seen. Exercise 5.7 requires calculation of the position of several colors for primary and secondary rainbows. When the rainbows are very pronounced, supernumerary rainbows often become visible. It is an interference phenomenon accompanying the refraction of the light in the drop and cannot be explained by the geometrical ray-tracing approach.

The above analyses only give the angles at which rainbows may be produced. However, the geometrical optics cannot provide the intensity of the rainbow. It is evident from Eq. (5.152) that when minimum deviations occur, the intensity approaches infinity. The geometrical optics approximation assumes that the wave fronts near any point are sufficiently characterized by their normals and by their local radii of curvature. It is clear that such an approximation breaks down near the rainbow. The next higher approximation is a cubic wave front, which leads to the Airy theory. It should be noted, however, that the intensity results derived from geometrical optics are accurate in the vicinity of the rainbow angles, and with appropriate extrapolations, intensity patterns at the rainbow angles may be approximately obtained. We shall demonstrate this when the comparison between the ray optics and Mie theory is made.

In this subsection we shall also introduce briefly the Airy theory for rainbows. Since a raindrop is spherical, it is sufficient to use only a single plane containing the center of the drop and the luminous object, and to trace rays incident on one quadrant of the intersection circle. Refer to Fig. 5.9b; let AB be the wave front of parallel incident rays above the ray that passes through the center of the drop (the axial ray), and consider rays which undergo only one internal reflection. (Note that the following discussions are general and applicable to internal reflections more than once.) The heavy line denotes the course of the ray of minimum deviation (the Descartes ray) for a water drop having an index of refraction of $\frac{4}{3}$. Since the deviations of the rays incident between the Descartes and the axial rays are greater than that of the Descartes ray, the exits of these rays must lie between those two rays. Likewise, rays which locate between the Descartes and the edge rays must also have more deviations than that of the Descartes ray. As a consequence, while they leave the drop beyond this ray, they eventually will come between it and the axial ray. Thus, the one internally reflected light rays are diffuse and weakened, except near the direction of minimum deviation, and are confined to the region between this direction and the axial ray. The wave front is now described by ACB''.

The outgoing wave front near the Descartes ray is in the form of $A'B'$. Through tedious geometrical analyses and numerous approximations [e.g., see Humphreys (1954)], it can be shown that the wave front in Cartesian coordinates is related by the cubic equation in the form

$$y = hx^3/(3a^2), \tag{5.157}$$

where a is the radius of the drop and

$$h = \frac{(\not{\!{p}}^2 + 2\not{\!{p}})^2}{(\not{\!{p}} + 1)^2(m^2 - 1)}\left[\frac{(\not{\!{p}} + 1)^2 - m^2}{m^2 - 1}\right]^{1/2}$$

Equation (5.157) represents a curve closely coincident with the portion of the wave front to which the rainbows are produced.

To evaluate the intensity and its variation with angular distance from the ray of minimum deviation, we consider the diagram depicted in Fig. 5.9b, and let 0 be the point of inflection of the outgoing cubic wave front near a drop. Let θ_0 and θ be the deviation angles for the Descartes ray and the neighboring rays, respectively, and let P be a distant point in the direction $\theta - \theta_0$ from the Descartes ray. We find that the phase difference of the disturbance at P for waves passing through points 0 and M is given by

$$\delta = k[x\sin(\theta - \theta_0) - y\cos(\theta - \theta_0)]$$
$$= k\left[x\sin(\theta - \theta_0) - \frac{h}{3a^2}x^3\cos(\theta - \theta_0)\right], \tag{5.158}$$

where $k = 2\pi/\lambda$. The amplitude of the wave disturbance u_p is then proportional to integration of all the possible vibrations due to phase differences along the X axis as

$$u_p \sim \int_{-\infty}^{\infty} \exp\left\{-ik\left[x\sin(\theta - \theta_0) - \frac{h}{3a^2}x^3\cos(\theta - \theta_0)\right]\right\}dx. \quad (5.159a)$$

It suffices to use the cosine representation, and if we let

$$(2h/3a^2\lambda)x^3\cos(\theta - \theta_0) = t^3/2, \qquad (2x/\lambda\sin(\theta - \theta_0)) = zt/2, \quad (5.160a)$$

we find that the amplitude is now given by

$$u_p \sim 2\left[\frac{3a^2\lambda}{4h\cos(\theta - \theta_0)}\right]^{1/3} f(z), \quad (5.159b)$$

and the intensity $I = u_p^2$, where the *rainbow integral* due to Airy is defined by

$$f(z) = \int_0^{\infty} \cos\frac{\pi}{2}(zt - t^3)\,dt. \quad (5.161)$$

From Eq. (5.160a), we get

$$z^3 = \frac{48a^2}{h\lambda^2}\frac{\sin^3(\theta - \theta_0)}{\cos(\theta - \theta_0)} \quad (6.160b)$$

which, for small values of $\theta - \theta_0$, is proportional to $(\theta - \theta_0)^3$. Thus, when one is dealing only with small angles of departure from the Descartes ray, we have

$$z \approx \left(\frac{48a^2}{h\lambda^2}\right)^{1/3}(\theta - \theta_0). \quad (5.160c)$$

Table 5.3 gives maximum and minimum values of z and $f^2(z)$ for a given wavelength and drops of a definite size. Note that the first maximum (main rainbow) does not coincide with $z = 0$, the geometrical position of the primary rainbow ($\theta = 138.0°$). Also note that the absolute intensity from the Airy theory may be obtained by comparing the result from the geometrical optics with the value of $f^2(z)$ for large z.

5.6.2.2 *Halos* The high cirrus clouds in the atmosphere consist of ice crystals having predominately hexagonal structures. Ice-crystal clouds normally have low concentrations and thus present a tenuous appearance in the sky. The solar or lunar disks may be clearly visible through the cloud cover. Owing to the great variety of shapes and orientations of ice particles, a large number of fascinating optical phenomena are noticeable under favorable atmospheric conditions, among which the halos around the sun or

TABLE 5.3   *Maxima and Minima of the Rainbow Integral*

	Maxima			Minima
Number	z	$f^2(z)$	Number	z
1	1.0845	1.005	1	2.4955
2	3.4669	0.615	2	4.3631
3	5.1446	0.510	3	5.8922
4	6.5782	0.450	4	7.2436
5	7.8685	0.412	5	8.4788
6	9.0599	0.384	6	9.6300
7	10.1774	0.362	7	10.7161
8	11.2364	0.345	8	11.7496
9	12.2475	0.330	9	12.7395
10	13.2185	0.318	10	13.6924

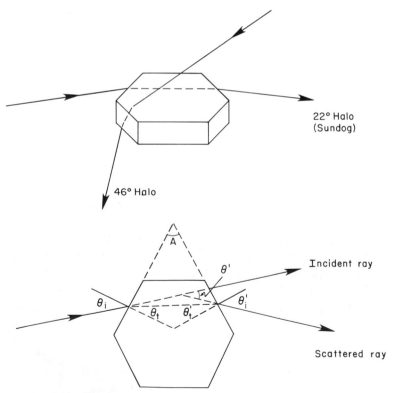

Fig. 5.10   Geometrical reflection and refraction by hexagonal crystals.

moon are the most commonly observed features. We may use the ray-tracing technique previously developed to locate the positions at which halos are produced. Owing to the plane crystal surface, consideration of the curvature effect is not required in the calculation of the scattered intensity.

In reference to Fig. 5.10, if a light ray passes through a prism of angle A in a plane at right angles to the refracting edge, the deviation angle is

$$\theta' = (\theta_i - \theta_t) + (\theta_i' - \theta_t') = 2\theta_i' - A. \tag{5.162}$$

The minimum deviation occurs when

$$\frac{d\theta'}{d\theta_i} = 0 = 1 + \frac{d\theta_i'}{d\theta_i}.$$

Since $A = \theta_t + \theta_t'$, we also have

$$1 + \frac{d\theta_t'}{d\theta_t} = 0.$$

An obvious solution to the last two equations is when $\theta_i' = \theta_i$ and $\theta_t' = \theta_t$. Thus, from Eq. (5.162), the incident angle at which minimum deviation occurs is

$$\theta_i = \tfrac{1}{2}(\theta' + A). \tag{5.163a}$$

Also, the angle of refraction at minimum deviation is given by $\theta_t = A/2$. When Snell's law is used, we have

$$\sin\left[\tfrac{1}{2}(\theta' + A)\right] = m \sin \tfrac{1}{2}A. \tag{5.163b}$$

This is the basic relationship for finding minimum deviation from the known index of refraction and prism angle. As illustrated in Fig. 5.10, the possible prism angles are 60°, 90°, and 120°. However, owing to the property of the sine function and the fact that the refractive index of ice is about 1.31 in the visible wavelengths, a prism angle of 120° cannot produce minimum deviation. The common halo has an angular radius of 22° indicating refraction by hexagonal prisms. The halo of 46° is produced by refraction of rectangular prisms. Since the index of refraction varies with wavelength, white light is dispersed into its component colors with red refracted least and blue refracted most. The color sequence is the reverse of that observed in the corona. This provides a means of distinction between ice and water clouds.

The fact that halos appear in the form of circles is due to the orientation property of hexagonal prisms. Observation shows that hexagonal plates and needles fall through the air having their major axes parallel to the ground. They are continuously spinning on the horizontal axis and thus produce random orientation in the horizontal plane. Light rays, which reach a sheet of ice columns or needles randomly oriented in a horizontal plane will then produce halos in directions along conical circles at the angle of minimum deviation surrounding the sun or the moon. When the sun is close to the horizon and thin cirrus clouds are present, colored streaks greater than 22°, at the same elevation as the sun, may sometimes be observed. This optical phenomenon is called parhelia of 22° or, commonly, sun dogs or mock suns. The sun dogs are only two bright spots since we are unable to see the other rays from the sun deviated by clouds higher above the horizon. External reflections from the flat horizontal faces of ice plates produce streaks of white light, which may be above, below, or both above and below the sun; they are called sun pillars. There are other colored arcs caused by the re-fraction and reflection by ice crystals. Identifications of their positions require tedious geometrical exercises.

Since ice crystals are nonspherical, it is obvious that the Mie scattering theory developed earlier cannot be used to calculate their scattering and extinction cross sections nor their phase functions. However, these para-meters are needed to understand the transfer of solar and infrared radiation through cirrus cloud layers. One approach to the difficult scattering problem involving ice crystals is to utilize the ray-optics technique described pre-viously. Another approach would be experimentally determining the scat-tering properties of ice-crystal clouds generated in the laboratory or in the atmosphere. We shall introduce the scattering behavior of ice crystals in the following section.

5.6.3   Comparison between Ray Optics and Mie Theory for Spheres

In order to compare the scattering results derived from ray optics with those from Mie scattering, we define the *gain* G relative to an isotropic scatterer. This gain is defined as the ratio of the scattered intensity to the intensity that would be found in any direction if the particle scattered the incoming energy isotropically. Thus, the averaged gain over the entire solid angle is unity such that

$$\frac{1}{4\pi} \int_{4\pi} G_{1,2}(\theta) d\Omega = 1. \qquad (5.164)$$

Isotropic scattering implies that the incident energy $I_0 \pi a^2$ to a sphere with radius a is uniformly distributed over the surface of a sphere $4\pi r^2$. Con-sequently,

$$I_{1,2}(\theta) = \frac{I_0 \pi a^2}{4\pi r^2} G_{1,2}(\theta). \qquad (5.165)$$

From Eq. (5.86), we find

$$G_{1,2} = 4i_{1,2}/x^2. \tag{5.166}$$

In a similar manner, the gain due to diffraction may be written as

$$G^f = 4i_p/x^2, \tag{5.167}$$

where i_p is given in Eq. (5.133). The gain due to diffraction is the same for the perpendicular and parallel components.

The total gain caused by diffraction and geometrical reflection and refraction can now be expressed by

$$G^t_{1,2} = G^f + \sum_{p=0}^{N} G^\rho_{1,2}, \tag{5.168}$$

where the geometrical reflection and refraction are represented by the index ρ. They include external reflection ($\rho = 0$), refraction ($\rho = 1$), and internal reflection ($\rho \geq 2$).

The foregoing treatment neglects the different contribution to the scattered intensity caused by the phase interference produced by various ray-optics components. In the case of large particles, the phase interferences give rise to rapidly oscillating intensities as a function of scattering as shown in Fig. 5.4. However, if the particles are randomly located and separated by distances much larger than the incident wavelength, the intensities from separate particles may be added without regard to the phase. For a sample of such particles, called *polydispersion*, the numerous maxima and minima are then lost in the integration over particle size. Hence it is reasonable to ignore the phase altogether in adding the intensities for diffraction, reflection, and refraction for a sample of large particles of various sizes. Figure 5.11 compares phase function P_{11} and degree of linear polarization from the Mie theory to the corresponding results from ray optics for the typical refractive indices of 1.33 and 1.50 for water drops and aerosols, respectively, in the visible spectrum. Mie calculations were made for a size parameter of 400. The size distribution employed in the Mie and diffraction calculations is the gamma function with its mode at x_m. It is given by

$$n(x) = cx^6 e^{-6x/x_m}, \tag{5.169}$$

where c is an arbitrary constant. There is close agreement between ray optics and the Mie theory when the size parameter is as large as 400. An exception is the glory feature for $m_r = 1.33$ which, as discussed earlier, does not occur in the ray-optics results. Most of the discrepancies and their variation with the size parameter can be qualitatively understood in terms

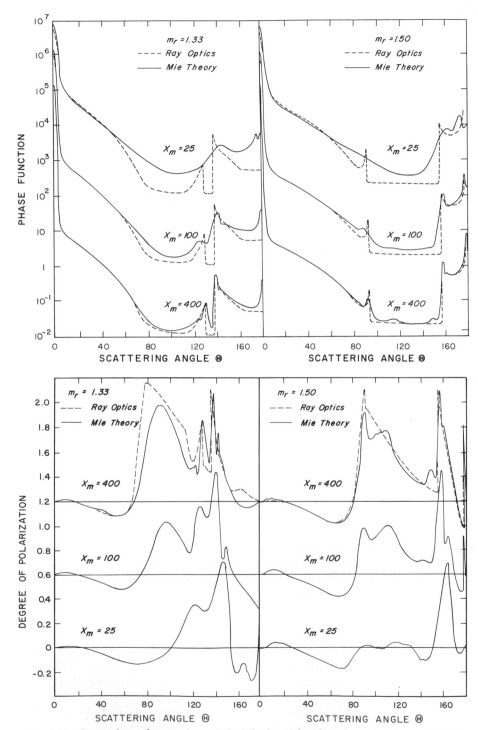

Fig. 5.11 Comparison of ray optics and the Mie theory for phase functions P_{11} and degree of linear polarization. Two refractive indices are shown along with three size distributions. The vertical scale applies to the lowermost curves. The other curves are successively displayed upward by factors of 10^2 and 0.6 for the phase function and the degree of linear polarization, respectively (after Liou and Hansen, 1971).

of the increasing inapplicability of the localization principle for decreasing size parameters. This causes the light in the individual features to be blurred over a wider range of angles than predicted by ray optics. The secondary rainbow is quite smooth at $x_m = 100$ and is lost at $x_m = 25$, while the primary rainbow is still easily visible. The number of rainbows visible in the intensity pattern thus give some indication of the particle size. The rainbow, in addition to being smoothed out, tends to move away from its ray-optics location as the size parameter decreases. For Mie scattering with $x_m = 400$, the small secondary peaks on the less steep side of the rainbows are supernumerary bows, which are caused by interference phenomena and hence are not rendered by ray optics in which the phase is neglected. There is also a small but noticeable discrepancy in the diffraction peak. The higher value for Mie scattering perhaps may arise from surface waves which scatter in the forward direction. The lower figure compares the degree of linear polarization defined in Eq. (3.75). The polarization patterns contain much stronger imprints of most of the features occurring in the scattered light such as the rainbows, the supernumerary bows, the glory, and the external reflections all of which produce positive polarization.

In Fig. 5.12, we also show comparisons of the theoretical and experimental scattering phase function and polarization patterns. These scattering patterns are derived from measurements of a rather dense water cloud in the cold chamber, utilizing He–Ne (0.6328 μm) laser light. The curves depicted in this figure are comprised of five successive nephelometer scans (10°–175°) which were each normalized at 10° scattering angles and then averaged, with the standard deviations of the averages shown as vertical bars. The measured cloud droplet size, using a continuous-impactor-replicator device, displayed a modal diameter of 2 μm and a maximum diameter of 10 μm. This size spectra was fitted with a zeroth order log-normal distribution (Kerker, 1969) in the form

$$n(a) = \exp\left[-(\log a - \log a_m)^2/(2\sigma_0)\right]/\left[\sqrt{2\pi}\sigma_0 a_m \exp(\sigma_0^2/2)\right], \quad (5.170)$$

where a_m (= 2 μm) denotes the modal diameter, and σ_0(= 0.275) the geometric mean standard deviation. Mie scattering calculations employing this size distribution were made, and the results were compared with the measured data. It is seen that the comparison of the experimental and theoretical patterns yields rather close agreement. Owing to the relatively small size of the cloud droplets (the modal size parameter $x_m \approx 20$), both theory and measurment show that the secondary cloudbow is absent, and that the primary cloudbow reaches a maximum at about 146°. Close agreement is found also for the linear polarization pattern, especially in the vicinity of the primary cloudbow where strongly positive polarization values are observed.

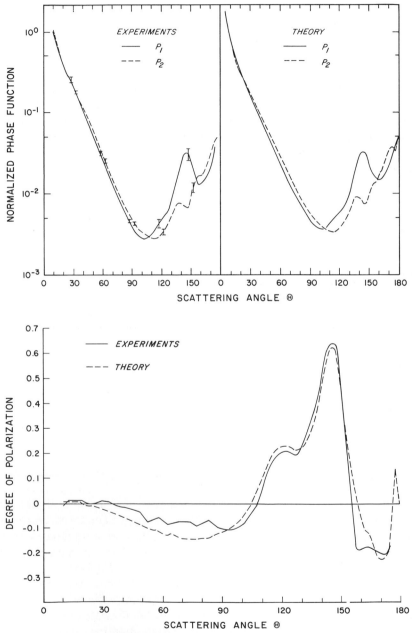

Fig. 5.12   Comparison of the normalized phase functions P_1 and P_2 and the degree of linear polarization from water cloud experiments and Mie scattering theory, using the same cloud droplet size distriubtion (after Sassen and Liou, 1979).

5.7 LIGHT SCATTERING BY NONSPHERICAL ICE CRYSTALS

The angular scattering behavior of water droplet clouds may be precisely described by the Mie scattering theory for a representative polydispersion of homogeneous water spheres. Based on the exact Mie theory, the optical properties of water droplets for any wavelength in the solar, infrared, and microwave spectra can be evaluated, provided that the droplet size distribution is given. With the resulting comprehensive numerical tables (Deirmendjian, 1969) and the existing Mie computer programs, the problem of polarized light scattered by water spheres seems to have been completely solved.

However, the changing atmosphere also contains micrometer-sized aerosol particles and large ice crystals, which are nonspherical. The determination of polarized light scattered by these irregular particles is made very difficult by their nonsphericity and the consequent problem of orientation. Knowledge of the scattering and absorbing behavior of atmospheric clouds and aerosols is of vital importance for remote sounding of cloud and aerosol compositions of the atmosphere by means of intensity and polarization techniques. Also it is closely relevant to the radiation budget and hence the climate and climatic changes of the earth–atmosphere sytem.

Clouds regularly cover about 50 % of the earth. In addition to absorbing and scattering the incoming solar radiation, clouds also trap the outgoing terrestrial radiation and produce the greenhouse effect. Thus, clouds represent the most important modulators of radiation in the earth–atmosphere system. However, effects of cirrus clouds on the radiation budget of the earth–atmosphere are less understood because of their high location in the troposphere and their semitransparent property with respect to both solar and terrestrial radiation. Moreover, cirrus clouds exclusively consist of nonspherical ice crystals whose light scattering properties in solar and infrared spectra are largely unknown. The nonspherical shapes of ice crystals depend upon such variables as temperature, saturation ratio, and atmospheric conditions. Under normal circumstances, ice crystals have the basic hexagonal structure. The fact that we see halos in cirrus cloudy atmospheres as illustrated in Section 5.6.2.2 proves that cirrus clouds must be composed of hexagonal crystals. Moreover, according to a number of *in situ* observations, the sizes of hexagonal crystals normally are on the order of several hundred micrometers. Thus, the ray-optics approach described in the previous section may be applicable for the scattering study. The scattering phase functions for hexagonal columns and plates have been reported by Jacobowitz (1971), Wendling *et al.* (1979), and Coleman (1979). In this section, we present and discuss physically some scattering characteristics of hexagonal ice crystals.

The general geometry of light rays incident on a hexagonal crystal is depicted in Fig. 5.13. The geometry of the crystal is defined by the length L and radius R, while the incident light rays are described by the ray plane. Normal to the Z axis, we define the principle plane, which lies on an XY plane. A hexagon has six equal sides along with the top and bottom faces. To describe the geometry of the hexagon with respect to the incident ray plane, seven variables are required: the length and radius of the hexagon and the position of the principal plane, the position of the incident ray on the ray axis, and three angles defining the orientation of the crystal with respect to the incident ray, i.e., the elevation angle ε, the rotation angle Ψ, and the azimuthal angle ϕ.

Having defined the variables involved, the ray tracing procedures may be outlined. We first find the position of the entry ray in the (X, Y, Z) co-

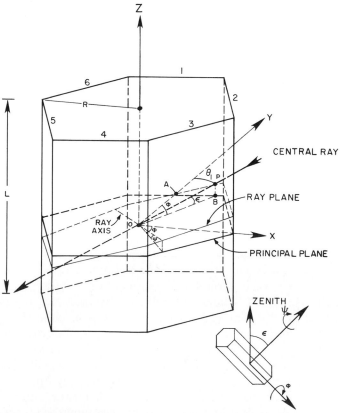

Fig. 5.13   General geometry of light rays incident on a hexagonal crystal: P is the point where the ray enters, θ_1 is the incident angle, A and B are points on the principle plane, O is the center point on the principal plane, and other symbols are explained in the text.

ordinates in terms of the seven geometrical variables. Through Snell's law, we find the refracted angles in terms of the incident angle mapped on the principal plane and the elevation angle. It follows that the position of the exit ray, the face that the ray will hit and the geometrical path length in the crystal can be determined through the procedures of the analytical geometry. The procedures are then repeated for internally reflected rays. Finally, we need to find the scattering angle with respect to the incident ray, to perform the summation of the refracted and reflected components, and to carry out the normalization of the energy pattern to get the scattering phase function.

The geometrical ray tracing equations for hexagons differ greatly from those for spheres. Spheres have a curvature effect, whereas hexagons do not. Also, a hexagon does not have the symmetry of the geometrical path length that a sphere inherently possesses. Basically, the general equation for the scattered energy per unit angle normalized with respect to the incident energy perpendicular to the X axis may be described by

$$E_1^s(\theta, \not{h}) = \sum_j E_1^s(\theta, \not{h}, \theta_j) \exp(-2km_i l_{\not{h}}) \Big/ \sum_j \cos\theta_{1j}$$

$$= \frac{\exp(-2km_i l_{\not{h}})}{\sum \cos\theta_{1j}} \begin{cases} \sum |R_1(\theta_{1j})|^2 \cos\theta_{1j}, \\ \quad \not{h} = 0 \quad \text{(external reflection)} \\ \sum [1 - |R_1(\theta_{1j})|^2][1 - |R_1(\theta_{2j})|^2] \cos\theta_{1j}, \\ \quad \not{h} = 1 \quad \text{(two refraction)} \\ \sum [1 - |R_1(\theta_{1j})|^2][1 - |R_1(\theta_{(\not{h}+1)j})|^2] \\ \quad \times \prod_{n=2}^{\not{h}} |R_1(\theta_{nj})|^2 \cos\theta_{1j}, \\ \quad \not{h} \geq 2 \quad \text{(internal reflection)} \end{cases} \tag{5.171}$$

where subscript $1(r)$ denotes the perpendicular polarization component. The equation is also valid for the parallel component $2(l)$. In this equation, θ is the scattering angle; j the index for the entry rays; $l_{\not{h}}$ the ray path length in the crystal ($l_{\not{h}} = 0$, when $\not{h} = 0$); θ_1 the incident angle, which normally has three different values; $\not{h}$ the index denoting the event of reflection and the refraction; and $\theta_2, \theta_3, \ldots$ are incident angles in the crystal. In this equation the effect of absorption is included so that absolute values need to be taken for the reflection and transmission components (see Exercise 5.8). Scattering energy patterns for two and three-dimensional orientations subsequently may be computed by noting the specific relation of the incident angle and the elevation and azimuthal angles. For horizontal orientation cases, the incident angle is the elevation angle. But for general cases, $\cos\theta_i = \cos\varepsilon_i \cos\phi_i$.

Snell's law governing the incident angles (ε_i, ϕ_i) and refracted angles (ε_t, ϕ_t) can be proven to be

$$m_r \sin \varepsilon_t = \sin \varepsilon_i, \qquad m_r(\cos \varepsilon_t / \cos \varepsilon_i) \sin \phi_t = \sin \phi_i, \qquad (5.172)$$

where m_r is the real index of refraction.

To complete the ray tracing exercise, we must include the diffraction pattern. The projection of a hexagonal column onto a horizontal plane clearly resembles a rectangle. The diffraction pattern for a rectangular aperture can be easily derived from the Franhofer diffraction theory. It is given by

$$E^d(\theta, \phi, L) = \frac{\sin^2(Rk \sin \theta \cos \phi)}{(Rk \sin \theta \cos \phi)^2} \frac{\sin^2[(L/2)k \sin \theta \sin \phi]}{[(L/2)k \sin \theta \sin \phi]^2}, \qquad (5.173)$$

where k is the wave number. Clearly, three parameters are required to define the position of a hexagon in reference to the incident ray; i.e., the scattering angle θ, the azimuthal angle ϕ, and the geometrical length L. For horizontally oriented hexagons the diffraction pattern can be obtained by performing integration in ϕ from 0 to π; i.e.,

$$E^d_{2D}(\theta, L) = \frac{1}{\pi} \int_0^\pi E^d(\theta, \phi, L) \, d\phi. \qquad (5.174)$$

For three-dimensional random orientation, integration with respect to the length of hexagons is required. Thus, we find

$$E^d_{3D}(\theta) = \begin{cases} \dfrac{1}{(L/2 - R)} \displaystyle\int_R^{L/2} E^d_{2D}(\theta, L') \, dL' & \text{for columns} \\[3mm] \dfrac{1}{(R - L/2)} \displaystyle\int_{L/2}^R E^d_{2D}(\theta, L') \, dL' & \text{for plates.} \end{cases} \qquad (5.175)$$

It should be noted that for plates, the approximate equation is less accurate because the major axis is on the plane of the hexagon.

Since the equations derived from the ray tracing procedure are presented in units of energy per degree, we must now perform normalization so that the scattering phase function can be derived. On the basis of the definition of gain with respect to isotropic scatterers, we find

$$\frac{G(\theta) 2\pi \sin \theta \, d\theta \, r^2}{4\pi r^2} = E(\theta) \, d\theta, \qquad (5.176)$$

where r denotes the distance, and the gain is normalized as in Eq. (5.164).

Figure 5.14 shows the scattering patterns due to geometrical reflection and refraction for horizontally oriented and randomly oriented columns with lengths and radii of 300 and 60 μm, respectively, incident by a visible wavelength of 0.55 μm. The index $\not{\iota}$ in the diagram denotes the contribution

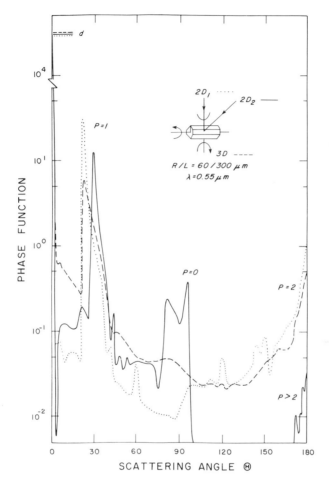

Fig. 5.14 Scattering phase functions for randomly oriented columns (3D) and for horizontally oriented columns (2D) with elevation angles ε of 0° and 42°.

of the scattering energy: $\not p = 0$, external reflection; $\not p = 1$, two refractions; and $\not p \geq 2$, internal reflection. The dashed and dashed–dot lines represent the scattering patterns for horizontally oriented columns with elevation angles of 0° (normal incidence) and 42°, respectively, while the solid curve denotes the scattering pattern for random orientation. The major features for these three cases are the strong forward scattering and halo in the region 20°–30°. For horizontal orientation, the halo feature shifts to a larger scattering angle when the incident angle increases. We see an 8° difference for incident angles of 0° and 42°. Owing to the shift of the halo features for different incident angles, the halo feature in the case of random orientation is broadened and smeared out. The less pronounced 46° halo features are also evident in the random orientation case. The strong peak at 96° in the case of horizontal orientation with an incident angle of 42° is strictly due to the external reflection. Note that the scattering pattern beyond 96° is caused by the end effects and internal reflections. For random orientation and horizontal orientation with normal incidence, the backscattering is primarily produced by one internal reflection. The less pronounced backscattering in the random orientation case is the result of the averaging over many oblique incidence cases. The diffraction peaks show about the same value for these orientation cases.

Comparison of the scattering phase functions for randomly oriented columns with lengths of 300 μm and radii of 60 μm, and plates with lengths of 25 μm and radii of 125 μm is illustrated in Fig. 5.15. The most significant scattering differences between plates and columns are the much lower

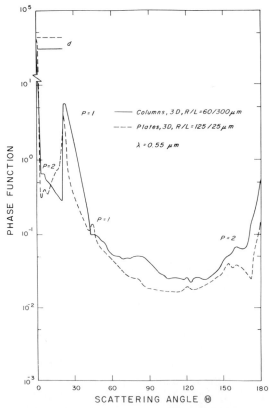

Fig. 5.15   Phase functions for randomly oriented (3D) columns and plates.

forward peak for columns, and the lower side scattering for plates. The well-defined 22° halo for columns is stronger than that for plates. Both scattering patterns depict the very narrow diffraction peak, strong 22° halo feature, and peak at backscattering. The 46° halo feature is less pronounced for columns. Note that columns and plates have about the same volume.

The degree of linear polarization is shown in Fig. 5.16 for randomly oriented columns and plates as well as for spheres based on the geometrical ray tracing for comparison purposes. The polarization pattern for plates remains negative from 0 to about 66°, whereas for columns negative polarization extends only from 0 to about 39°. The strong polarization maximum for plates at about 136° is caused by external reflection ($p = 0$). Such a maximum occurs at about 70° for columns. The positive polarization peaks at about 156 and 178° for columns are associated with one internal reflection ($p = 2$). There is a slight negative polarization for plates in the

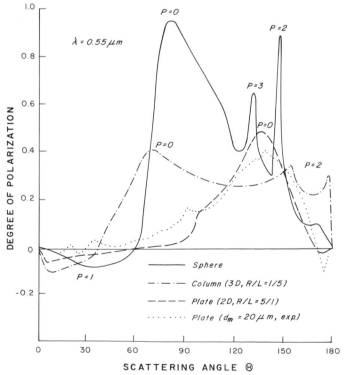

Fig. 5.16 The degree of linear polarization for randomly oriented columns and plates and for spheres. Also shown is the polarization pattern derived from experimental data for plates having a modal diameter of 20 μm.

backscattering direction from about 165 to 180°. The polarization patterns for nonspherical plates and columns differ significantly from the polarization produced by spheres. Large spheres (i.e., in geometrical regions) generate strong polarization at about 80° due to the external reflection, and at the first ($\sim 138°$) and second ($\sim 126°$) rainbow angles caused by one and two internal reflections, respectively. The apparent and significant differences in polarization patterns caused by the shape factor may provide a practical and feasible means for the identification of spheres, plate- and column-like particles in clouds. Also shown in this figure is the polarization pattern derived from experimental data for plates having a modal diameter of 20 μm. There is general agreement between the measured and calculated polarization patterns for plates despite the size difference.

Finally, it should be noted that information and physical understanding of the basic scattering parameters for oriented columns and plates are required to perform radiative transfer calculations for cirrus clouds and to develop active remote sensing techniques for the identification of the phase, shape, and size of cloud particles.

EXERCISES

5.1   Based on the definitions of $\mathbf{M}_\psi$ and $\mathbf{N}_\psi$ in Eqs. (5.23) and (5.24), show that

$$\nabla \times \mathbf{N}_\psi = mk\mathbf{M}_\psi$$

and prove that

$$\nabla^2 \mathbf{N}_\psi + k^2 m^2 \mathbf{N}_\psi = 0, \qquad \nabla^2 \mathbf{M}_\psi + k^2 m^2 \mathbf{M}_\psi = 0.$$

5.2   From the radial component of the magnetic vector

$$H_r^i = e^{-ik\cos\theta}\sin\theta\sin\phi = \frac{i}{k}\left[\frac{\partial^2(rv^i)}{\partial r^2} + k^2(rv^i)\right].$$

show that

$$rv^i = \frac{1}{k}\sum_{n=1}^{\infty}(-i)^n\frac{2n+1}{n(n+1)}\psi_n(kr)P_n^1(\cos\theta)\sin\phi.$$

5.3   The electric and magnetic field vectors in a homogeneous medium satisfy the following vector wave equation:

$$\nabla^2 \mathbf{A} + k^2 m^2 \mathbf{A} = 0.$$

If ψ satisfies the scalar wave equation

$$\nabla^2 \psi + k^2 m^2 \psi = 0,$$

(a)   show that vectors $\mathbf{M}_\psi$ and $\mathbf{N}_\psi$ in cylindrical coordinates (r, ϕ, z) defined by

$$\mathbf{M}_\psi = \nabla \times (\mathbf{a}_z \psi), \qquad mk\mathbf{N}_\psi = \nabla \times \mathbf{M}_\psi$$

satisfy the vector wave equation, where $\mathbf{a}_z$ is a unit vector in Z direction

(b)   Also prove that

$$\mathbf{E} = \mathbf{M}_v + i\mathbf{N}_u, \qquad \mathbf{H} = m(-\mathbf{M}_u + i\mathbf{N}_v)$$

satisfy the Maxwell equations, where u and v are solutions of the scalar wave equation. Write out the expressions for $\mathbf{E}$ and $\mathbf{H}$ in terms of u and v.

5.4   The scalar wave equation in cylindrical coordinates is given by

$$\frac{1}{r}\frac{\partial}{\partial r}\left(r\frac{\partial \psi}{\partial r}\right) + \frac{1}{r^2}\frac{\partial^2 \psi}{\partial \phi^2} + \frac{\partial^2 \psi}{\partial z^2} + m^2 k^2 \psi = 0.$$

Utilizing the method of separation of variables, show that the solution can be written as

$$\psi_n(r, \phi, z) = e^{i\omega t} Z_n(jr) e^{in\phi} e^{-ihz},$$

where h is an arbitrary constant, n is an integer, $j = (m^2 k^2 - h^2)^{1/2}$, and Z_n is any Bessel function of order n.

5.5   The electric fields scattered by nonspherical particles, in general, may be expressed by

$$\begin{bmatrix} E_l^s \\ E_r^s \end{bmatrix} = \frac{e^{-ikr+ikz}}{ikr} \begin{bmatrix} S_2 & S_3 \\ S_4 & S_1 \end{bmatrix} \begin{bmatrix} E_l^i \\ E_r^i \end{bmatrix}.$$

Derive the explicit form of the transformation matrix $\mathbf{M}$ associated with Stokes parameters in terms of S_j ($j = 1, 2, 3, 4$).

5.6   Using the Fresnel formulas, show explicitly that the transmitted and reflected portions of the energy for the two polarization components are conserved. Also compute the incident angle θ_i at which $R_2 = 0$. This angle is called the polarizing or *Brewster* angle, under which the electric vector of the reflected light has no component in the plane of incidence.

5.7   The wavelengths of red and violet light and the corresponding refractive indices of ice and water at these wavelengths are given as follows:

λ (μm)	m (ice)	m (water)
0.656	1.307	1.332
0.405	1.317	1.344

(a)   Describe the color sequence of the corona. Find the radius of cloud particles which produce a secondary white corona of $10°$ angular radius about the sun.

(b) Describe the mechanism of cloudbow formation. Find the scattering angles for the primary and secondary cloudbows at these two wavelengths.

(c) Find the angular radii of the rings of the halos formed by prism angles of 60 and 90°. What will be the widths of the rings?

(d) Derive Eq. (5.172) and find the angular distance between the sun and sun dogs when the elevation angle of the sun is 30°. Sketch a diagram of sun dogs indicating the angular and azimuthal distances and the width of two colors.

From these cloud optics, what may be concluded about the shape, size distribution, and orientation of cloud particles?

5.8   When absorption is involved, the refractive index $m = m_r - im_i$, prove that the Fresnel reflection coefficients are given by

$$|R_1|^2 = \frac{(\cos \theta_i - u)^2 + v^2}{(\cos \theta_i + u)^2 + v^2},$$

$$|R_2|^2 = \frac{[(m_r^2 - m_i^2)\cos \theta_i - u]^2 + (2m_r m_i \cos \theta_i - v)^2}{[(m_r^2 - m_i^2)\cos \theta_i + u]^2 + (2m_r m_i \cos \theta_i + v)^2},$$

$$u^2 = \tfrac{1}{2}\{m_r^2 - m_i^2 - \sin^2 \theta_i + [(m_r^2 - m_i^2 - \sin^2 \theta_i)^2 + 4m_r^2 m_i^2]^{1/2}\},$$

$$v^2 = \tfrac{1}{2}\{-(m_r^2 - m_i^2 - \sin^2 \theta_i) + [(m_r^2 - m_i^2 - \sin^2 \theta_i)^2 + 4m_r^2 m_i^2]^{1/2}\}.$$

To derive these equations, let $m\cos \theta_t = u + iv$ in Eq. (5.145) and use the law of refraction $\sin \theta_i = m \sin \theta_t$.

5.9   Neglecting the proportionality constant, derive Eq. (5.173) from Eq. (5.125).

SUGGESTED REFERENCES

Born, M., and Wolf, E. (1975). *Principles of Optics*. Pergamon Press, New York. Chapter 13, Section 13.5, gives a tractable approach to the solution of the vector wave equation for a homogeneous sphere utilizing the Debye potentials up to the point of the derivation of the scattered electric and magnetic fields. Basic and advanced materials regarding the electromagnetic scattering, geometrical optics, and diffraction are treated elegantly in the book.

Bullrich, K. (1964). Scattered radiation in the atmosphere and the natural aerosol. *Adv. Geophys.* **10**, 99–260. This paper comprehensively discusses the scattering and polarization properties of spherical aerosol particles.

Deirmendjian, D. (1969). *Electromagnetic Scattering on Spherical Polydispersions*. Elsevier, New York. The book provides detailed discussions on the numerical techniques for the computations of Mie theory and presents helpful scattering numerical results in terms of tables for various aerosol, cloud, and rain models covering the electromagnetic spectrum.

Hansen, J. E., and Travis, L. D. (1974). Light scattering in planetary atmospheres. *Space Sci. Rev.*, **16**, 527–610. The first part of this review paper gives a concise description of light scattering by spherical particles and also provides a number of illustrative numerical examples.

Kerker, M. (1969). *The Scattering of Light and Other Electromagnetic Radiation.* Academic Press, New York. Chapter 3 comprehensively gives the history of the solution for scattering by a homogeneous sphere. This book also provides practical discussions on the scattering of light by spheres and cylinders in conjunction with the particle-size determination from scattering information.

Liou, K. N. (1977). A complementary theory of light scattering by homogeneous spheres. *Appl. Math. Comput.* **3**, 331–358. This review paper presents, in a coherent manner, the Mie scattering theory.

Stratton, J. A. (1941). *Electromagnetic Theory.* McGraw-Hill, New York. Chapters 7 and 9 contain advanced and organized materials for the first time for the solution of vector wave equations in spherical coordinates. Basic Bessel, Neumann, and Hankel functions and Legendre polynomials are also discussed in some detail in the book.

Tricker, R. A. R. (1970). *Introduction to Meteorological Optics.* Elsevier, New York. The book offers delightful illustrations of several optical phenomena produced by atmospheric particulates, especially the ice crystals.

van de Hulst, H. C. (1957). *Light Scattering by Small Particles.* Wiley, New York. Chapters 8, 9, and 12 provide authoritative analyses of diffraction, Mie scattering theory, and geometrical optics, respectively. In Chapters 2–4, fundamental scattering equations are derived by means of physical insight and postulations.

Chapter 6

PRINCIPLE OF MULTIPLE SCATTERING IN PLANE-PARALLEL ATMOSPHERES

In Section 3.7 and Chapter 5, single scattering processes involving molecules and particulates in planetary atmospheres were discussed. In a realistic atmosphere, molecules and particulates not only undergo single scattering but also multiple scattering which is responsible for the transfer of energy within the atmosphere. We have introduced the concept of multiple scattering in Section 1.1.4 and have presented the basic equation of transfer for plane-parallel atmospheres in Section 1.4.4. The source function associated with multiple scattering, however, was not specifically defined. In this chapter, we discuss the fundamentals and various approximations and methods dealing with multiple scattering problems utilizing the single-scattering parameters defined in Section 3.7 and Chapter 5. We begin with formulating the basic radiative transfer equation for the scattering of sunlight in plane-parallel atmospheres and discuss the representation of the phase function. Simplified approximations including the order of scattering, two stream and Eddington's methods for multiple scattering are then described. Presentation of the classical discrete-ordinates method for radiative transfer, especially for cases of isotropic scattering, and description of the principles of invariance for infinite and finite atmospheres leading to the H-function and X and Y functions, respectively, are followed. The inclusion of surface reflection in multiple scattering problems is also outlined. We then introduce the adding method for multiple scattering and illustrate the equivalence of the adding method and the principles of invariance for finite atmospheres. The following section is concerned with the use of the Stokes parameters in multiple scattering. i.e., the inclusion of polarization. The final section deals with the fundamental transfer equations for multiple scattering by oriented nonspherical particles and for multiple scattering in three-dimensional space. Subject matters including the diffusion approximation, the spherical harmonics method, and the invariant imbedding are also introduced in the exercise section.

6.1   FORMULATION OF THE SCATTERING OF SUNLIGHT IN PLANE-PARALLEL ATMOSPHERES

Consider a plane-parallel atmosphere illuminated by flux of radiation πF_0 emitted from the sun. Assuming that the diffuse intensity is from below, we then have the reduction of the differential diffuse intensity caused by events of single scattering and absorption by particles. This is expressed by

$$dI(z, \mathbf{\Omega}) = -\bar{\sigma}_e N I(z, \mathbf{\Omega}) \, dz / \cos \theta, \qquad (6.1)$$

where dz is the differential thickness, $\bar{\sigma}_e$ the mean extinction cross section of a sample of particles, N the total number of particles per volume, and $\mathbf{\Omega}$ the directional element of solid angle that represents the pencil of radiation (see Fig. 6.1).

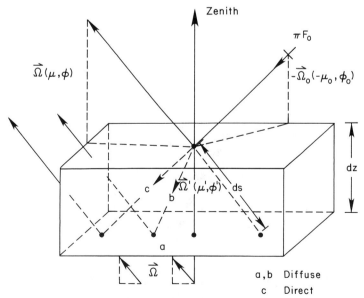

Fig. 6.1 Transfer of solar radiation in plane-parallel layers, illustrating attenuation by extinction, a; multiple scattering, b; and single scattering of the unscattered solar flux, c.

Meanwhile, the differential diffuse intensity in the direction $\boldsymbol{\Omega}$ may be increased by multiple scattering, arising from the scattering of a pencil of radiation of solid angle $d\boldsymbol{\Omega}'$ in the direction $\boldsymbol{\Omega}'$. This is given by

$$dI(z, \boldsymbol{\Omega}) = \bar{\sigma}_s N \, dz / \cos\theta \int_{4\pi} I(z, \boldsymbol{\Omega}') \frac{P(\boldsymbol{\Omega}, \boldsymbol{\Omega}')}{4\pi} d\boldsymbol{\Omega}', \qquad (6.2)$$

where $\bar{\sigma}_s$ denotes the mean scattering cross section, and the nondimensional phase function introduces the appropriate radiation stream from $\boldsymbol{\Omega}'$ to $\boldsymbol{\Omega}$. Thus, integration over the entire solid angle gives all the possible contributions of multiply scattered radiant energy from $\boldsymbol{\Omega}'$ to $\boldsymbol{\Omega}$.

In addition, the differential diffuse intensity in the direction $\boldsymbol{\Omega}$ also may be increased due to single scattering of the direct solar radiation whose direction is represented by $-\boldsymbol{\Omega}_0$ (where the minus sign denotes that the direct solar radiation is always downward). If the direct solar flux at level z is $F(z)$, then

$$dI(z, \boldsymbol{\Omega}) = \frac{\bar{\sigma}_s N \, dz}{\cos\theta} F(z) \frac{P(\boldsymbol{\Omega}, -\boldsymbol{\Omega}_0)}{4\pi}, \qquad (6.3)$$

From the Beer–Bouguer–Lambert law discussed in Section 1.3.2, $F(z)$ is simply

$$F(z) = \pi F_0 \exp\left\{ -\frac{1}{\cos\theta_0} \int_z^\infty \bar{\sigma}_e(z') N(z') \, dz' \right\}. \qquad (6.4)$$

Upon combining Eqs. (6.1)–(6.4), and introducing the optical depth defined in Section 1.4.4 and the single-scattering albedo defined in Eq. (5.117), we obtain the basic equation for scattering of solar radiation in plane-parallel atmospheres as

$$\mu \frac{dI(\tau, \boldsymbol{\Omega})}{d\tau} = I(\tau, \boldsymbol{\Omega}) - \frac{\tilde{\omega}}{4\pi} \int_{4\pi} I(\tau, \boldsymbol{\Omega}') P(\boldsymbol{\Omega}, \boldsymbol{\Omega}') \, d\boldsymbol{\Omega}'$$

$$- \frac{\tilde{\omega}}{4\pi} \pi F_0 P(\boldsymbol{\Omega}, -\boldsymbol{\Omega}_0) e^{-\tau/\mu_0}. \qquad (6.5)$$

Note that $\mu = \cos\theta$, $\mu_0 = \cos\theta_0$, $d\boldsymbol{\Omega} = d\mu \, d\phi$, $\boldsymbol{\Omega} = (\mu, \phi)$, and ϕ represent the azimuthal angle.

In reference to Section 1.4.4, it is clear that the source function in the solar spectral region is

$$J(\tau; \mu, \phi) = \frac{\tilde{\omega}}{4\pi} \int_0^{2\pi} \int_{-1}^1 I(\tau; \mu', \phi') P(\mu, \phi; \mu', \phi') \, d\mu' \, d\phi'$$

$$+ \frac{\tilde{\omega}}{4\pi} \pi F_0 P(\mu, \phi; -\mu_0, \phi_0) e^{-\tau/\mu_0}. \qquad (6.6)$$

The scattering geometry is shown in Fig. 6.2. As noted in Section 3.7.1, the scattering angle is defined as the angle between the incident and scattered beams. Lines AO and BO in the figure denote the incident and scattered beams, respectively. Based on the spherical geometry as shown in Appendix F, the cosine of the scattering angle can be expressed by

$$\cos \Theta = \cos \theta \cos \theta' + \sin \theta \sin \theta' \cos(\phi' - \phi)$$
$$= \mu\mu' + (1 - \mu^2)^{1/2}(1 - \mu'^2)^{1/2} \cos(\phi' - \phi). \tag{6.7}$$

The phase function defined in Eq. (3.63) may be numerically expanded in Legendre polynomials with a finite number of terms, N. Thus,

$$P(\cos \Theta) = \sum_{l=0}^{N} \tilde{\omega}_l P_l(\cos \Theta), \tag{6.8}$$

where $\tilde{\omega}_l$ are a set of $N + 1$ constants and $\tilde{\omega}_0 = 1$. In view of Eq. (6.7), we have

$$P(\mu, \phi; \mu', \phi') = \sum_{l=0}^{N} \tilde{\omega}_l P_l[\mu\mu' + (1 - \mu^2)^{1/2}(1 - \mu'^2)^{1/2} \cos(\phi' - \phi)].$$
$$\tag{6.9}$$

The Legendre polynomials for the argument shown in Eq. (6.9) can be expanded by the addition theorem for spherical harmonics (see Appendix G) to give

$$P(\mu, \phi; \mu', \phi') = \sum_{m=0}^{N} \sum_{l=m}^{N} \tilde{\omega}_l^m P_l^m(\mu) P_l^m(\mu') \cos m(\phi' - \phi), \tag{6.10}$$

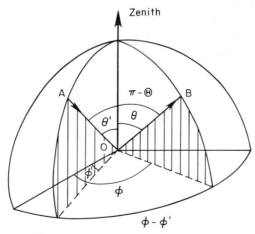

Fig. 6.2 Relation of scattering, zenith, and azimuthal angles.

where

$$\tilde{\omega}_l^m = (2 - \delta_{0,m})\tilde{\omega}_l \frac{(l - m)!}{(l + m)!} \quad (l = m, \ldots, N, \quad 0 \le m \le N), \quad (6.11)$$

$$\delta_{0,m} = \begin{cases} 1 & \text{if } m = 0, \\ 0 & \text{otherwise,} \end{cases} \quad (6.12)$$

and p_l^m denote the associated Legendre polynomials.

In view of the phase function expansion, we also may expand the intensity in the form

$$I(\tau; \mu, \phi) = \sum_{m=0}^{N} I^m(\tau, \mu) \cos m(\phi_0 - \phi). \quad (6.13)$$

Upon inserting Eqs. (6.10) and (6.13) into Eq. (6.5) and noting the orthogonality of the associated Legendre polynomial, Eq. (6.5) splits up into $(N + 1)$ independent equations:

$$\mu \frac{dI^m(\tau, \mu)}{d\tau} = I^m(\tau, \mu) - (1 + \delta_{0,m}) \frac{\tilde{\omega}}{4} \sum_{l=m}^{N} \tilde{\omega}_l^m P_l^m(\mu) \int_{-1}^{1} P_l^m(\mu') I^m(\tau; \mu') du'$$

$$- \frac{\tilde{\omega}}{4\pi} \sum_{l=m}^{N} \tilde{\omega}_l^m P_l^m(\mu) P_l^m(-\mu_0) \pi F_0 e^{-\tau/\mu_0} \quad (m = 0, 1, \ldots, N). \quad (6.14)$$

Each equation may be solved independently for I^m, and from Eq. (6.13), I may be determined.

For $m = 0$, the intensity expressed in Eq. (6.13) corresponds to the azimuthal independence case. We neglect the superscript 0 for simplicity and rewrite Eq. (6.14) to yield

$$\mu \frac{dI(\tau, \mu)}{d\tau} = I(\tau, \mu) - \frac{\tilde{\omega}}{2} \sum_{l=0}^{N} \tilde{\omega}_l P_l(\mu) \int_{-1}^{1} P_l(\mu') I(\tau; \mu') d\mu'$$

$$- \frac{\tilde{\omega}}{4\pi} \sum_{l=0}^{N} \tilde{\omega}_l P_l(\mu) P_l(-\mu_0) \pi F_0 e^{-\tau/\mu_0}. \quad (6.15)$$

Equation (6.15) is particularly useful for flux calculations as will be seen in the following discussions.

For scattering atmospheres, the diffuse upward and downward flux densities for any given τ are given, respectively, by [see the definition of flux density in Eq. (1.9)]

$$F_{\text{dif}}^{\uparrow}(\tau) = \int_{0}^{2\pi} \int_{0}^{1} I(\tau; \mu, \phi) \mu \, d\mu \, d\phi, \qquad \mu \ge 0, \quad (6.16)$$

$$F_{\text{dif}}^{\downarrow}(\tau) = \int_{0}^{2\pi} \int_{0}^{-1} I(\tau; \mu, \phi) \mu \, d\mu \, d\phi, \qquad \mu \le 0. \quad (6.17)$$

Thus, by noting that

$$\int_0^{2\pi} \cos m(\phi_0 - \phi)\, d\phi = 0, \qquad m \neq 0, \tag{6.18}$$

in Eq. (6.13), we obtain the upward and downward flux densities

$$F_{\text{dif}}^{\uparrow\downarrow}(\tau) = 2\pi \int_0^{\pm 1} I(\tau; \mu)\mu\, d\mu. \tag{6.19}$$

Consequently, for flux calculations in the atmosphere, the azimuthal dependence of the intensity expansion can be ignored, and Eq. (6.15) is sufficient for radiation studies.

Moreover, for azimuthal independent cases, we may define the phase function

$$P(\mu, \mu') = \frac{1}{2\pi} \int_0^{2\pi} P(\mu, \phi; \mu', \phi')\, d\phi'. \tag{6.20}$$

In view of the phase function expansion represented by Eq. (6.10), we shall have then

$$P(\mu, \mu') = \sum_{l=0}^{N} \tilde{\omega}_l P_l(\mu) P_l(\mu'). \tag{6.21}$$

By virtue of this equation, the azimuthally independent transfer equation for diffuse radiation expressed in Eq. (6.15) can be rewritten as

$$\mu \frac{dI(\tau, \mu)}{d\tau} = I(\tau, \mu) - \frac{\tilde{\omega}}{2} \int_{-1}^{1} I(\tau, \mu')P(\mu, \mu')\, d\mu'$$

$$- \frac{\tilde{\omega}}{4\pi} \pi F_0 P(\mu, -\mu_0) e^{-\tau/\mu_0}. \tag{6.22}$$

It should be noted that we use the positive μ to denote the upward radiation, whereas the negative μ denotes the downward radiation; this is evident in Eqs. (6.16) and (6.17). Hence, the μ_0s that denote the direct solar radiation component are negative values. However, a positive μ_0 has been employed for convenience, and $-\mu_0$ represents the fact that the direct solar radiation is downward. For the transfer of terrestrial infrared radiation in scattering atmospheres, which are in local thermodynamic equilibrium, we simply replace the last term in Eq. (6.22) by the thermal emission contribution $(1 - \tilde{\omega}_v)B_v[T(\tau)]$, where B_v is the Planck function of temperature T.

Since the equation of transfer only describes the diffuse component of radiation, i.e., radiation scattered more than once, we must include the direct component of radiation to account for the downward radiation. It is given by the simple Beer–Bouguer–Lambert law of extinction

$$F^{\downarrow}_{\text{dir}}(\tau) = \mu_0 \pi F_0 e^{-\tau/\mu_0}. \tag{6.23}$$

The total upward and downward flux densities at a given τ are, respectively,

$$F^{\uparrow}(\tau) = F^{\uparrow}_{\text{dif}}(\tau) = 2\pi \int_0^1 I(\tau, \mu)\mu \, d\mu, \tag{6.24}$$

$$F^{\downarrow}(\tau) = F^{\downarrow}_{\text{dif}}(\tau) + F^{\downarrow}_{\text{dir}}(\tau) = 2\pi \int_0^{-1} I(\tau, \mu)\mu \, d\mu + \mu_0 \pi F_0 e^{-\tau/\mu_0}. \tag{6.25}$$

The net flux density for a given level is therefore

$$F(\tau) = F^{\uparrow}(\tau) - F^{\downarrow}(\tau). \tag{6.26}$$

It follows that the heating rate in the atmosphere, due to solar radiation, can be evaluated according to discussions made in Section 3.5.

6.2   APPROXIMATIONS FOR RADIATIVE TRANSFER

6.2.1   Single Scattering and Order of Scattering Approximations

Consider the emergent radiation as consisting of light which has been scattered only once. Then, the radiation source function is simply

$$J(\tau; \mu, \phi) = \frac{\tilde{\omega}}{4\pi} \pi F_0 P(\mu, \phi; -\mu_0, \phi_0) e^{-\tau/\mu_0}. \tag{6.27}$$

On the basis of Eqs. (1.64) and (1.65), the upward (reflected) and downward (transmitted) intensities for a finite atmosphere bounded on two sides at $\tau = 0$ and $\tau = \tau_1$ are

$$I(\tau; \mu, \phi) = I(\tau_1; \mu, \phi)e^{-(\tau_1 - \tau)/\mu} + \frac{\tilde{\omega}}{4\pi} \pi F_0 P(\mu, \phi; -\mu_0, \phi_0)$$

$$\times \int_\tau^{\tau_1} \exp\{-[(\tau' - \tau)/\mu + \tau'/\mu_0]\} \frac{d\tau'}{\mu}, \tag{6.28}$$

$$I(\tau; -\mu, \phi) = I(0; -\mu, \phi)e^{-\tau/\mu} + \frac{\tilde{\omega}}{4\pi} \pi F_0 P(-\mu, \phi; -\mu_0, \phi_0)$$

$$\times \int_0^\tau \exp\{-[(\tau - \tau')/\mu + \tau'/\mu_0]\} \frac{d\tau'}{\mu}. \tag{6.29}$$

When we assume that there is no diffuse downward and upward radiation at the top and the base of the finite atmosphere, i.e.,

$$I(0; -\mu, \phi) = 0,$$
$$I(\tau_1; \mu, \phi) = 0,$$

(6.30)

then the reflected and transmitted intensities for a finite atmosphere with an optical depth τ_1 are

$$I(0; \mu, \phi) = \frac{\tilde{\omega}\mu_0 F_0}{4(\mu + \mu_0)} P(\mu, \phi; -\mu_0, \phi_0)$$

$$\times \left\{ 1 - \exp\left[-\tau_1 \left(\frac{1}{\mu} + \frac{1}{\mu_0} \right) \right] \right\},$$

(6.31)

$$I(\tau_1; -\mu, \phi) = \begin{cases} \dfrac{\tilde{\omega}\mu_0 F_0}{4(\mu - \mu_0)} P(-\mu, \phi; -\mu_0, \phi_0)(e^{-\tau_1/\mu} - e^{-\tau_1/\mu_0}), \\[6pt] \hspace{5cm} \mu \neq \mu_0 \\[6pt] \dfrac{\tilde{\omega}\tau_1 F_0}{4\mu_0} P(-\mu_0, \phi_0; -\mu_0, \phi_0)e^{-\tau_1/\mu_0}, \hspace{0.5cm} \mu = \mu_0. \end{cases}$$

(6.32)

It is clear that for the single-scattering approximation, intensities are directly proportional to the phase function.

The method of successive orders of scattering is one in which the intensity is computed individually for photons scattered once, twice, three times, and so forth, with the total intensity obtained as the sum over all orders. Hence for diffuse reflected and transmitted intensities we may write, respectively,

$$I(\tau; \mu, \phi) = \sum_{n=1}^{\infty} I_n(\tau; \mu, \phi),$$

$$I(\tau; -\mu, \phi) = \sum_{n=1}^{\infty} I_n(\tau; -\mu, \phi),$$

(6.33)

where n denotes the order of scattering.

Subject to the boundary conditions denoted in Eq. (6.30), the formal solution of the equation of transfer is given by

$$I(\tau; \mu, \phi) = \int_{\tau}^{\tau_1} J(\tau'; \mu, \phi) \exp[-(\tau' - \tau)/\mu] \frac{d\tau'}{\mu},$$

$$I(\tau; -\mu, \phi) = \int_{0}^{\tau} J(\tau'; \mu, \phi) \exp[-(\tau - \tau')/\mu] \frac{d\tau'}{\mu}.$$

(6.34)

The source function for the first-order scattering of the incident radiation is given by Eq. (6.27). Inserting it into the formal solution of the equation of transfer [Eq. (6.34)] and integrating over the appropriate optical depths, we obtain the intensity due to photons scattered once. It follows that the source functions and intensities may be derived successively by means of the recursion relationships

$$J_{n+1}(\tau; \mu, \phi) = \frac{\tilde{\omega}}{4\pi} \int_0^{2\pi} \int_{-1}^1 P(\mu, \phi; \mu', \phi') I_n(\tau; \mu', \phi')\, d\mu'\, d\phi',$$

$$I_n(\tau; \mu, \phi) = \int_\tau^{\tau_1} J_n(\tau'; \mu, \phi) \exp[-(\tau' - \tau)/\mu] \frac{d\tau'}{\mu}, \qquad n \geq 1, \quad (6.35)$$

$$I_n(\tau; -\mu, \phi) = \int_0^\tau J_n(\tau'; -\mu, \phi) \exp[-(\tau - \tau')/\mu] \frac{d\tau'}{\mu}, \qquad n \geq 1,$$

where the zero-order intensity is given by Dirac's δ function

$$I_0(\tau; \mu', \phi') = \pi F_0 e^{-\tau/\mu_0} \delta(\mu' - \mu_0)\, \delta(\phi' - \phi_0). \qquad (6.36)$$

Numerical techniques may be devised to carry out the integrations for a finite interval of τ in Eq. (6.35) to obtain the intensity distribution.

6.2.2   Two-Stream and Eddington Approximations

In order to solve Eq. (6.15) analytically, the integral has to be replaced by summation over a finite number of quadrature points. It has been found that Gauss's formula is superior to other formulas for quadratures in the interval $(-1, 1)$. For any function $f(\mu)$, Gauss's formula is expressed by

$$\int_{-1}^1 f(\mu)\, d\mu \approx \sum_{j=-n}^n a_j f(\mu_j), \qquad (6.37)$$

where the weights are

$$a_j = \frac{1}{P'_{2n}(\mu_j)} \int_{-1}^1 \frac{P_{2n}(\mu)}{\mu - \mu_j}\, d\mu, \qquad (6.38)$$

and μ_j are the zeros of the even-order Legendre polynomials $P_{2n}(\mu)$. We also find that

$$a_{-j} = a_j, \qquad \mu_{-j} = -\mu_j, \qquad \sum_{j=-n}^n a_j = 2. \qquad (6.39)$$

Table 6.1 lists the Gaussian points and weights for the first four approximations.

TABLE 6.1 *The Gaussian Points and Weights*

n	$2n$	$\pm \mu_n$	a_n
1	2	$\mu_1 = 0.5773503$	$a_1 = 1$
2	4	$\mu_1 = 0.3399810$	$a_1 = 0.6521452$
		$\mu_2 = 0.8611363$	$a_2 = 0.3478548$
3	6	$\mu_1 = 0.2386192$	$a_1 = 0.4679139$
		$\mu_2 = 0.6612094$	$a_2 = 0.3607616$
		$\mu_3 = 0.9324695$	$a_3 = 0.1713245$
4	8	$\mu_1 = 0.1834346$	$a_1 = 0.3626838$
		$\mu_2 = 0.5255324$	$a_2 = 0.3137066$
		$\mu_3 = 0.7966665$	$a_3 = 0.2223810$
		$\mu_4 = 0.9602899$	$a_4 = 0.1012285$

On utilizing Gauss's formula, Eq. (6.15) can be written as

$$\mu_i \frac{dI(\tau; \mu_i)}{d\tau} = I(\tau; \mu_i) - \frac{\tilde{\omega}}{2} \sum_{l=0}^{N} \tilde{\omega}_l P_l(\mu_i) \sum_{j=-n}^{n} a_j P_l(\mu_j) I(\tau; \mu_j)$$

$$- \frac{\tilde{\omega}}{4} F_0 \left[\sum_{l=0}^{N} (-1)^l \tilde{\omega}_l P_l(\mu_i) P_l(\mu_0) \right] e^{-\tau/\mu_0}, \quad i = -n, n, \quad (6.40)$$

where $\mu_i(-n, n)$ represent the directions of radiation streams.

For simplicity of solving Eq. (6.40), we take two radiation streams, i.e., $j = -1$ and 1, and $N = 1$. Note that $\mu_1 = 1/\sqrt{3}$ and $a_1 = a_{-1} = 1$. After rearranging terms and denoting $I^\uparrow = I(\tau; \mu_1)$ and $I^\downarrow = I(\tau; -\mu_1)$, two simultaneous equations are derived. They are

$$\mu_1 \frac{dI^\uparrow}{d\tau} = I^\uparrow - \tilde{\omega}(1 - b)I^\uparrow - \tilde{\omega}bI^\downarrow - S^- e^{-\tau/\mu_0}, \quad (6.41a)$$

$$-\mu_1 \frac{dI^\downarrow}{d\tau} = I^\downarrow - \tilde{\omega}(1 - b)I^\downarrow - \tilde{\omega}bI^\uparrow - S^+ e^{-\tau/\mu_0}, \quad (6.41b)$$

where

$$g = \frac{\tilde{\omega}_1}{3} = \frac{1}{2} \int_{-1}^{1} P(\cos \Theta) \cos \Theta \, d\cos \Theta = \langle \cos \Theta \rangle,$$

$$b = \frac{1 - g}{2} = \frac{1}{2} \int_{-1}^{1} P(\cos \Theta) \frac{1 - \cos \Theta}{2} d\cos \Theta, \quad (6.42)$$

$$S^\pm = \frac{F_0 \tilde{\omega}}{4} (1 \pm 3g\mu_1\mu_0).$$

The parameter g is called the *asymmetry factor* and is the first moment of the phase function. It is derived from Eq. (6.8) by using the orthogonal property of the Legendre polynomials. Note that the zero moment of the phase function is simply equal to $\tilde{\omega}_0 (=1)$. For isotropic scattering, g is zero as it is for Rayleigh scattering. The asymmetry factor increases as the diffraction peak of the phase function becomes increasingly sharpened. Conceivably, the asymmetry factor may be negative if the phase function is peaked at the backward directions (90–180°). For Mie particles whose phase function has a general sharp peak at 0° scattering angle (e.g., see Fig. 5.11), the asymmetry factor denotes the relative strength of forward scattering. The parameters b and $(1 - b)$ can be interpreted as the integrated fraction of energy backscattered and forward scattered, respectively. Thus, it is apparent in Eq. (6.41) that the multiple scattering contribution in the two-stream approximation is represented by the upward and downward intensities weighed by the appropriate fraction of the forward or backward phase function. The upward intensity is strengthened by its coupling with the forward fraction (0–90°) of the phase function plus the downward intensity which appears in backward fraction (90–180°) of the phase. A similar argument holds for the downward intensity.

Equations (6.41a) and (6.41b) represent two first-order inhomogeneous differential equations. Let $M = I^\uparrow + I^\downarrow$ and $N = I^\uparrow - I^\downarrow$, and note that $(1 - 2b) = g$; then by subtracting and adding, Eq. (6.41) becomes

$$\mu_1 \frac{dM}{d\tau} = (1 - \tilde{\omega}g)N - (S^- - S^+)e^{-\tau/\mu_0}, \tag{6.43a}$$

$$\mu_1 \frac{dN}{d\tau} = (1 - \tilde{\omega})M - (S^- + S^+)e^{-\tau/\mu_0}. \tag{6.43b}$$

Further, by differentiating both equations with respect to τ, we obtain

$$\mu_1 \frac{d^2M}{d\tau^2} = (1 - \tilde{\omega}g)\frac{dN}{d\tau} + \frac{(S^- - S^+)}{\mu_0}e^{-\tau/\mu_0}, \tag{6.44a}$$

$$\mu_1 \frac{d^2N}{d\tau^2} = (1 - \tilde{\omega})\frac{dM}{d\tau} + \frac{(S^- + S^+)}{\mu_0}e^{-\tau/\mu_0}. \tag{6.44b}$$

Upon inserting Eqs. (6.43b) and (6.43a) into Eqs. (6.44a) and (6.44b), respectively, we find

$$\frac{d^2M}{d\tau^2} = k^2M + Z_1e^{-\tau/\mu_0}, \tag{6.45a}$$

$$\frac{d^2N}{d\tau^2} = k^2N + Z_2e^{-\tau/\mu_0}, \tag{6.45b}$$

where the eigenvalue is given by

$$k^2 = (1 - \tilde{\omega})(1 - \tilde{\omega}g)/\mu_1^2 \qquad (6.46)$$

and

$$Z_1 = -\frac{(1 - \tilde{\omega}g)(S^- + S^+)}{\mu_1^2} + \frac{S^- + S^+}{\mu_1 \mu_0},$$

$$Z_2 = -\frac{(1 - \tilde{\omega})(S^- - S^+)}{\mu_1^2} + \frac{S^- + S^+}{\mu_1 \mu_0}, \qquad (6.47)$$

Equation (6.45) represents a set of second-order differential equations, which can be solved by seeking first the homogeneous part and then by adding a particular solution. In seeking the homogeneous solution, the homogeneous parts of Eq. (6.43) need to be satisfied so that there are only two unknown constants involved. Straightforward analyses yield the solutions

$$I^\uparrow = I(\tau, \mu_1) = Kve^{k\tau} + Hue^{-k\tau} + \varepsilon e^{-\tau/\mu_0}, \qquad (6.48)$$

$$I^\downarrow = I(\tau, -\mu_1) = Kue^{k\tau} + Hve^{-k\tau} + \gamma e^{-\tau/\mu_0}, \qquad (6.49)$$

where

$$v = (1 + a)/2, \qquad u = (1 - a)/2, \qquad (6.50)$$

$$a^2 = (1 - \tilde{\omega})/(1 - \tilde{\omega}g), \qquad (6.51)$$

$$\varepsilon = (\alpha + \beta)/2, \qquad \gamma = (\alpha - \beta)/2, \qquad (6.52)$$

$$\alpha = Z_1\mu_0^2/(1 - \mu_0^2 k^2), \qquad \beta = Z_2\mu_0^2/(1 - \mu_0^2 k^2). \qquad (6.53)$$

The constants K and H are to be determined from the diffuse radiation boundary conditions at the top and bottom of the scattering layer. For boundary conditions given in Eq. (6.30), we get

$$K = -(\varepsilon v e^{-\tau_1/\mu_0} - \gamma u e^{-k\tau_1})/(v^2 e^{k\tau_1} - u^2 e^{-k\tau_1}), \qquad (6.54)$$

$$H = -(\varepsilon u e^{-\tau_1/\mu_0} - \gamma v e^{-k\tau_1})/(v^2 e^{k\tau_1} - \mu^2 e^{-k\tau_1}). \qquad (6.55)$$

Once the upward and downward intensities have been evaluated, the upward and downward diffuse flux densities are given simply by

$$F^\uparrow(\tau) = 2\pi\mu_1 I^\uparrow, \qquad F^\downarrow(\tau) = 2\pi\mu_1 I^\downarrow. \qquad (6.56)$$

This is the two-stream approximation for radiative transfer.

In a similar approach, we may expand the intensity component in Eq. (6.15) [see also Eq. (6.40)] as

$$I(\tau, \mu) = I_0(\tau) + I_1(\tau)\mu \qquad (-1 \le \mu \le 1). \qquad (6.57)$$

And by letting $N = 1$, we have

$$\mu \frac{d(I_0 + \mu I_1)}{d\tau} = (I_0 + I_1\mu) - \tilde{\omega}(I_0 + g\mu I_1) - \frac{\tilde{\omega}}{4} F_0 (1 - 3g\mu\mu_0)e^{-\tau/\mu_0}. \quad (6.58)$$

Upon integrating Eq. (6.58) and Eq. (6.58) times μ both over μ, the following two first-order differential equations are derived. They are

$$\frac{dI_1}{d\tau} = 3(1 - \tilde{\omega})I_0 - \tfrac{3}{4}\tilde{\omega}F_0 e^{-\tau/\mu_0}, \quad (6.59)$$

$$\frac{dI_0}{d\tau} = (1 - \tilde{\omega}g)I_1 + \tfrac{3}{4}\tilde{\omega}g\,\mu_0 F_0 e^{-\tau/\mu_0}. \quad (6.60)$$

The reader is invited to derive the solutions for I_0 and I_1, subject to radiation boundary conditions given by Eq. (6.30).

Once the I_0 and I_1 have been obtained, then the upward and downward diffuse flux densities are

$$F^{\uparrow}(\tau) = 2\pi \int_0^1 [I_0(\tau) + \mu I_1(\tau)]\mu\,d\mu = \pi[I_0(\tau) + \tfrac{2}{3}I_1(\tau)],$$
$$F^{\downarrow}(\tau) = 2\pi \int_0^{-1} [I_0(\tau) + \mu I_1(\tau)]\mu\,d\mu = \pi[I_0(\tau) - \tfrac{2}{3}I_1(\tau)]. \quad (6.61)$$

The foregoing analysis constitutes the so-called Eddington's approximation. It is evident that a two-term expansion in the intensity component is utilized, and its concept of simplification is similar to that of the two-stream approximation.

Now returning to Eq. (6.40), we may consider the approximation for Eq. (6.40) a step further and assume that $j = -1, -2, 1, 2$ and $l = 0, 1, 2, 3$. Consequently, four first-order, inhomogeneous differential equations can be derived. Thus, the solution of the four-stream approximation for radiative transfer may also be derived analytically (Liou, 1974).

Tables 6.2–6.4 present numerical results of reflection and total transmission (diffuse plus direct) for two-stream, four-stream, and high-order discrete stream for anisotropic scattering discussed in Section 6.3.3. The reflection r, and the diffuse (t_{dif}) and direct (t_{dir}) transmission for solar radiation are defined by [also see Eqs. (6.109) and (6.110)]

$$r = F^{\uparrow}(0)/(\pi\mu_0 F_0), \qquad t_{\text{dif}} = F^{\downarrow}(\tau_1)/(\pi\mu_0 F_0), \qquad t_{\text{dir}} = \exp(-\tau_1/\mu_0).$$

The computations were all made for the analytic Henyey–Greenstein phase function

$$P(\Theta) = (1 - g^2)/(1 + g^2 - 2g\cos\Theta)^{3/2}$$

TABLE 6.2 *Comparisons of Reflection and Transmission (Direct + Diffuse) as Computed by the Discrete-Ordinates Method (DOM) with Discrete Streams of 2, 4, 8, and 16 and by the Doubling Method for Conservative Scattering;* $\tilde{\omega} = 1$

τ_1	Method	μ_0:	Reflection					Transmission				
			0.1	0.3	0.5	0.7	0.9	0.1	0.3	0.5	0.7	0.9
0.25	DOM, 2		0.41133	0.18018	0.07635	0.02109	−0.01294	0.58867	0.81982	0.92365	0.97891	1.01294
	4		0.40339	0.15250	0.05743	0.02457	0.03221	0.59661	0.84750	0.94257	0.97542	0.97593
	8		0.40985	0.15185	0.06937	0.03884	0.02114	0.59015	0.84815	0.93063	0.96116	0.97888
	16		0.41768	0.15776	0.07165	0.03797	0.02246	0.58239	0.84229	0.92842	0.96206	0.97741
	Doubling		0.41610	0.15795	0.07179	0.03801	0.02250	0.58390	0.84205	0.92821	0.96200	0.97751
1	DOM, 2		0.51962	0.36837	0.22559	0.11158	0.02389	0.48222	0.63163	0.77441	0.88842	0.97611
	4		0.56631	0.37569	0.22498	0.13468	0.09826	0.43368	0.62431	0.77501	0.86531	0.90173
	8		0.58967	0.38728	0.24037	0.15066	0.09582	0.41034	0.61273	0.75963	0.84934	0.90424
	16		0.58567	0.38676	0.24068	0.15019	0.09654	0.41440	0.61337	0.75951	0.84991	0.90249
	Doubling		0.58148	0.38571	0.24048	0.15019	0.09672	0.41852	0.61430	0.75952	0.84981	0.90328
4	DOM, 2		0.68564	0.59282	0.49999	0.40737	0.31612	0.31436	0.40718	0.50001	0.59263	0.68388
	4		0.73722	0.62493	0.52120	0.42825	0.34962	0.26278	0.37507	0.47880	0.57174	0.65036
	8		0.73877	0.61974	0.52046	0.43052	0.34806	0.26124	0.38026	0.47953	0.56948	0.65203
	16		0.73541	0.61830	0.51977	0.42964	0.34776	0.26469	0.38190	0.48053	0.57051	0.65077
	Doubling		0.73254	0.61732	0.51932	0.42945	0.34823	0.26746	0.38269	0.48069	0.57055	0.65178
16	DOM, 2		0.86860	0.82980	0.79100	0.75220	0.71340	0.13140	0.17020	0.20900	0.24780	0.28660
	4		0.88407	0.83459	0.78881	0.74721	0.71005	0.11592	0.16540	0.21118	0.25278	0.28993
	8		0.88397	0.83127	0.78659	0.74699	0.70755	0.11604	0.16873	0.21260	0.25300	0.29254
	16		0.88240	0.83055	0.78702	0.74642	0.70627	0.11770	0.16965	0.21329	0.25373	0.29225
	Doubling		0.88103	0.82995	0.78659	0.74618	0.70722	0.11897	0.17005	0.21342	0.25382	0.29279

TABLE 6.3　*Same as Table 6.2, except for the Single-scattering Albedo $\tilde{\omega}$ of 0.95*

τ_1	Method	μ_0:	Reflection					Transmission				
			0.1	0.3	0.5	0.7	0.9	0.1	0.3	0.5	0.7	0.9
0.25	DOM, 2		0.38739	0.16913	0.07140	0.01942	−0.01258	0.55674	0.79643	0.90463	0.96228	0.99780
	4		0.37699	0.14149	0.05281	0.02242	0.02231	0.56058	0.82043	0.92134	0.95860	0.96318
	8		0.37921	0.13924	0.06352	0.03568	0.01940	0.55067	0.81933	0.90969	0.94519	0.96576
	16		0.38384	0.14389	0.06531	0.03466	0.02056	0.54098	0.81302	0.90728	0.94583	0.96451
	Doubling		0.38124	0.14369	0.06530	0.03464	0.02055	0.54162	0.81249	0.90699	0.94571	0.96439
1	DOM, 2		0.47866	0.33481	0.20147	0.09620	0.01565	0.42640	0.56941	0.71177	0.82725	0.91672
	4		0.51064	0.32845	0.19197	0.11336	0.08455	0.36991	0.54933	0.70281	0.80215	0.85086
	8		0.52254	0.33114	0.20309	0.12721	0.08069	0.34357	0.53305	0.68700	0.78768	0.85154
	16		0.51657	0.32962	0.20270	0.12612	0.08129	0.34647	0.53302	0.68625	0.78751	0.85072
	Doubling		0.51164	0.32828	0.20233	0.12560	0.08123	0.34963	0.53370	0.68629	0.78743	0.85052
4	DOM, 2		0.58357	0.47900	0.38080	0.28822	0.20120	0.20988	0.27631	0.34604	0.41952	0.49574
	4		0.60735	0.47440	0.36883	0.28929	0.23531	0.16090	0.23578	0.31347	0.39447	0.47611
	8		0.60317	0.46306	0.36689	0.29228	0.22966	0.15970	0.23803	0.31335	0.39268	0.47448
	16		0.59728	0.46062	0.36550	0.29058	0.22963	0.16138	0.23855	0.31346	0.39286	0.47412
	Doubling		0.59296	0.45923	0.36491	0.29027	0.22942	0.16294	0.23902	0.31360	0.39289	0.47396
16	DOM, 2		0.62095	0.52820	0.44243	0.36286	0.28886	0.01681	0.02213	0.02772	0.03363	0.03994
	4		0.63151	0.50984	0.41595	0.34830	0.30531	0.01426	0.02088	0.02776	0.03509	0.04312
	8		0.62675	0.49829	0.41336	0.35026	0.29829	0.01416	0.02104	0.02764	0.03479	0.04308
	16		0.62102	0.49581	0.41182	0.34839	0.29799	0.01431	0.02110	0.02765	0.03481	0.04305
	Doubling		0.61691	0.49445	0.41122	0.34804	0.29771	0.01445	0.02114	0.02767	0.03482	0.04304

TABLE 6.4 *Same as Table 6.2, except for the Single-scattering Albedo ω̃ of 0.8*

τ₁	Method μ₀:	Reflection					Transmission				
		0.1	0.3	0.5	0.7	0.9	0.1	0.3	0.5	0.7	0.9
0.25	DOM, 2	0.31802	0.13747	0.05739	0.01489	−0.01125	0.46566	0.72916	0.84979	0.91426	0.95403
	4	0.30269	0.11116	0.04040	0.01674	0.01746	0.46032	0.74403	0.86090	0.91033	0.92623
	8	0.29599	0.10578	0.04799	0.02714	0.01473	0.44354	0.73879	0.84949	0.89808	0.92728
	16	0.29406	0.10781	0.04888	0.02608	0.01558	0.43120	0.73272	0.84795	0.89947	0.92679
	Doubling	0.28961	0.10686	0.04855	0.02590	0.01547	0.43017	0.73172	0.84756	0.89938	0.92669
1	DOM, 2	0.37519	0.24977	0.14279	0.06101	−0.00064	0.29023	0.41377	0.55267	0.67036	0.76333
	4	0.37646	0.22124	0.12003	0.06785	0.05425	0.22724	0.37500	0.52936	0.64559	0.72003
	8	0.36938	0.21105	0.12471	0.07775	0.04901	0.20192	0.35487	0.51471	0.63379	0.71702
	16	0.36071	0.20875	0.12396	0.07644	0.04942	0.20416	0.35597	0.51601	0.63571	0.71784
	Doubling	0.35487	0.20714	0.12342	0.07622	0.04929	0.20556	0.35621	0.51606	0.63580	0.71772
4	DOM, 2	0.40411	0.29377	0.20057	0.12070	0.05152	0.06605	0.09165	0.12281	0.16164	0.20828
	4	0.39835	0.25840	0.16792	0.11604	0.09563	0.04412	0.07045	0.10648	0.15660	0.22130
	8	0.38582	0.24266	0.16755	0.12180	0.08907	0.04453	0.07115	0.10625	0.15573	0.21918
	16	0.37725	0.24028	0.16677	0.12063	0.08944	0.04505	0.07167	0.10710	0.15690	0.21959
	Doubling	0.37148	0.23862	0.16615	0.12036	0.08925	0.04539	0.07179	0.10718	0.15697	0.21953
16	DOM, 2	0.40571	0.29599	0.20354	0.12458	0.05636	0.00018	0.00026	0.00034	0.00045	0.00060
	4	0.39914	0.25968	0.16985	0.11886	0.09945	0.00026	0.00042	0.00063	0.00094	0.00138
	8	0.38661	0.24394	0.16947	0.12458	0.09280	0.00026	0.00042	0.00062	0.00091	0.00138
	16	0.37805	0.24157	0.16870	0.12343	0.09316	0.00027	0.00042	0.00062	0.00092	0.00139
	Doubling	0.37229	0.23990	0.16808	0.12315	0.09297	0.00027	0.00042	0.00062	0.00092	0.00139

with an asymmetry factor $g = 0.75$. Single-scattering albedos of 1, 0.95, and 0.8 were used. Also shown in these tables are results computed from the adding method discussed in Section 6.6 for comparison purposes. The accuracy of the doubling method as described by van de Hulst and Grossman (1968) is about 5 decimal points. For discrete streams of 16, the values of reflection and transmission generally are accurate up to ~ 3–4 digits for all three cases, as compared to the doubling computations. The accuracy decreases to ~ 2–3 decimal points for discrete streams of 8. For discrete streams of 4, small deviations occur, with the absolute differences, in general, on the order of ~ 0.01. The two-stream approximation appears to show fairly good accuracy for $\tau_1 = 4$ and 16 for both conservative and nonconservative scattering. However, the two-stream approximation tends to produce larger errors as the optical thickness descreases, especially for reflection because the values are generally small. At near-normal incidence, negative values are seen at $\tau_1 = 0.25$ for both cases.

The accuracy of Eddington's approximation, as illustrated by Shettle and Weinman (1970), is about the same as the two-stream approximation. Thus, both approximations should be used with care for optically thin cases and when large absorption is involved.

Because of the unsatisfactory accuracy of the Eddington approximation, Joseph *et al.* (1976) developed an empirical adjustment in which the fractional forward peak of the scattering phase function was taken into account by redefining the asymmetry factor, the single-scattering albedo, and the optical depth in the forms

$$ g' = g/(1 + g), \qquad \tilde{\omega}' = (1 - g^2)\tilde{\omega}/(1 - g^2\tilde{\omega}), \qquad \tau' = (1 - \tilde{\omega}g^2)\tau. $$

These empirical adjustments lead to a considerable improvement in the reflection and transmission computation.

6.3 DISCRETE-ORDINATES METHOD FOR RADIATIVE TRANSFER

6.3.1 General Solution for Isotropic Scattering

For simplicity in introducing the discrete-ordinates method for radiative transfer, we assume isotropic scattering, i.e., the scattering phase function $P(\mu, \phi; \mu', \phi') = 1$. We further define the azimuthally independent intensity in the form

$$ I(\tau, \mu) = \frac{1}{2\pi} \int_0^{2\pi} I(\tau; \mu, \phi)\, d\phi. \tag{6.62} $$

Hence the equation of transfer shown in Eq. (6.22) becomes

$$\mu \frac{dI(\tau, \mu)}{d\tau} = I(\tau, \mu) - \frac{\tilde{\omega}}{2} \int_{-1}^{1} I(\tau, \mu') d\mu' - \frac{\tilde{\omega} F_0}{4} e^{-\tau/\mu_0}. \qquad (6.63)$$

Now replacing the integral by a summation, according to Gauss's formula discussed in Section 6.2.2, and letting $I_i = I(\tau, \mu_i)$, we obtain

$$\mu_i \frac{dI_i}{d\tau} = I_i - \frac{\tilde{\omega}}{2} \sum_j I_j a_j - \frac{\tilde{\omega} F_0}{4} e^{-\tau/\mu_0}, \qquad i = -n, \dots, n, \qquad (6.64)$$

where Σ_j denotes summation from $-n$ to n, i.e., $2n$ terms.

The solution of Eq. (6.64) may be derived by seeking first the general solution for the homogeneous part of the differential equation and then by adding a particular solution. For the homogeneous part of the differential equation, we set

$$I_i = g_i e^{-k\tau}, \qquad (6.65)$$

where g_i and k are constants. Substituting Eq. (6.65) into the homogeneous part of Eq. (6.64), we find

$$g_i(1 + \mu_i k) = \frac{\tilde{\omega}}{2} \sum_j a_j g_j. \qquad (6.66)$$

This implies that g_i must be in the following form with a constant L:

$$g_i = L/(1 + \mu_i k). \qquad (6.67)$$

With the expression of g_i given by Eq. (6.67), we obtain the characteristic equation for the determination of the eigenvalue k as

$$1 = \frac{\tilde{\omega}}{2} \sum_j \frac{a_j}{1 + \mu_j k} = \tilde{\omega} \sum_{j=1}^{n} \frac{a_j}{1 - \mu_j^2 k^2}. \qquad (6.68)$$

For $\tilde{\omega} < 1$, it is apparent that Eq. (6.68) admits $2n$ distinct nonzero eigenvalues which occur in pairs as $\pm k_j (j = 1, \dots, n)$. Thus the general solution for the homogeneous part is

$$I_i = \sum_j \frac{L_j}{1 + \mu_i k_j} e^{-k_j \tau}. \qquad (6.69)$$

For the particular solution, we assume

$$I_i = (\tilde{\omega} F_0/4) h_i e^{-\tau/\mu_0}, \qquad (6.70)$$

where h_i are constants. Inserting Eq. (6.70) into Eq. (6.64), we find

$$h_i(1 + \mu_i/\mu_0) = (\tilde{\omega}/2) \sum_j a_j h_j + 1. \qquad (6.71)$$

Hence, the constants h_i must be in the form

$$h_i = \gamma/(1 + \mu_i/\mu_0) \qquad (6.72)$$

with γ to be determined from Eq. (6.71), i.e.,

$$\gamma = 1 \Big/ \left[1 - \tilde{\omega} \sum_{j=1}^{n} a_j/(1 - \mu_j^2/\mu_0^2) \right]. \qquad (6.73)$$

Adding the general and particular solution, we get

$$I_i = \sum_j \frac{L_j}{1 + \mu_i k_j} e^{-k_j \tau} + \frac{\tilde{\omega} F_0 \gamma}{4(1 + \mu_i/\mu_0)} e^{-\tau/\mu_0}, \qquad i = -n, \ldots, n. \quad (6.74)$$

The unknown coefficients of proportionality L_j are to be determined from the boundary conditions imposed [e.g., see Eq. (6.30)].

The next step is to introduce Chandrasekhar's H function to replace the constant γ. We consider the function

$$T(z) = 1 - \frac{\tilde{\omega} z^2}{2} \sum_j \frac{a_j}{z + \mu_j} = 1 - \tilde{\omega} z^2 \sum_{j=1}^{n} \frac{a_j}{z^2 - \mu_j^2}. \qquad (6.75)$$

This function is a polynomial of degree $2n$ in z, and if we compare it with the characteristic equation (6.68), we find that $z = \pm 1/k_j$ for $T(z) = 0$. Thus we must have

$$\prod_{j=1}^{n} (z^2 - \mu_j^2) T(z) = \text{const} \prod_{j=1}^{n} (1 - k_j^2 z^2) \qquad (6.76)$$

since the two polynomials of degree $2n$ have the same zeros. For $z = 0$ we find that

$$\text{const} = \prod_{j=1}^{n} (-\mu_j^2).$$

Thus,

$$T(z) = (-1)^n \mu_1^2 \cdots \mu_n^2 \prod_{j=1}^{n} (1 - k_j^2 z^2) \Big/ \prod_{j=1}^{n} (z^2 - \mu_j^2). \qquad (6.77)$$

We now introduce the H function as

$$H(\mu) = \frac{1}{\mu_1 \cdots \mu_n} \frac{\prod_{j=1}^{n}(\mu + \mu_j)}{\prod_{j=1}^{n}(1 + k_j \mu)}. \qquad (6.78)$$

In terms of the H function,

$$\gamma = 1/T(z) = H(z)H(-z). \qquad (6.79)$$

So, the complete solution to the isotropic, nonconservative radiative transfer equation in the nth approximation is now given by

$$I_i = \sum_j \frac{L_j}{1 + \mu_i k_j} e^{-k_j \tau} + \frac{\tilde{\omega} F_0 H(\mu_0) H(-\mu_0)}{4(1 + \mu_i/\mu_0)} e^{-\tau/\mu_0}. \tag{6.80}$$

Figure 6.3 illustrates a distribution of eigenvalues for isotropic scattering with a single-scattering albedo $\tilde{\omega} = 0.9$ using four discrete streams tabulated in Table 6.1. The characteristic equation denoted in Eq. (6.68) may be written in the form

$$f(k) = 1 - \tilde{\omega} \sum_{j=1}^{n} \frac{a_j}{1 - \mu_j^2 k^2}. \tag{6.68a}$$

It is clear that for $\tilde{\omega} \neq 0$, $f(k_j) \to \pm\infty$, as $k_j \to \mu_j^{-1}$. In the figure, the same intervals between each μ_j^{-1} were divided so that lines across the zeros can be clearly seen. The eigenvalues occur in pairs and there exists one, and only one, eigenvalue in each interval, which can be proven mathematically. The eigenvalue in the discrete-ordinates method for radiative transfer may be physically interpreted as an effective extinction coefficient that when multiplied by the normal optical depth represents an effective optical path length in each discrete stream.

For conservative scattering $\tilde{\omega} = 1$, we note that the characteristic equation (6.68) admits two zero eigenvalues, namely, $k^2 = 0$. Upon using the relation

$$\sum_{j=-n}^{n} a_j \mu_j^l = \int_{-1}^{1} \mu^l \, d\mu = 2\delta_l/(2l + 1), \qquad \delta_l = \begin{cases} 1 & \text{even,} \\ 0 & \text{odd,} \end{cases}$$

it can be shown that

$$I_i = b(\tau + \mu_i + Q) \tag{6.81}$$

satisfies the homogeneous part of the differential equation, with b and Q as two arbitrary constants of integration. Thus, the complete solution to the isotropic radiative transfer equation in the nth approximation may be written

$$I_i = \sum_{j=-(n-1)}^{n-1} \frac{L_j}{1 + \mu_i k_j} e^{-k_j \tau} + (\tau + \mu_i)L_{-n} + L_n$$
$$+ \frac{F_0 H(\mu_0) H(-\mu_0)}{4(1 + \mu_i/\mu_0)} e^{-\tau/\mu_0}. \tag{6.82}$$

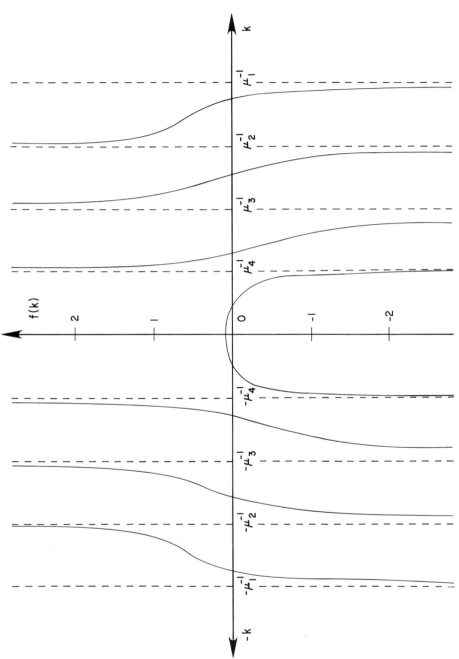

Fig. 6.3   A typical distribution of eigenvalues for isotropic scattering with a single-scattering albedo of 0.95 for equal intervals between each μ_j^{-1} ($j = \pm1, \pm2, \pm3, \pm4$).

6.3.2   The law of Diffuse Reflection for Semi-Infinite Isotropic Scattering Atmospheres

Assume that there is no diffuse downward and upward radiation at the top and base of a semi-infinite atmosphere so that

$$I(0, -\mu_i) = 0, \qquad I(\tau_1, +\mu_i) = 0, \qquad \tau_1 \to \infty. \qquad (6.83)$$

Inserting the second boundary condition into the solution to the isotropic radiative transfer equation in the nth approximation denoted in Eq. (6.80), we obtain

$$I(\tau_1, +\mu_i) = 0 = \sum_{j=-n}^{-1} \frac{L_j}{1 - \mu_i k_j} e^{k_j \tau_1}, \qquad i = 1, \ldots, n. \qquad (6.84)$$

It is apparent that $L_j = 0 \, (j = -n, \ldots, -1)$ in order to satisfy the boundary condition for a semi-infinite atmosphere. Thus,

$$I(\tau, \mu_i) = \sum_{j=1}^{n} \frac{L_j}{1 + \mu_i k_j} e^{-k_j \tau} + \frac{\tilde{\omega} F_0 H(\mu_0) H(-\mu_0)}{4(1 + \mu_i/\mu_0)} e^{-\tau/\mu_0}. \qquad (6.85)$$

We define

$$S(\mu) = \sum_{j=1}^{n} \frac{L_j}{1 - k_j \mu} + \frac{\tilde{\omega} F_0 H(\mu_0) H(-\mu_0)}{4(1 - \mu/\mu_0)}. \qquad (6.86)$$

Thus,

$$S(\mu_i) = I(0, -\mu_i) = 0, \qquad i = 1, \ldots, n, \qquad (6.87)$$

and the reflected intensity

$$I(0, \mu) = S(-\mu). \qquad (6.88)$$

Moreover, we consider the function

$$(1 - \mu/\mu_0) \prod_{j=1}^{n} (1 - k_j \mu) S(\mu), \qquad (6.89)$$

which is a polynomial of degree n in μ and vanishes for $\mu = \mu_i, \, i = 1, \ldots, n$. Hence, this function must be equal to $\prod_{j=1}^{n} (\mu - \mu_j)$ apart from a constant value. We may write

$$(1 - \mu/\mu_0) \prod_{j=1}^{n} (1 - k_j \mu) S(\mu) = \text{const} \frac{(-1)^n}{\mu_1 \cdots \mu_n} \prod_{j=1}^{n} (\mu - \mu_j). \qquad (6.90)$$

Upon employing the definition of the H function defined in Eq. (6.78), we obtain

$$S(\mu) = \text{const} \, H(-\mu)/(1 - \mu/\mu_0). \qquad (6.91)$$

To obtain the value of the constant, we observe

$$\lim_{\mu \to \mu_0} (1 - \mu/\mu_0)S(\mu) = \text{const } H(-\mu_0). \tag{6.92}$$

But from Eq. (6.86) we have

$$\lim_{\mu \to \mu_0} (1 - \mu/\mu_0)S(\mu) = \tfrac{1}{4}\tilde{\omega}F_0H(\mu_0)H(-\mu_0). \tag{6.93}$$

Comparing Eq. (6.92) and Eq. (6.93), we find

$$\text{const} = \tfrac{1}{4}\tilde{\omega}F_0H(\mu_0), \tag{6.94}$$

and

$$S(\mu) = \frac{\tilde{\omega}F_0H(\mu_0)H(-\mu)}{4(1 - \mu/\mu_0)}. \tag{6.95}$$

The reflected intensity for a semi-infinite, isotropic scattering atmosphere is then given by

$$I(0, \mu) = S(-\mu) = \tfrac{1}{4}\tilde{\omega}F_0 \frac{\mu_0}{\mu + \mu_0} H(\mu_0)H(\mu). \tag{6.96}$$

Thus the diffuse reflection can be expressed in terms of the H function. This simple expression has been used to interpret the absorption line formation in cloudy atmospheres of other planets. Exercise 6.7 illustrates the applicability of the law of diffuse reflection.

6.3.3 General Solution for Anisotropic Scattering

To solve Eq. (6.14), we first seek the solution for the homogeneous part of the differential equation and then add a particular solution for the inhomogeneous part. After some mathematical manipulation, it is given in the form

$$I^m(\tau, \mu_i) = \sum_j L_j^m \phi_j^m(\mu_i)e^{-k_j^m\tau} + Z^m(\mu_i)e^{-\tau/\mu_0}, \tag{6.97}$$

where the eigenfunction derived from the associated homogeneous system is

$$\phi_j^m(\mu_i) = \frac{1}{1 + \mu_i k_j^m} \sum_{l=0}^{N} \tilde{\omega}_l^m \xi_l^m P_l^m(\mu_i), \tag{6.98}$$

and the Z function is

$$Z^m(\mu_i) = \tfrac{1}{4}\tilde{\omega}F_0P_m^m(-\mu_0)\frac{H^m(\mu_0)H^m(-\mu_0)}{1 + \mu_i/\mu_0} \sum_{l=0}^{N} \tilde{\omega}_l^m \xi_l^m \left(\frac{1}{\mu_0}\right) P_l^m(\mu_i). \tag{6.99}$$

The ξ function has the recursion form

$$\xi_{l+1}^m = -\frac{2l+1-\tilde{\omega}_l}{k(l-m+1)}\xi_l^m - \frac{l+m}{l-m+1}\xi_{l-1}^m. \qquad (6.100)$$

Finally, the eigenvalues k_j^m can be determined from the characteristic equation described by

$$1 = \frac{\tilde{\omega}}{2}\sum_j \frac{a_j}{1+\mu_j k}\left[\sum_{\lambda=m}^N \tilde{\omega}_\lambda^m \xi_\lambda^m(k)P_\lambda^m(\mu_j)P_m^m(\mu_j)\right]. \qquad (6.101)$$

Equation (6.101) is of order n in k^2 and admits, in general, $2n$ distinct nonvanishing eigenvalues which must occur in pairs. For strong anisotropic scattering, having a sharp phase function, a number of eigenvalues are normally contained in the interval $(0, \mu_n^{-1})$, and the eigenvalue pattern is unsymmetric, differing considerably from that in Fig. 6.3.

Now the unknown coefficients L_j^m are to be determined, subject to the radiation boundary conditions. For the simple boundary conditions given by Eq. (6.30), and in view of the intensity expansion in Eq. (6.13), we shall have

$$\left.\begin{array}{l} I^m(0, -\mu_i) = 0 \\ I^m(\tau_1, \mu_i) = 0 \end{array}\right\} \quad \text{for} \quad i = 1, \ldots, n \quad \text{and} \quad m = 0, \ldots, N. \qquad (6.102)$$

We may then determine L_j^m m times independently with the final result given by Eq. (6.13). At this point, the analytic solution for Eq. (6.14) is therefore complete.

The solution expressed in Eq. (6.97) is valid only for nonconservative scattering because when $\tilde{\omega} = 1$, $k^2 = 0$ will satisfy the characteristic equation for $m = 0$, and $\xi_1^0(k)$ becomes indefinite. Thus, a different solution has to be derived. In conservative cases since there is no absorption, the flux of radiation normal to the plane of stratification is constant. It can be shown that the transfer equation admits a solution of the form for $m = 0$:

$$I^0(\tau; \mu_i) = \sum_{j=-(n-1)}^{n-1} L_j^0 \phi_j^0(\mu_i)e^{-k_j^0\tau}$$

$$+ \left[(1 - \tilde{\omega}_1/3) + \mu_i\right]L_{-n}^0 + L_n^0 + Z^0(\mu_i)e^{-\tau/\mu_0}. \qquad (6.103)$$

The discrete-ordinates method gives the analytical solution of the radiative transfer equation, and therefore the internal radiation field, as well as the reflected and transmitted intensities, can be evaluated without extra computational effort. For the azimuthally independent term, the method yields rather accurate results even for $n = 2$ or 4 (Liou, 1973). These may be sufficiently accurate for computations of the flux density in many applications. Numerical exercises using the discrete-ordinates method for the azimuthally dependent case, however, have not been carried out for general anisotropic phase functions.

6.4 PRINCIPLES OF INVARIANCE

6.4.1 Definitions of Various Scattering Parameters

The method of principles of invariance for radiative transfer seeks certain physical statements and mathematical formulations regarding the fields of reflection and transmission. The radiation field from this method is not derived directly from the equation of transfer, and so it differs from the approximate and exact solutions for radiative transfer discussed in the previous sections.

It is necessary to define and clarify a number of parameters which have been used in literature in order to introduce the principles of invariance and other multiple-scattering problems. We find it convenient to express the solutions to the multiple-scattering problem in terms of *reflection function R* and *transmission function T* in the forms [see Eqs. (6.31) and (6.32)]

$$I_r(0, \mu, \phi) = \frac{1}{\pi} \int_0^{2\pi} \int_0^1 R(\mu, \phi; \mu', \phi') I_0(-\mu', \phi') \mu' \, d\mu' \, d\phi', \quad (6.104)$$

$$I_t(\tau_1, -\mu, \phi) = \frac{1}{\pi} \int_0^{2\pi} \int_0^1 T(\mu, \phi; \mu', \phi') I_0(-\mu', \phi') \mu' \, d\mu' \, d\phi', \quad (6.105)$$

where $I_0(-\mu, \phi)$ represents the intensity of sunlight incident on the top of the scattering layer. It suffices for most practical problems to approximate it as monodirectional in the form

$$I_0(-\mu, \phi) = \delta(\mu - \mu_0)\delta(\phi - \phi_0)\pi F_0, \quad (6.106)$$

where δ is the Dirac delta function, and πF_0 denotes the incident solar flux density perpendicular to the incident beam. Thus, we find from Eqs. (6.104) and (6.105) the definitions of reflection and transmission functions

$$R(\mu, \phi; \mu_0, \phi_0) = I_r(0, \mu, \phi)/(\mu_0 F_0), \quad (6.107)$$

$$T(\mu, \phi; \mu_0, \phi_0) = I_t(\tau_1, -\mu, \phi)/(\mu_0 F_0). \quad (6.108)$$

Note here that $I_t(\tau_1, -\mu, \phi)$ represents the diffusely transmitted intensity, which does not include the directly transmitted solar intensity $\pi F_0 e^{-\tau_1/\mu_0}$. This quantity represents the reduced incident radiation, which penetrates to the level τ_1 without suffering any extinction process. In cases where polarization is included, i.e., where the four Stokes parameters are considered for the incident intensity, R and T are composed of four rows and four columns and are called *reflection* and *transmission matrices*. The reflection and transmission functions also have been referred to as reflection and transmission coefficients by Ambartsumyan (1958) and Sobolev (1975). In satellite meteorology, a parameter called *bidirectional reflectance*, which is analogous to the reflection function, is widely used.

On the basis of Eqs. (6.107) and (6.108), we may define the *reflection r* (also called *local* or *planetary albedo*) and *transmission* (diffuse) *t* associated with reflected (upward) and transmitted (downward) flux densities in the forms

$$r(\mu_0) = \frac{F_{\mathrm{dif}}^{\uparrow}(0)}{\pi\mu_0 F_0} = \frac{1}{\pi}\int_0^{2\pi}\int_0^1 R(\mu,\phi;\mu_0,\phi_0)\mu\,d\mu\,d\phi, \qquad (6.109)$$

$$t(\mu_0) = \frac{F_{\mathrm{dif}}^{\downarrow}(\tau_1)}{\pi\mu_0 F_0} = \frac{1}{\pi}\int_0^{2\pi}\int_0^1 T(\mu,\phi;\mu_0,\phi_0)\mu\,d\mu\,d\phi, \qquad (6.110)$$

where the definitions of the upward and downward diffuse flux densities are given by Eqs. (6.16) and (6.17). Note that the direct transmission is simply $e^{-\tau_1/\mu_0}$. In a similar manner, the *absorption* of the atmosphere, bounded by the optical depths of 0 and τ_1, may be obtained from the net flux density divergence (including the direct transmission component) at levels of 0 and τ_1, normalized with respect to $\pi\mu_0 F_0$. For the azimuthally independent case, Eqs. (6.109) and (6.110) reduce to

$$r(\mu_0) = 2\int_0^1 R(\mu,\mu_0)\mu\,d\mu, \qquad (6.111)$$

$$t(\mu_0) = 2\int_0^1 T(\mu,\mu_0)\mu\,d\mu. \qquad (6.112)$$

The total amount of radiant energy per unit time incident on the planet having a radius a is $\pi a^2 \pi F_0$. To find the flux of energy reflected by the planet, we consider on the disk a ring with radius a' and width da', where a' is the projected distance from the center of the disk (see Fig. 6.4). Hence, the

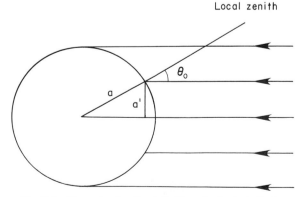

Fig. 6.4 Geometry for the definition of spherical albedo.

flux of energy reflected by this ring would be $r(\mu_0)\pi F_0 2\pi a'\, da'$. But $a' = a$ sin θ_0, and $da' = a$ cos $\theta_0\, d\theta_0$, so we may write this flux of energy as $2\pi a^2 \pi F_0 r(\mu_0)\mu_0\, d\mu_0$. The flux of energy reflected by the entire planet is therefore

$$f^\uparrow(0) = 2\pi a^2 \pi F_0 \int_0^1 r(\mu_0)\mu_0\, d\mu_0. \qquad (6.113)$$

The *spherical* (or *global*) *albedo*, which represents the ratio of the energy reflected by the entire planet to the energy incident on it, is then given by

$$\bar{r} = \frac{f^\uparrow(0)}{\pi a^2 \pi F_0} = 2 \int_0^1 r(\mu_0)\mu_0\, d\mu_0. \qquad (6.114)$$

Similarly, the *global diffuse transmission* is

$$\bar{t} = \frac{f^\downarrow(\tau_1)}{\pi a^2 \pi F_0} = 2 \int_0^1 t(\mu_0)\mu_0\, d\mu_0, \qquad (6.115)$$

and the global direct transmission is simply $2\int_0^1 e^{-\tau_1/\mu_0}\mu_0\, d\mu_0$. Likewise, the *global absorption* may also be expressed in terms of absorption.

Chandrasekhar (1950) expresses the resulting laws of diffuse reflection and transmission for a finite atmosphere with an optical depth τ_1 in terms of a *scattering function S* and a *transmission function T_c* (diffuse), which differ from the parameters defined in Eqs. (6.104) and (6.105), in the forms

$$I_r(0, \mu, \phi) = \frac{1}{4\pi\mu} \int_0^{2\pi} \int_0^1 S(\mu, \phi; \mu', \phi')I_0(-\mu', \phi')\, d\mu'\, d\phi', \quad (6.116)$$

$$I_t(\tau_1, -\mu, \phi) = \frac{1}{4\pi\mu} \int_0^{2\pi} \int_0^1 T_c(\mu, \phi; \mu', \phi')I_0(-\mu', \phi')\, d\mu'\, d\phi', \quad (6.117)$$

where T_c is utilized to distinguish from T described previously. Upon substituting Eq. (6.106) into Eqs. (6.116) and (6.117), we obtain the definitions of Chandrasekhar's scattering and transmission functions as

$$S(\mu, \phi; \mu_0, \phi_0) = (4\mu/F_0)I_r(0, \mu, \phi), \qquad (6.118)$$

$$T_c(\mu, \phi; \mu_0, \phi_0) = (4\mu/F_0)I_t(\tau_1, -\mu, \phi). \qquad (6.119)$$

The introduction of the factor μ in the intensity parameters gives the required symmetry of S and T_c in the pair of variable (μ, ϕ) and (μ_0, ϕ_0) such that

$$S(\mu, \phi; \mu_0, \phi_0) = S(\mu_0, \phi_0; \mu, \phi), \qquad (6.120)$$

$$T_c(\mu, \phi; \mu_0, \phi_0) = T_c(\mu_0, \phi_0; \mu, \phi). \qquad (6.121)$$

This is the so-called Helmholtz's principle of *reciprocity*, which is stated here without proof.

6.4.2 Principles of Invariance for Semi-Infinite Atmospheres

Consider a flux of parallel solar radiation πF_0 in a direction defined by $(-\mu_0, \phi_0)$ ($-\mu_0$ denotes that the light beam is downward) incident on the outer boundary of a semi-infinite, plane-parallel atmosphere. The principles of invariance originally introduced by Ambartsumyan (1942, 1958) state that the diffuse reflected intensity from such an atmosphere can not be changed if a plane layer of finite optical depth, having the same optical properties as those of the original atmosphere, is added. Let the optical depth of the added layer be $\Delta\tau$. This is so small that $(\Delta\tau)^2$ can be neglected when it is compared with $\Delta\tau$ itself. For simplicity in presenting the principles of invariance, we neglect the azimuthal dependence of the diffuse reflected intensity and define the *reflection function* in terms of the diffuse reflected intensity at the top of a semi-infinite atmosphere $I(0, \mu)$ in the form [see Eq. (6.107)]

$$R(\mu, \mu_0) = I(0, \mu)/(\mu_0 F_0). \tag{6.122}$$

In reference to Fig. 6.5, we find that the reduction or increase of the reflection function, after the addition of an infinitesimal layer, follows these principles:

(1) The differential attenuation of the reflection function in passing through $\Delta\tau$ downward, based on Eq. (6.1) $(\Delta\tau = \bar{\sigma}_e N \Delta z)$ is

$$\Delta R_1' = -R(\mu, \mu_0)\,\Delta\tau/\mu_0. \tag{6.123}$$

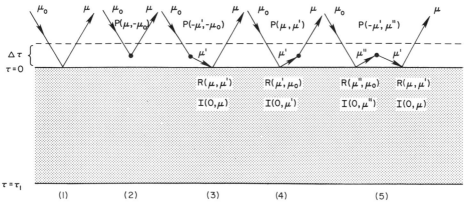

Fig. 6.5 The principles of invariance for a semi-infinite, plane-parallel atmosphere. The μ and $-\mu$ denote the upward and downward directions, respectively. Black dots show that scattering events take place in which the scattering phase function is required. The directional representation in the argument is such that the emergent angle is stated first; then it is followed by the incident angle. A similar rule governs the argument of the reflection function.

The reflection function at $\tau = 0$ is now $(R + \Delta R_1')$, which is again attenuated in passing through $\Delta\tau$ upward. Thus,

$$\Delta R_1'' = -[R(\mu, \mu_0) + \Delta R_1']\Delta\tau/\mu. \tag{6.124}$$

The total attenuation is therefore

$$\Delta R_1 = \Delta R_1' + \Delta R_1'' = -R(\mu, \mu_0)\left[\Delta\tau\left(\frac{1}{\mu_1} + \frac{1}{\mu_0}\right) - \frac{\Delta\tau^2}{\mu\mu_0}\right]$$

$$\approx -R(\mu, \mu_0)\Delta\tau\left(\frac{1}{\mu_1} + \frac{1}{\mu_0}\right). \tag{6.125}$$

(2)   However, $\Delta\tau$ may scatter directly in the direction μ a part of solar flux πF_0 incident on it, and according to Eq. (6.3), we find the additional reflection

$$\Delta R_2 = \frac{1}{\mu_0 F_0}\frac{\tilde{\omega}}{4\pi}\pi F_0 P(\mu, -\mu_0)\Delta\tau/\mu = \frac{\tilde{\omega}}{4}P(\mu, -\mu_0)\Delta\tau/(\mu\mu_0). \tag{6.126}$$

(3)   In addition, $\Delta\tau$ may scatter a part of the solar flux in the direction μ' onto the boundary $\tau = 0$. The diffuse light beam then undergoes reflection from this surface, and this additional reflection, analogous to Eq. (6.2), is given by

$$\Delta R_3 = \frac{1}{\mu_0 F_0}\frac{\tilde{\omega}}{4\pi}\int_0^{2\pi} d\phi' \int_0^1 I(0, \mu)P(-\mu', -\mu_0)\, d\mu' \frac{\Delta\tau}{\mu'}$$

$$= \frac{\tilde{\omega}}{2}\frac{\Delta\tau}{\mu_0}\int_0^1 R(\mu, \mu')P(-\mu', -\mu_0)\, d\mu'. \tag{6.127}$$

(4)   Moreover, $\Delta\tau$, after attenuating a fraction of the light beam diffusely reflected from the boundary $\tau = 0$ in the direction μ', may scatter a part of it in the direction μ. This incremental reflection is given by

$$\Delta R_4 = \frac{1}{\mu_0 F_0}\frac{\tilde{\omega}}{4\pi}\int_0^{2\pi} d\phi' \int_0^1 P(\mu, \mu')I(0, \mu')\, d\mu' \frac{\Delta\tau}{\mu}$$

$$= \frac{\tilde{\omega}}{2}\frac{\Delta\tau}{\mu}\int_0^1 P(\mu, \mu')R(\mu', \mu_0)\, d\mu'. \tag{6.128}$$

(5)   Finally, the unscattered component of the solar flux πF_0, which is reflected from the boundary $\tau = 0$ in the direct μ'', is scattered by $\Delta\tau$ back to $\tau = 0$ in the different direction μ', and again is reflected from the surface $\tau = 0$ in the direction μ. This additional contribution may be expressed by

$$\Delta R_5 = \frac{1}{\mu_0 F_0} \frac{\tilde{\omega}}{4\pi} \int_0^{2\pi} d\phi' \int_0^1 \frac{I(0,\mu)}{\pi F_0} d\mu'$$

$$\times \left[\int_0^{2\pi} d\phi'' \int_0^1 P(-\mu',\mu'')I(0,\mu'')\,d\mu'' \right] \frac{\Delta\tau}{\mu'}$$

$$= \tilde{\omega}\Delta\tau \int_0^1 R(\mu,\mu')\,d\mu' \left[\int_0^1 P(-\mu',\mu'')R(\mu'',\mu_0)\,d\mu'' \right]. \qquad (6.129)$$

On the basis of the principles of invariance already stated, we shall have

$$\Delta R_1 + \Delta R_2 + \Delta R_3 + \Delta R_4 + \Delta R_5 = 0. \qquad (6.130)$$

Thus,

$$R(\mu,\mu_0)\left(\frac{1}{\mu} + \frac{1}{\mu_0}\right) = \frac{\tilde{\omega}}{4\mu\mu_0}\left\{ P(\mu,-\mu_0) + 2\mu \int_0^1 R(\mu,\mu')P(-\mu',-\mu_0)\,d\mu' \right.$$

$$+ 2\mu_0 \int_0^1 P(\mu,\mu')R(\mu',\mu_0)\,d\mu' + 4\mu\mu_0 \int_0^1 R(\mu,\mu')\,d\mu'$$

$$\left. \times \left[\int_0^1 P(-\mu',\mu'')R(\mu'',\mu_0)\,d\mu'' \right] \right\}. \qquad (6.131)$$

For a simple case of isotropic scattering, Eq. (6.131) becomes

$$R(\mu,\mu_0)(\mu + \mu_0) = \frac{\tilde{\omega}}{4}\left[1 + 2\mu \int_0^1 R(\mu,\mu')\,d\mu' + 2\mu_0 \int_0^1 R(\mu',\mu_0)\,d\mu' \right.$$

$$\left. + 4\mu\mu_0 \int_0^1 R(\mu',\mu_0)\,d\mu' \int_0^1 R(\mu,\mu'')\,d\mu'' \right]$$

$$= \frac{\tilde{\omega}}{4}\left[1 + 2\mu \int_0^1 R(\mu,\mu')\,d\mu' \right]\left[1 + 2\mu_0 \int_0^1 R(\mu',\mu_0)\,d\mu' \right].$$

$$(6.132)$$

Inspection of Eq. (6.132) reveals that if it is satisfied by the function $R(\mu,\mu_0)$, it must also be satisfied by the function $R(\mu_0,\mu)$. And since Eq. (6.132) can have only one solution, $R(\mu,\mu_0)$ must be symmetrical, i.e.,

$$R(\mu,\mu_0) = R(\mu_0,\mu). \qquad (6.133)$$

With this relationship, which we state here without rigorous mathematical proof, we may define

$$H(\mu) = 1 + 2\mu \int_0^1 R(\mu,\mu')\,d\mu', \qquad (6.134)$$

such that

$$R(\mu,\mu_0) = \frac{\tilde{\omega}}{4} \frac{H(\mu)H(\mu_0)}{\mu + \mu_0}. \qquad (6.135)$$

If we now review the analyses in Section 6.3.2, we find this expression is exactly the same as that in Eq. (6.96). It is indeed an exact solution for the semi-infinite atmosphere. To examine the H function we insert Eq. (6.135) into (6.134) to obtain

$$H(\mu) = 1 + \frac{\tilde{\omega}}{2} \mu H(\mu) \int_0^1 \frac{H(\mu') d\mu'}{\mu + \mu'}. \tag{6.136}$$

It is now clear that the solution of Eq. (6.132) is reduced to solving the H function given in Eq. (6.136). Its exact value in this case can be obtained by selecting an approximate value and then carrying out appropriate iterations. We first seek the mean value of H in the form

$$H_0 = \int_0^1 H(\mu) d\mu. \tag{6.137}$$

From Eq. (6.136) we have

$$\int_0^1 H(\mu) d\mu = 1 + \frac{\tilde{\omega}}{2} \int_0^1 \int_0^1 \frac{H(\mu)H(\mu')\mu}{\mu + \mu'} d\mu \, d\mu'. \tag{6.138}$$

Upon interchanging μ with μ', we find that Eq. (6.138) does not vary. Thus, we may write

$$\int_0^1 H(\mu) d\mu = 1 + \frac{\tilde{\omega}}{4} \int_0^1 \int_0^1 \frac{H(\mu)H(\mu')\mu}{\mu + \mu'} d\mu \, d\mu' + \frac{\tilde{\omega}}{4} \int_0^1 \int_0^1 \frac{H(\mu)H(\mu')\mu'}{\mu + \mu'} d\mu \, d\mu'$$

$$= 1 + \frac{\tilde{\omega}}{4} \int_0^1 H(\mu) d\mu \int_0^1 H(\mu') d\mu'. \tag{6.139}$$

It is apparent that

$$H_0 = 1 + (\tilde{\omega}/4)H_0^2. \tag{6.140}$$

This gives the solution of H_0 in the form

$$H_0 \equiv \int_0^1 H(\mu) d\mu = (2/\tilde{\omega})(1 - \sqrt{1 - \tilde{\omega}}), \tag{6.141}$$

where the positive root is found to be unrealistic because the albedo values become greater than unity, as evident in Exercise 6.8. To find $H(\mu)$ in Eq. (6.136), we may insert this zero-order approximation into the right-hand side of Eq. (6.136) to obtain the first approximation for $H(\mu)$. The procedure can be continued until the desirable accuracy is achieved.

6.4.3   Principles of Invariance for Finite Atmospheres

In the last subsection, we described the principles of invariance for a semi-infinite atmosphere in which only the reflection function is involved. We now introduce the general principles of invariance for a finite atmosphere developed by Chandrasekhar (1950). To be consistent with our previous discussions, we neglect the azimuthal dependence of the scattering parameters and use the reflection and transmission functions defined in Eqs. (6.107) and (6.108), instead of the scattering and transmission functions proposed originally by Chandrasekhar, as defined in Eqs. (6.118) and (6.119). We note, however, the relationships

$$S(\mu, \mu_0) = 4\mu\mu_0 R(\mu, \mu_0), \tag{6.142}$$

$$T_c(\mu, \mu_0) = 4\mu\mu_0 T(\mu, \mu_0). \tag{6.143}$$

In reference to Fig. 6.6, we find the following four principles governing the reflection and transmission of a light beam:

(1) The reflected (upward) intensity at level τ is caused by the reflection of the attenuated incident solar flux density $\pi F_0 e^{-\tau/\mu_0}$, and the downward

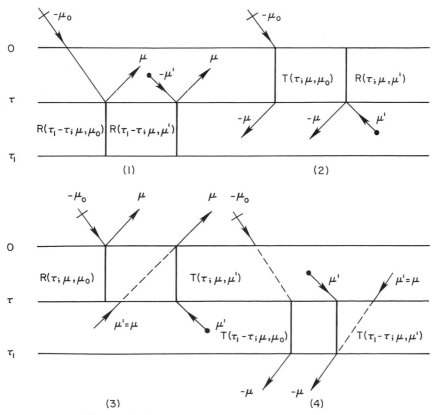

Fig. 6.6 Principles of invariance for a finite atmosphere.

diffuse intensity incident on the surface τ from the finite optical depth $(\tau_1 - \tau)$ below [see Eqs. (6.104) and (6.105)]. Thus, we find

$$I(\tau, \mu) = \mu_0 F_0 e^{-\tau/\mu_0} R(\tau_1 - \tau; \mu, \mu_0) + 2 \int_0^1 R(\tau_1 - \tau; \mu, \mu') I(\tau, -\mu') \mu' \, d\mu'.$$

(2)   The diffusely transmitted (downward) intensity at level τ is due to the transmission of the incident solar flux density by the optical depth τ above, and the reflection of the upward diffuse intensity incident on the surface τ from below. Thus, we find

$$I(\tau, -\mu) = \mu_0 F_0 T(\tau; \mu, \mu_0) + 2 \int_0^1 R(\tau; \mu, \mu') I(\tau, \mu') \mu' \, d\mu'. \qquad (6.145)$$

(3)   The reflected (upward) intensity at the top of the finite atmosphere $(\tau = 0)$ can be taken as the reflection by the optical depth τ of the atmosphere plus the transmission of this atmosphere, the upward diffuse and direct intensities incident on the surface τ from below. Thus, we find

$$I(0, \mu) = \mu_0 F_0 R(\tau; \mu, \mu_0) + 2 \int_0^1 T(\tau; \mu, \mu') I(\tau, \mu') \mu' \, d\mu' + e^{-\tau/\mu} I(\tau, \mu). \qquad (6.146)$$

(4)   The diffusely transmitted (downward) intensity at the bottom of the finite atmosphere $(\tau = \tau_1)$ is equivalent to the transmission of the attenuated incident solar flux density plus the transmission of the downward diffuse and direct intensities incident on the surface τ from above. Thus, we find

$$I(\tau_1, -\mu) = \mu_0 F_0 e^{-\tau/\mu_0} T(\tau_1 - \tau; \mu, \mu_0) + 2 \int_0^1 T(\tau_1 - \tau; \mu, \mu') I(\tau_1 - \mu') \mu' \, d\mu'$$

$$+ e^{-(\tau_1 - \tau)/\mu} I(\tau, -\mu). \qquad (6.147)$$

To obtain the reflection and transmission functions of a finite atmosphere with an optical depth of τ_1, we first differentiate Eqs. (6.144)–(6.147) with respect to τ and evaluate the values at $\tau = 0$ and τ_1, where the boundary conditions stated in Eq. (6.30) can be applied. (For azimuthally independent cases, we have $I(0, -\mu) = 0$ and $I(\tau_1, \mu) = 0$.) After differentiation with respect to τ, we set $\tau = 0$ and $\tau = \tau_1$ for principles (1) and (4) and for principles (2) and (3), respectively, to obtain the equations

$$\left. \frac{dI(\tau, \mu)}{d\tau} \right|_{\tau = 0} = \mu_0 F_0 \left[-\frac{\partial R(\tau_1; \mu, \mu_0)}{\partial \tau_1} - \frac{1}{\mu_0} R(\tau_1; \mu, \mu_0) \right]$$

$$+ 2 \int_0^1 \mu' \, d\mu' R(\tau_1; \mu, \mu') \left. \frac{dI(\tau, -\mu')}{d\tau} \right|_{\tau = 0}, \qquad (6.148)$$

$$\left. \frac{dI(\tau, -\mu)}{d\tau} \right|_{\tau = \tau_1} = \mu_0 F_0 \frac{\partial T(\tau_1; \mu, \mu_0)}{\partial \tau_1} + 2 \int_0^1 \mu' \, d\mu' R(\tau_1; \mu, \mu') \left. \frac{dI(\tau, \mu')}{d\tau} \right|_{\tau = \tau_1},$$

$$(6.149)$$

$$\frac{dI(0,\mu)}{d\tau}\bigg|_{\tau=\tau_1} = 0 = \mu_0 F_0 \frac{\partial R(\tau_1;\mu,\mu_0)}{\partial \tau_1} + 2\int_0^1 \mu' \, d\mu' T(\tau_1;\mu,\mu') \frac{dI(\tau,\mu')}{d\tau}\bigg|_{\tau=\tau_1}$$

$$+ e^{-\tau_1/\mu} \frac{dI(\tau,\mu)}{d\tau}\bigg|_{\tau=\tau_1}, \tag{6.150}$$

$$\frac{dI(\tau_1,-\mu)}{d\tau}\bigg|_{\tau=0} = 0 = \mu_0 F_0 \left[-\frac{\partial T(\tau_1;\mu,\mu_0)}{\partial \tau_1} - \frac{1}{\mu_0} T(\tau_1;\mu,\mu_0) \right]$$

$$+ 2\int_0^1 \mu' \, d\mu' T(\tau_1;\mu,\mu') \frac{dI(\tau,-\mu')}{d\tau}\bigg|_{\tau=0}$$

$$+ e^{-\tau_1/\mu} \frac{dI(\tau,-\mu)}{d\tau}\bigg|_{\tau=0}. \tag{6.151}$$

To eliminate the derivatives of the intensity, we utilize the azimuthally independent transfer equation [Eq. (6.22)] to find

$$\mu \frac{dI(\tau,\mu)}{d\tau}\bigg|_{\tau=0} = \mu_0 F_0 R(\tau_1;\mu,\mu_0) - \frac{\tilde{\omega}}{2}\int_0^1 P(\mu,\mu'')I(0,\mu'') \, d\mu''$$

$$- \frac{\tilde{\omega}}{4} F_0 P(\mu,-\mu_0), \tag{6.152}$$

$$-\mu \frac{dI(\tau,-\mu)}{d\tau}\bigg|_{\tau=0} = 0 - \frac{\tilde{\omega}}{2}\int_0^1 P(-\mu,\mu'')I(0,\mu'') \, d\mu''$$

$$- \frac{\tilde{\omega}}{4} F_0 P(-\mu,-\mu_0), \tag{6.153}$$

$$\mu \frac{dI(\tau,\mu)}{d\tau}\bigg|_{\tau=\tau_1} = 0 - \frac{\tilde{\omega}}{2}\int_0^1 P(\mu,-\mu'')I(\tau_1,-\mu'') \, d\mu''$$

$$- \frac{\tilde{\omega}}{4} F_0 P(\mu,-\mu_0)e^{-\tau_1/\mu_0}, \tag{6.154}$$

$$-\mu \frac{dI(\tau,-\mu)}{d\tau}\bigg|_{\tau=\tau_1} = \mu_0 F_0 T(\tau_1;\mu,\mu_0) - \frac{\tilde{\omega}}{2}\int_0^1 P(-\mu,-\mu'')I(\tau_1,-\mu'') \, d\mu''$$

$$- \frac{\tilde{\omega}}{4} F_0 P(-\mu,-\mu_0)e^{-\tau_1/\mu_0}. \tag{6.155}$$

In these four equations, $\mu \geq 0$. We also note that $I(0,\mu) = \mu_0 F_0 R(\tau_1;\mu,\mu_0)$, and $I(\tau_1,-\mu) = \mu_0 F_0 T(\tau_1;\mu,\mu_0)$. Upon substituting Eqs. (6.152) and (6.153), (6.154) and (6.155), (6.154) and (6.153) into Eqs. (6.148)–(6.151), respectively, and rearranging the terms, we obtain

$$\frac{\partial R(\tau_1; \mu, \mu_0)}{\partial \tau_1} = -\left(\frac{1}{\mu} + \frac{1}{\mu_0}\right) R(\tau_1; \mu, \mu_0) + \frac{\tilde{\omega}}{4\mu\mu_0} P(\mu, -\mu_0)$$

$$+ \frac{\tilde{\omega}}{2\mu} \int_0^1 P(\mu, \mu'') R(\tau_1; \mu'', \mu_0)\, d\mu''$$

$$+ \frac{\tilde{\omega}}{2\mu_0} \int_0^1 R(\tau_1; \mu, \mu') P(-\mu', -\mu_0)\, d\mu'$$

$$+ \tilde{\omega} \int_0^1 R(\tau_1; \mu, \mu')\, d\mu' \left[\int_0^1 P(-\mu', \mu'') R(\tau_1; \mu'', \mu_0)\, d\mu'' \right],$$

$$(6.156)$$

$$\frac{\partial T(\tau_1; \mu, \mu_0)}{\partial \tau_1} = -\frac{1}{\mu} T(\tau_1; \mu, \mu_0) + \frac{\tilde{\omega}}{4\mu\mu_0} e^{-\tau_1/\mu_0} P(-\mu, -\mu_0)$$

$$+ \frac{\tilde{\omega}}{2\mu} \int_0^1 P(-\mu, -\mu'') T(\tau_1; \mu'', \mu_0)\, d\mu''$$

$$+ \frac{\tilde{\omega}}{2\mu_0} e^{-\tau_1/\mu_0} \int_0^1 R(\tau_1; \mu, \mu') P(\mu', -\mu_0)\, d\mu'$$

$$+ \tilde{\omega} \int_0^1 R(\tau_1; \mu, \mu')\, d\mu' \left[\int_0^1 P(\mu', -\mu'') T(\tau_1; \mu'', \mu_0)\, d\mu'' \right],$$

$$(6.157)$$

$$\frac{\partial R(\tau_1; \mu, \mu_0)}{\partial \tau_1} = \frac{\tilde{\omega}}{4\mu\mu_0} \exp\left[-\tau\left(\frac{1}{\mu} + \frac{1}{\mu_0}\right) \right] P(\mu, -\mu_0)$$

$$+ \frac{\tilde{\omega}}{2\mu} e^{-\tau_1/\mu} \int_0^1 P(\mu, -\mu'') T(\tau_1; \mu'', \mu_0)\, d\mu''$$

$$+ \frac{\tilde{\omega}}{2\mu_0} e^{-\tau_1/\mu_0} \int_0^1 T(\tau_1; \mu, \mu') P(\mu', -\mu_0)\, d\mu'$$

$$+ \tilde{\omega} \int_0^1 T(\tau_1; \mu, \mu')\, d\mu' \left[\int_0^1 P(\mu', -\mu'') T(\tau_1; \mu'', \mu_0)\, d\mu'' \right],$$

$$(6.158)$$

$$\frac{\partial T(\tau_1; \mu, \mu_0)}{\partial \tau_1} = -\frac{1}{\mu_0} T(\tau_1; \mu, \mu_0) + \frac{\tilde{\omega}}{4\mu\mu_0} e^{-\tau_1/\mu} P(-\mu, -\mu_0)$$

$$+ \frac{\tilde{\omega}}{2\mu} e^{-\tau_1/\mu} \int_0^1 P(-\mu, \mu'') R(\tau_1; \mu'', \mu_0) d\mu''$$

$$+ \frac{\tilde{\omega}}{2\mu_0} \int_0^1 T(\tau_1; \mu, \mu') P(-\mu', -\mu_0) d\mu'$$

$$+ \tilde{\omega} \int_0^1 T(\tau_1; \mu, \mu') d\mu' \left[\int_0^1 P(-\mu', \mu'') R(\tau_1; \mu'', \mu_0) d\mu'' \right].$$

$$(6.159)$$

Equations (6.156)–(6.159) represent four nonlinear integral equations governing the complete field of radiation at $\tau = 0$ and $\tau = \tau_1$ for plane-parallel atmospheres of finite optical depths. It should be noted that for simplicity we have neglected the azimuthal dependence in the reflection, transmission, and phase functions in the derivation of these four integral equations. However, it is straightforward to write down the four integral equations for the azimuthal-dependent case. We further note that as $\tau_1 \to \infty$, we have $\partial R/\partial \tau_1 \to 0$ and Eq. (6.156) reduces to Eq. (6.131), formulated for a semi-infinite atmosphere. Equations (6.156) and (6.159) may be obtained by adding a thin layer ($\Delta \tau \ll 1$) to the top of a finite atmosphere following the principles outlined in Section 6.4.2. Moreover, by adding a thin layer to the bottom of a finite atmosphere, Eqs. (6.157) and (6.158) may be derived. The adding method is referred to as *invariant imbedding* (Bellman *et al.*, 1963).

6.4.4   The X and Y Functions

In Section 6.4.2, we showed that the reflection function of a semi-infinite atmosphere for isotropic scattering is given by the H function. Here, we shall demonstrate that the reflection and transmission functions of a finite atmosphere for isotropic scattering are governed by the X and Y functions. In the case of isotropic scattering, Eqs. (6.156)–(6.159) become

$$\frac{\partial R(\tau_1; \mu, \mu_0)}{\partial \tau_1} + \left(\frac{1}{\mu} + \frac{1}{\mu_0} \right) R(\tau_1; \mu, \mu_0)$$

$$= \frac{\tilde{\omega}}{4\mu\mu_0} \left[1 + 2\mu_0 \int_0^1 R(\tau_1; \mu'', \mu_0) d\mu'' + 2\mu \int_0^1 R(\tau_1; \mu, \mu') d\mu' \right.$$

$$\left. + 4\mu\mu_0 \int_0^1 R(\tau_1; \mu, \mu') \int_0^1 R(\tau_1; \mu'', \mu_0) d\mu'' \right], \qquad (6.160)$$

$$\frac{\partial R(\tau_1; \mu, \mu_0)}{\partial \tau_1}$$

$$= \frac{\tilde{\omega}}{4\mu\mu_0} \left[\exp\left\{ -\tau_1\left(\frac{1}{\mu} + \frac{1}{\mu_0} \right) \right\} + 2\mu_0 e^{-\tau_1/\mu} \int_0^1 T(\tau_1; \mu'', \mu_0) \, d\mu'' \right.$$

$$+ 2\mu e^{-\tau_1/\mu_0} \int_0^1 T(\tau_1; \mu, \mu') \, d\mu'$$

$$\left. + 4\mu\mu_0 \int_0^1 T(\tau_1; \mu, \mu') \, d\mu' \int_0^1 T(\tau_1; \mu'', \mu_0) \, d\mu'' \right], \qquad (6.161)$$

$$\frac{\partial T(\tau_1; \mu, \mu_0)}{\partial \tau_1} + \frac{1}{\mu} T(\tau_1; \mu, \mu_0)$$

$$= \frac{\tilde{\omega}}{4\mu\mu_0} \left[e^{-\tau_1/\mu_0} + 2\mu_0 \int_0^1 T(\tau_1; \mu'', \mu_0) \, d\mu'' + 2\mu e^{-\tau_1/\mu_0} \int_0^1 R(\tau_1; \mu, \mu') \, d\mu' \right.$$

$$\left. + 4\mu\mu_0 \int_0^1 R(\tau_1; \mu, \mu') \, d\mu' \int_0^1 T(\tau_1; \mu'', \mu_0) \, d\mu'' \right], \qquad (6.162)$$

$$\frac{\partial T(\tau_1; \mu, \mu_0)}{\partial \tau_1} + \frac{1}{\mu_0} T(\tau_1; \mu, \mu_0)$$

$$= \frac{\tilde{\omega}}{4\mu\mu_0} \left[e^{-\tau_1/\mu} + 2\mu \int_0^1 T(\tau_1; \mu, \mu') \, d\mu' + 2\mu_0 e^{-\tau_1/\mu} \int_0^1 R(\tau_1; \mu'', \mu_0) \, d\mu'' \right.$$

$$\left. + 4\mu\mu_0 \int_0^1 T(\tau_1; \mu, \mu') \int_0^1 R(\tau_1; \mu'', \mu_0) \, d\mu'' \right]. \qquad (6.163)$$

Upon inspection of Eqs. (6.160)–(6.163), we define Chandrasekhar's X and Y functions in the forms

$$X(\mu) = 1 + 2\mu \int_0^1 R(\tau_1; \mu, \mu') \, d\mu', \qquad (6.164)$$

$$Y(\mu) = e^{-\tau_1/\mu} + 2\mu \int_0^1 T(\tau_1; \mu, \mu') \, d\mu'. \qquad (6.165)$$

Equations (6.160)–(6.163) now may be expressed by

$$\frac{\partial R(\tau_1; \mu, \mu_0)}{\partial \tau_1} + \left(\frac{1}{\mu} + \frac{1}{\mu_0} \right) R(\tau_1; \mu, \mu_0) = \frac{\tilde{\omega}}{4\mu\mu_0} X(\mu) X(\mu_0), \quad (6.166)$$

$$\frac{\partial R(\tau_1; \mu, \mu_0)}{\partial \tau_1} = \frac{\tilde{\omega}}{4\mu\mu_0} Y(\mu) Y(\mu_0), \quad (6.167)$$

$$\frac{\partial T(\tau_1; \mu, \mu_0)}{\partial \tau_1} + \frac{1}{\mu} T(\tau_1; \mu, \mu_0) = \frac{\tilde{\omega}}{4\mu\mu_0} X(\mu) Y(\mu_0), \quad (6.168)$$

$$\frac{\partial T(\tau_1; \mu, \mu_0)}{\partial \tau_1} + \frac{1}{\mu_0} T(\tau_1; \mu, \mu_0) = \frac{\tilde{\omega}}{4\mu\mu_0} X(\mu_0)Y(\mu). \quad (6.169)$$

It is evident that by eliminating $\partial R/\partial \tau_1$ from Eqs. (6.166) and (6.167), we obtain

$$\left(\frac{1}{\mu} + \frac{1}{\mu_0}\right) R(\tau_1; \mu, \mu_0) = \frac{\tilde{\omega}}{4\mu\mu_0} [X(\mu)X(\mu_0) - Y(\mu)Y(\mu_0)], \quad (6.170)$$

and by eliminating $\partial T/\partial \tau_1$ from Eqs. (6.168) and (6.169), we have

$$\left(\frac{1}{\mu} - \frac{1}{\mu_0}\right) T(\tau_1; \mu, \mu_0) = \frac{\tilde{\omega}}{4\mu\mu_0} [X(\mu)Y(\mu_0) - X(\mu_0)Y(\mu)]. \quad (6.171)$$

Inserting Eqs. (6.170) and (6.171) into Eqs. (6.164) and (6.165), we find

$$X(\mu) = 1 + \mu \int_0^1 \frac{\psi(\mu')}{\mu + \mu'} [X(\mu)X(\mu') - Y(\mu)Y(\mu')] \, d\mu', \quad (6.172)$$

$$Y(\mu) = e^{-\tau_1/\mu} + \mu \int_0^1 \frac{\psi(\mu')}{\mu' - \mu} [X(\mu)Y(\mu') - X(\mu')Y(\mu)] \, d\mu', \quad (6.173)$$

where the characteristic function $\psi(\mu') = \tilde{\omega}/2$. Thus, the exact solutions of the reflection and transmission functions are now given by the X and Y functions, which are solutions of the nonlinear integral equations. It is also clear that for a semi-infinite atmosphere $Y(\mu) = 0$, and the X function defined in Eqs. (6.164) and (6.172) is equivalent to the H function introduced in Eqs. (6.134) and (6.136). The characteristic function $\psi(\mu')$ differs from problem to problem, but it has a simple algebraic form for the Rayleigh scattering phase function. For a more general case involving the Mie scattering phase function, however, the analytic characteristic functions $\psi(\mu')$ have not been derived. The iteration procedure may be utilized to solve the above nonlinear integral equations for X and Y functions, and extensive tables of these two functions for conservative and nonconservative isotropic scattering, as well as anisotropic phase functions, with as many as three terms are available.

6.5 THE INCLUSION OF SURFACE REFLECTION

For planetary applications, the surface reflection plays an important role for the reflected and transmitted sunlight. In this section, we introduce the inclusion of surface reflection for the scattered intensity and flux density. It is assumed that the ground reflects according to Lambert's law, with a reflectivity (or surface albedo) of r_s. Under this condition the diffuse upward intensity denoted in Eq. (6.30) is

$$I(\tau_1; \mu, \phi) = I_s = \text{const} \quad (6.173)$$

Let $I^*(0; \mu, \phi)$ represent the reflected intensity including the contribution of surface reflection, and in reference to Fig. 6.7a, we find

$$I^*(0; \mu, \phi) = I(0; \mu, \phi) + \frac{1}{\pi} \int_0^{2\pi} \int_0^1 T(\mu, \phi; \mu', \phi') I_s \mu' \, d\mu' \, d\phi' + I_s e^{-\tau_1/\mu}. \quad (6.174)$$

The last two terms represent, respectively, the diffuse and direct transmission of the upward isotropic intensity I_s.

Equation (6.174) can be rewritten in terms of the reflection function and the direct and diffuse transmission defined in Section 6.4.1 in the form

$$I^*(0; \mu, \phi) = \mu_0 F_0 R(\mu, \phi; \mu_0, \phi_0) + I_s \gamma(\mu), \quad (6.175)$$

where

$$\gamma(\mu) = e^{-\tau_1/\mu} + t(\mu), \quad (6.176)$$

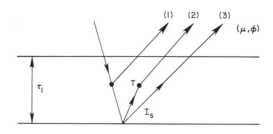

(a) Upward diffuse intensity

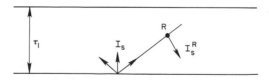

(b) Reflection of upward isotropic intensity

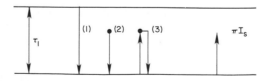

(c)   Downward flux density   Upward flux density

Fig. 6.7 Scattering configuration for the inclusion of surface reflection: (a) upward diffuse intensity, (b) reflection of upward isotropic intensity, and (c) downward flux density and upward flux density.

and $t(\mu)$ is defined in Eq. (6.110). Here we note that the principle of reciprocity, i.e., $T(\mu, \phi; \mu', \phi') = T(\mu', \phi'; \mu, \phi)$, is used to obtain the diffuse transmission $t(\mu)$.

The upward isotropic intensity from the surface also will be reflected by the atmosphere and will contribute to the downward intensity in the additional amount (see Fig. 6.7b)

$$I_s^R(-\mu) = \frac{1}{\pi} \int_0^{2\pi} \int_0^1 R(\mu, \phi; \mu', \phi') I_s \mu' \, d\mu' \, d\phi' = I_s r(\mu), \qquad (6.177)$$

where again the principle of reciprocity, $R(\mu, \phi; \mu', \phi') = R(\mu', \phi'; \mu, \phi)$, is used. Thus the total transmitted intensity, including the ground contribution, is given by

$$I^*(\tau_1; -\mu, \phi) = I(\tau_1; -\mu, \phi) + I_s^R(-\mu)$$
$$= \mu_0 F_0 T(\mu, \phi; \mu_0, \phi_0) + I_s r(\mu). \qquad (6.178)$$

It now requires an equation to determine I_s. Since the upward flux density has to be equal to the downward flux multiplied by the surface albedo r_s, we have

$$\pi I_s = r_s \times \text{downward flux density.} \qquad (6.179)$$

The downward flux density includes three components, as evident from Fig. 6.7c.

(1) Direct transmission component:

$$\pi \mu_0 F_0 e^{-\tau_1/\mu_0}.$$

(2) Diffuse transmission component:

$$\int_0^{2\pi} \int_0^1 I(\tau_1; -\mu, \phi) \mu \, d\mu \, d\phi = \int_0^{2\pi} \int_0^1 \mu_0 F_0 T(\mu, \phi; \mu_0, \phi_0) \mu \, d\mu \, d\phi$$
$$= \pi \mu_0 F_0 t(\mu_0).$$

(3) The component of I_s reflected by the atmosphere:

$$\int_0^{2\pi} \int_0^1 I_s^R(-\mu) \mu \, d\mu \, d\phi = \pi I_s \bar{r}.$$

From Eq. (6.179), we have the equality at $\tau = \tau_1$

$$\pi I_s = r_s [\pi \mu_0 F_0 e^{-\tau_1/\mu_0} + \pi \mu_0 F_0 t(\mu_0) + \pi I_s \bar{r}]. \qquad (6.180)$$

We then rearrange these terms to yield

$$I_s = \frac{r_s}{1 - r_s \bar{r}} \mu_0 F_0 \gamma(\mu_0). \qquad (6.181)$$

It follows from Eqs. (6.175) and (6.178) that the reflected and transmitted intensities, including the ground reflection, are respectively,

$$I^*(0; \mu, \phi) = I(0; \mu, \phi) + \frac{r_s}{1 - r_s \bar{r}} \mu_0 F_0 \gamma(\mu) \gamma(\mu_0), \qquad (6.182)$$

$$I^*(\tau_1; -\mu, \phi) = I(\tau_1; -\mu, \phi) + \frac{r_s}{1 - r_s \bar{r}} \mu_0 F_0 \gamma(\mu_0) r(\mu). \qquad (6.183)$$

To obtain the reflected and transmitted flux densities, we perform the integration of the intensity over the solid angle, according to Eqs. (6.16) and (6.17), to yield

$$F^*(0) = F(0) + \frac{r_s}{1 - r_s \bar{r}} \pi \mu_0 F_0 \gamma(\mu_0) \bar{\gamma}, \qquad (6.184)$$

$$F^*(\tau_1) = F(\tau_1) + \frac{r_s}{1 - r_s \bar{r}} \pi \mu_0 F_0 \gamma(\mu_0) \bar{r}, \qquad (6.185)$$

where

$$\bar{\gamma} = \bar{t} + 2 \int_0^1 e^{-\tau_1/\mu_0} \mu_0 \, d\mu_0, \qquad (6.186)$$

and $\bar{t}$ and $\bar{r}$ are defined in Eqs. (6.114) and (6.115). Further, by dividing $\pi \mu_0 F_0$ and adding $e^{-\tau_1/\mu_0}$ to both sides in Eq. (6.185), the preceding two equations become

$$r^*(\mu_0) = r(\mu_0) + f(\mu_0) \bar{\gamma}, \qquad (6.187)$$

$$\gamma^*(\mu_0) = \gamma(\mu_0) + f(\mu_0) \bar{r}, \qquad (6.188)$$

where

$$f(\mu_0) = \frac{r_s}{1 - r_s \bar{r}} \gamma(\mu_0). \qquad (6.189)$$

In Exercise 6.10, the reader is invited to derive Eqs. (6.187) and (6.188) by means of the ray-tracing technique.

6.6 ADDING METHOD FOR MULTIPLE SCATTERING

In essence, the adding method uses a straightforward geometrical ray-tracing technique. If the reflection and transmission properties of two individual layers are known, then the reflection and transmission properties of the combined layer may be obtained by computing the successive reflections back and forth between the two layers. When the two layers have the same optical depth, the adding method is referred to as the *doubling method*. The adding method for radiative transfer provides traceable mathematical and physical deductions of the reflection and transmission of light. The principle involved establishes the foundation for a number of numerical methods. In this section we introduce the basic procedures of the adding method developed by van de Hulst (1963).

Consider Fig. 6.8 and assume that radiation comes from the top of the layer. Let R_1 and $\tilde{T}_1$ denote the reflection and total (direct plus diffuse) transmission functions for the first layer and R_2 and $\tilde{T}_2$ for the second layer, and define $\tilde{D}$ and U for the combined total transmission and reflection functions between layers 1 and 2. In principle, photons may undergo one to an infinite number of scattering events. Inspection of this diagram makes it evident that the combined reflection and transmission functions are given by

$$\begin{aligned}
R_{12} &= R_1 + \tilde{T}_1 R_2 \tilde{T}_1 + \tilde{T}_1 R_2 R_1 R_2 \tilde{T}_1 + \tilde{T}_1 R_2 R_1 R_2 R_1 R_2 \tilde{T}_1 + \cdots \\
&= R_1 + \tilde{T}_1 R_2 \tilde{T}_1 [1 + R_1 R_2 + (R_1 R_2)^2 + \cdots] \\
&= R_1 + R_2 \tilde{T}_1^2 (1 - R_1 R_2)^{-1}
\end{aligned} \tag{6.190}$$

$$\begin{aligned}
\tilde{T}_{12} &= \tilde{T}_1 \tilde{T}_2 + \tilde{T}_1 R_2 R_1 \tilde{T}_2 + \tilde{T}_1 R_2 R_1 R_2 R_1 \tilde{T}_2 + \cdots \\
&= \tilde{T}_1 \tilde{T}_2 [1 + R_1 R_2 + (R_1 R_2)^2 + \cdots] \\
&= \tilde{T}_1 \tilde{T}_2 (1 - R_1 R_2)^{-1}.
\end{aligned} \tag{6.191}$$

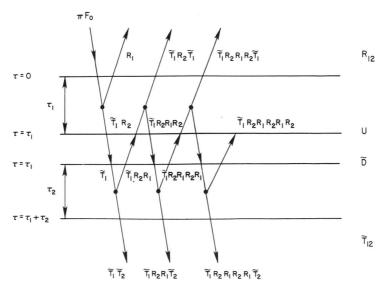

Fig. 6.8   Configuration of the adding method. The two layers of optical depths τ_1 and τ_2 are for convenient illustration, as if they were physically separated.

We also find the expressions for U and $\tilde{D}$ in the forms

$$
\begin{aligned}
U &= \tilde{T}_1 R_2 + \tilde{T}_1 R_2 R_1 R_2 + \tilde{T}_1 R_2 R_1 R_2 R_1 R_2 + \cdots \\
&= \tilde{T}_1 R_2 [1 + R_1 R_2 + (R_1 R_2)^2 + \cdots] = \tilde{T}_1 R_2 (1 - R_1 R_2)^{-1}, \quad (6.192)
\end{aligned}
$$

$$
\begin{aligned}
\tilde{D} &= \tilde{T}_1 + \tilde{T}_1 R_2 R_1 + \tilde{T}_1 R_2 R_1 R_2 R_1 + \cdots \\
&= \tilde{T}_1 [1 + R_1 R_2 + (R_1 R_2)^2 + \cdots] = \tilde{T}_1 (1 - R_1 R_2)^{-1}, \quad (6.193)
\end{aligned}
$$

where we have replaced the infinite series by a single inverse function. On the basis of these equations, we find

$$
R_{12} = R_1 + \tilde{T}_1 U, \qquad \tilde{T}_{12} = \tilde{T}_2 \tilde{D}, \qquad U = R_2 \tilde{D}. \qquad (6.194)
$$

But the total transmission function arises from the diffuse as well as direct components and can be expressed by

$$
\tilde{T} = T + e^{-\tau/\mu'}, \qquad (6.195)
$$

where $\mu' = \mu_0$ when transmission is associated with the incident solar beam and $\mu' = \mu$ when transmission is related to the emergent beam in the direction μ. With this understanding, we now examine the parameters $\tilde{D}$ and $\tilde{T}_{12}$ and let

$$
S = R_1 R_2 (1 - R_1 R_2)^{-1}. \qquad (6.196)
$$

Thus, from Eqs. (6.193) and (6.194) we have

$$
\begin{aligned}
\tilde{D} &= D + e^{-\tau_1/\mu_0} = (T_1 + e^{-\tau_1/\mu_0})(1 + S) \\
&= (1 + S)T_1 + S e^{-\tau_1/\mu_0} + e^{-\tau_1/\mu_0}, \qquad (6.197)
\end{aligned}
$$

$$
\begin{aligned}
\tilde{T}_{12} &= (T_2 + e^{-\tau_2/\mu})(D + e^{-\tau_1/\mu_0}) = e^{-\tau_2/\mu} D \\
&\quad + T_2 e^{-\tau_1/\mu_0} + T_2 D + \exp\left[-\left(\frac{\tau_1}{\mu_0} + \frac{\tau_2}{\mu}\right)\right]\delta(\mu - \mu_0), \quad (6.198)
\end{aligned}
$$

where parameters without tilt notation (~) denote the diffuse component only. We add a delta function to the pure exponential terms to physically represent the direct transmission for the combined layer. At this point, we now may write a set of simultaneous equations governing the diffuse reflection and transmission functions for the two layers as

$$
\begin{aligned}
D &= T_1 + ST_1 + S e^{-\tau_1/\mu_0}, \\
U &= R_2 D + R_2 e^{-\tau_1/\mu_0}, \\
R_{12} &= R_1 + e^{-\tau_1/\mu} U + T_1 U, \\
T_{12} &= e^{-\tau_2/\mu} D + T_2 e^{-\tau_1/\mu_0} + T_2 D.
\end{aligned} \qquad (6.199)
$$

Note that the direct transmission function is simply $e^{-(\tau_1 + \tau_2)/\mu_0}$.

In Eqs. (6.196) and (6.199), the product of any two parameters implies that integration over the solid angle is to be performed so as to take into account all the possible multiple scattering contributions. Let the operators

$$A_1 B_2 = 2 \int_0^1 A(\tau_1; \mu, \mu') B(\tau_2; \mu', \mu_0) \mu' \, d\mu' \qquad (6.200)$$

in which A and B can be any of the parameters R, T, U, and D. In practice, one may begin with the computations for initial layers of such small optical depths that the single scattering approximation for R and T given in Eqs. (6.31) and (6.32) may be sufficiently accurate. Subsequent computations may be carried out employing Eqs. (6.196) and (6.199) to get R_{12} and T_{12}. The procedures may be repeated to evaluate the diffuse reflection and transmission functions for two thicker layers having optical depths of, say, $\tau_3 = \tau_1 + \tau_2$, and so on, until the desirable optical depth is reached. Numerical procedures referred to as the matrix formulation, matrix operator, or star product method are in essence the same as the adding method, so far as the principle and actual computations are concerned. Moreover, we shall also show that the adding method is equivalent to the principles of invariance for finite atmospheres introduced in Section 6.4.3.

In reference to the principles of invariance for finite atmospheres described in Section 6.4.3, replacing τ by τ_1, and τ_1 by $\tau_1 + \tau_2$ (see Fig. 6.8), we have

$$U(\mu, \mu_0) = R(\tau_2; \mu, \mu_0) e^{-\tau_1/\mu_0}$$
$$+ 2 \int_0^1 R(\tau_2; \mu, \mu') D(\mu', \mu_0) \mu' \, d\mu' \qquad (6.201)$$

$$D(\mu, \mu_0) = T(\tau_1; \mu, \mu_0) = 2 \int_0^1 R(\tau_1; \mu, \mu') U(\mu', \mu_0) \mu' \, d\mu', \qquad (6.202)$$

$$R(\tau_1 + \tau_2; \mu, \mu_0) = R(\tau_1; \mu, \mu_0) + e^{-\tau_1/\mu} U(\mu, \mu_0)$$
$$+ 2 \int_0^1 T(\tau_1; \mu, \mu') U(\mu', \mu_0) \mu' \, d\mu', \qquad (6.203)$$

$$T(\tau_1 + \tau_2; \mu, \mu_0) = T(\tau_2; \mu, \mu_0) e^{-\tau_1/\mu_0} + e^{-\tau_2/\mu} D(\mu, \mu_0)$$
$$+ 2 \int_0^1 T(\tau_2; \mu, \mu') D(\mu', \mu_0) \mu' \, d\mu', \qquad (6.204)$$

where we define

$$U(\mu, \mu_0) = I(\tau_1, \mu)/(\mu_0 F_0),$$
$$D(\mu, \mu_0) = I(\tau_1, -\mu)/(\mu_0 F_0). \qquad (6.205)$$

Upon utilizing the operators defined in Eq. (6.200), Eqs. (6.201)–(6.204) may be rewritten in the forms

$$U = R_2 e^{-\tau_1/\mu_0} + R_2 D,$$
$$D = T_1 + R_1 U,$$
$$R_{12} = R_1 + e^{-\tau_1/\mu} U + T_1 U, \qquad (6.206)$$
$$T_{12} = T_2 e^{-\tau_1/\mu_0} + e^{-\tau_2/\mu} D + T_2 D.$$

Employing these simplified operators and recognizing the iterative relationships between U and D, we find

$$D = T_1 + R_1(R_2 e^{-\tau_1/\mu_0} + R_2 D),$$ (6.207)

$$D(1 - R_1 R_2) = T_1 + R_1 R_2 e^{-\tau_1/\mu_0}.$$ (6.208)

But $(1 - R_1 R_2)^{-1} = 1 + R_1 R_2 + (R_1 R_2)^2 + \cdots = 1 + S$, so Eq. (6.208) becomes

$$D = (1 + S)T_1 + S e^{-\tau_1/\mu_0}.$$ (6.209)

At this point, it is evident that we have derived the adding (or doubling) equations depicted in Eqs. (6.196)–(6.199) by means of the principles of invariance.

6.7   MULTIPLE SCATTERING INCLUDING POLARIZATION

We first introduce the transformation matrix associated with the Stokes parameters (I, Q, U, V) for the rotation of the axes through an angle in the clockwise direction. Consider two orthogonal electric components, E_l and E_r, and rotate E_l clockwise through an angle χ. Let the two new orthogonal electric components be E_l' and E_r'. Referring to Fig. 6.9, we find

$$E_l' = \cos \chi E_l + \sin \chi E_r,$$
$$E_r' = -\sin \chi E_l + \cos \chi E_r.$$ (6.210)

Let the linear transformation matrix for the electric field be

$$\mathbf{L}_e(\chi) = \begin{bmatrix} \cos \chi & \sin \chi \\ -\sin \chi & \cos \chi \end{bmatrix}.$$ (6.211)

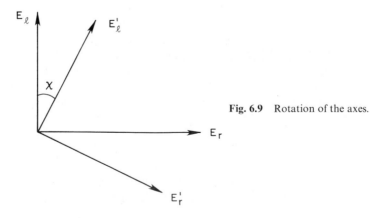

Fig. 6.9   Rotation of the axes.

It follows that

$$
\begin{bmatrix} E'_l \\ E'_r \end{bmatrix} = \mathbf{L}_e(\chi) \begin{bmatrix} E_l \\ E_r \end{bmatrix}.
\tag{6.212}
$$

Upon inserting Eq. (6.210) into the definition of the Stokes parameters for the prime system, i.e., (I', Q', U', V'), in Eq. (3.42), we find after some straightforward analysis

$$
\begin{bmatrix} I' \\ Q' \\ U' \\ V' \end{bmatrix} = \mathbf{L}(\chi) \begin{bmatrix} I \\ Q \\ U \\ V \end{bmatrix},
\tag{6.213}
$$

where the transformation matrix for the Stokes parameters is given by

$$
\mathbf{L}(\chi) = \begin{bmatrix} 1 & 0 & 0 & 0 \\ 0 & \cos 2\chi & \sin 2\chi & 0 \\ 0 & -\sin 2\chi & \cos 2\chi & 0 \\ 0 & 0 & 0 & 1 \end{bmatrix}.
\tag{6.214}
$$

On the basis of Eqs. (6.213) and (6.214), we see that I and V are invariant in the rotation process. We also find that $\mathbf{L}(\chi_1)\mathbf{L}(\chi_2) = \mathbf{L}(\chi_1 + \chi_2)$, and the inverse matrix $\mathbf{L}^{-1}(\chi) = \mathbf{L}(-\chi)$.

Having the transformation matrix defined, we should now proceed to formulate the equation of transfer in which polarization is included. In Section 6.1, the equation of transfer for plane-parallel atmospheres was formulated, based on the intensity quantity alone, without taking into account the effect of polarization. To represent the polarization property of the light wave, we have introduced in Section 3.6 a set of four parameters called Stokes parameters, defined in Eq. (3.42). To describe the radiation field at each point in space including polarization, we replace the scalar intensity I by the vector intensity $\mathbf{I} = (I, Q, U, V)$. The four Stokes parameters give, respectively, the intensity, the degree of polarization, the plane of polarization, and the ellipticity of the light waves as functions of the incoming and outgoing directions.

The equation of transfer given in Eq. (1.63) now may be written in the vector form as

$$
\mu \frac{d\mathbf{I}(\tau; \mu, \phi)}{d\tau} = \mathbf{I}(\tau; \mu, \phi) - \mathbf{J}(\tau; \mu, \phi),
\tag{6.215}
$$

where the source function is a vector consisting of four elements. To obtain the expression for $\mathbf{J}(\tau; \mu, \phi)$, we consider the differential increment

$d\mathbf{J}(\tau; \mu, \phi; \mu', \phi')$ due to multiple scattering of a pencil of radiation of solid angle $d\Omega'$ in the direction (μ', ϕ'). The diffuse intensity vector $\mathbf{I}(\tau; \mu', \phi')$, which generates the source term, is in reference to the meridian plane OP_1Z (see Fig. 6.10). However, the scattering phase matrix derived from the scattering theory [e.g., see Eq. (5.113)] refers to the plane of scattering OP_1P_2 containing the incident and scattered waves. Thus, we must first transform $\mathbf{I}(\tau; \mu', \phi')$ to the plane of scattering in order to obtain the proper source function. In view of the transformation concept, we may transform $\mathbf{I}(\tau; \mu', \phi')$ to the plane of scattering by applying the transformation matrix $\mathbf{L}(-i_1)$, where i_1 denotes the angle between the meridian plane OP_1Z and the plane of scattering OP_1P_2, and the minus sign indicates the rotation of the plane is counterclockwise. Thus, we have the contribution to the source function with reference to the plane of scattering at P_2 in the form

$$\tilde{\omega}\mathbf{P}(\Theta)\,\mathbf{L}(-i_1)\mathbf{I}(\tau; \mu', \phi')\,d\Omega'/(4\pi). \qquad (6.216)$$

To transform it to the scattering direction (μ, ϕ), i.e., the meridian plane OP_2Z, we must again apply the transformation matrix $\mathbf{L}(\pi - i_2)$ through the angle $(\pi - i_2)$ clockwise, where i_2 denotes the angle between the meridian plane OP_2Z and the plane of scattering OP_1P_2. Consequently, the desirable differential source function due to the diffuse component is

$$d\mathbf{J}(\tau; \mu, \phi; \mu', \phi') = \tilde{\omega}\mathbf{L}(\pi - i_2)\,\mathbf{P}(\Theta)\mathbf{L}(-i_1)\mathbf{I}(\tau; \mu', \phi')\,d\Omega'/(4\pi). \quad (6.217)$$

Thus, by performing the integration over all directions (μ', ϕ'), we obtain the source function vector for multiple scattering

$$\mathbf{J}(\tau; \mu, \phi; \mu', \phi') = \frac{\tilde{\omega}}{4\pi} \int_0^{2\pi} \int_{-1}^1 \mathbf{P}(\mu, \phi; \mu', \phi')\mathbf{I}(\tau; \mu', \phi')\,d\mu'\,d\phi', \quad (6.218)$$

where the phase matrix

$$\mathbf{P}(\mu, \phi; \mu', \phi') = \mathbf{L}(\pi - i_2)\mathbf{P}(\Theta)\mathbf{L}(-i_1). \qquad (6.219)$$

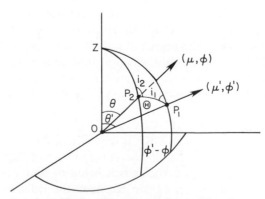

Fig. 6.10 Scattering plane OP_1P_2 with respect to the meridian planes OP_1Z and OP_2Z.

From the sperical trigonometry, as illustrated in Appendix F, the angles i_1 and i_2 can be expressed by

$$\cos i_1 = \frac{-\mu + \mu' \cos \Theta}{\pm (1 - \cos^2 \Theta)^{1/2} (1 - \mu'^2)^{1/2}}, \tag{6.220}$$

$$\cos i_2 = \frac{-\mu' + \mu \cos \Theta}{\pm (1 - \cos^2 \Theta)^{1/2} (1 - \mu^2)^{1/2}}, \tag{6.221}$$

where the plus sign is to be taken when $\pi < \phi - \phi' < 2\pi$ and the minus sign is to be taken when $0 < \phi - \phi' < \pi$. From Eq. (6.7), we also find $\cos \Theta = \mu\mu' + (1 - \mu^2)^{1/2}(1 - \mu'^2)^{1/2} \cos(\phi - \phi')$.

On following the same procedures as already outlined, the direct component of the source function due to the point source $\mathbf{I}_0(-\mu, \phi) = \delta(\mu - \mu_0) \delta(\phi - \phi_0) \pi \mathbf{F}_0$ [see Eq. (6.106)] is

$$\mathbf{J}(\tau; \mu, \phi) = \frac{\tilde{\omega}}{4\pi} \mathbf{P}(\mu, \phi; -\mu_0, \phi_0) \pi \mathbf{F}_0 e^{-\tau/\mu_0}. \tag{6.222}$$

Thus, the equation of transfer of sunlight including polarization can be written as

$$\mu \frac{d\mathbf{I}(\tau; \mu, \phi)}{d\tau} = \mathbf{I}(\tau; \mu, \phi) - \frac{\tilde{\omega}}{4\pi} \int_0^{2\pi} \int_{-1}^{1} \mathbf{P}(\mu, \phi; \mu', \phi') \mathbf{I}(\tau; \mu', \phi') \, d\mu' \, d\phi'$$

$$- \frac{\tilde{\omega}}{4\pi} \mathbf{P}(\mu, \phi; -\mu_0, \phi_0) \pi \mathbf{F}_0 e^{-\tau/\mu_0}. \tag{6.223}$$

Comparing with Eq. (6.5), we see that the scalar intensity now is replaced by a vector intensity consisting of four elements. Obviously, numerical computations of the intensity fields become much more involved and complicated in the case when polarization is included.

6.8   EQUATIONS FOR MULTIPLE SCATTERING BY ORIENTED NONSPHERICAL PARTICLES

Scattering of light by a nonspherical particle depends on the directions of the incoming and outgoing radiation, and the orientation of the particle with respect to the incoming beam. To formulate the transfer of solar radiation in a medium composed of oriented nonspherical particles, we begin by assuming that such a medium is plane-parallel so that the variation of the intensity is only in the Z direction. In reference to Fig. 6.11, we select a fixed coordinate system XYZ in such a manner that the Z axis is in the zenith direction. Also, we let $X'Y'Z'$ represent a coordinate system referring to the incoming light beam, which is placed on the Z' axis. Angles ϕ', ϕ, γ', and γ are azimuthal angles corresponding to θ', θ, α', and α denoted in

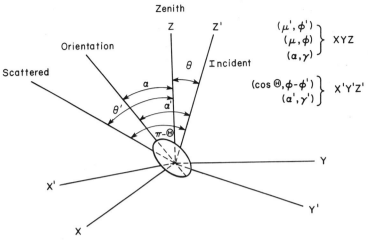

Fig.6.11   Single scattering configuration for a nonspherical particle.

the figure, and Θ is the scattering angle. The scattering parameters for a nonspherical particle, including the phase function, and the extinction and scattering cross sections, may be expressed with respect to either of these two coordinate systems. Thus, we may write symbolically

$$P(\alpha, \gamma; \mu', \phi'; \mu, \phi) = P(\alpha', \gamma'; \cos \Theta, \phi - \phi'),$$
$$\sigma_{e,s}(\alpha, \gamma; \mu', \phi') = \sigma_{e,s}(\alpha', \gamma').$$
$$(6.224a)$$

Here we note that the phase function depends on the directions of the incident and scattered beams as well as the orientation of the nonspherical particle. On the other hand, the extinction and scattering cross sections depend only on the direction of the incident beam and the orientation of the particle.

For a sample of nonspherical particles randomly oriented in space, average scattering properties may be expressed in the forms

$$P(\cos \Theta, \phi - \phi') = \frac{1}{\pi^2} \int_0^\pi \int_0^\pi P(\alpha', \gamma'; \cos \Theta, \phi - \phi')\, d\alpha'\, d\gamma',$$
$$(6.224b)$$

$$\sigma_{e,s} = \frac{1}{\pi^2} \int_0^\pi \int_0^\pi \sigma_{e,s}(\alpha', \gamma')\, d\alpha'\, d\gamma'.$$

Clearly, the extinction and scattering cross sections for randomly oriented nonspherical particles are directionally independent. Moreover, since $\cos \Theta$ can be expressed in terms of μ, ϕ and μ', ϕ' [see Eq. (6.7)], the source function in this case has the same form as that defined in Eq. (6.6). If all of the nonspherical particles have rotational symmetry (e.g., circular cylinders), then the scattering phase function is independent of the azimuthal angle $\phi - \phi'$. Consequently, formulations for multiple scattering in randomly oriented, symmetrical nonspherical particles involving the intensity follow the procedures discussed in previous sections. To include the effect of polarization, a proper phase matrix $\mathbf{P}(\Theta)$ is required. Generally, if no assumptions are made concerning the physical positions of nonspherical particles in space, the phase matrix contains 16 independent parameters (see Exercise 5.5). However, if these particles are randomly oriented in space and have a plane of symmetry, the 16 elements in the phase matrix may be reduced to only six independent elements as Perrin (1942) showed. We note that randomly oriented circular cylinders are symmetrical with respect to the incident beam, regardless of its direction. Thus, the reference plane of the incident beam may be used as a plane of symmetry for these particles. Furthermore, for hexagonal cylinders, if they rotate randomly with respect to their central axes, then their phase matrix should have the same number of elements as that of circular cylinders.

Generally, individual-falling hexagonal crystals with shapes of cylinders and disks tend to orient with their major axis parallel to the surface. It is highly probable that their orientations are random in a horizontal plane. In this case, then, $\alpha = \pi/2$ and we find

$$P(\mu', \phi'; \mu, \phi) = \frac{1}{\pi} \int_0^\pi P(\pi/2, \gamma; \mu', \phi'; \mu, \phi) \, d\gamma,$$

$$\sigma_{e,s}(\mu') = \frac{1}{\pi^2} \int_0^\pi \int_0^\pi \sigma_{e,s}(\pi/2, \gamma; \mu', \phi') \, d\gamma \, d\phi'.$$

(6.224c)

In Section 5.7 the importance of cirrus clouds in remote sensing and radiative transfer was pointed out. Thus, we wish to formulate the basic equation describing the transfer of solar radiation in cirrus cloud layers. It is apparent that since the scattering and extinction cross sections denoted in Eq. (6.236) depend on the direction of the incoming beam, we need to reformulate the source function due to multiple scattering. Let the number density of a sample of horizontally oriented hexagonal crystals be $N(s)$ and follow the procedures outlined in Section 6.1, we find

$$dI(s; \mu, \phi) = -N(s) \, ds \, I(s; \mu, \phi) \sigma_e(\mu)$$

$$+ N(s) \, ds \int_0^{2\pi} \int_{-1}^1 \sigma_s(\mu') \frac{P(\mu, \phi; \mu', \phi')}{4\pi} I(s; \mu', \phi') \, d\mu' \, d\phi'$$

$$+ N(s) \, ds \, \sigma_s(-\mu_0) \frac{P(\mu, \phi; -\mu_0, \phi_0)}{4\pi} \pi F_0$$

$$\times \exp\left[-\int_s^\infty \sigma_e(-\mu_0) N(s') \, ds' \right].$$

(6.225)

Upon defining the vertical path length

$$u = \int_0^z N(z')\,dz',\qquad(6.226)$$

and assuming that the particle number density varies only in the Z direction, Eq. (6.225) may be rewritten in the form

$$\mu\frac{dI(u;\mu,\phi)}{du} = -I(u;\mu,\phi)\sigma_e(\mu)$$

$$+\frac{1}{4\pi}\int_0^{2\pi}\int_{-1}^1 \sigma_s(\mu')P(\mu,\phi;\mu',\phi')I(u;\mu',\phi')\,d\mu'\,d\phi'$$

$$+\frac{1}{4\pi}\sigma_s(-\mu_0)P(\mu,\phi;-\mu_0,\phi_0)\pi F_0$$

$$\times\exp[-\sigma_e(-\mu_0)(u_1-u)/\mu_0],\qquad(6.227)$$

where u_1 corresponds to $z = \infty$. This basic equation differs from the conventional transfer equation in that the extinction and scattering cross sections are functions of the cosine of the zenith angles, and that the scattering phase function depends on the directions of the incoming and outgoing beams, which can be expressed in terms of the scattering angle and the zenith angle for the incoming beam. If the single-scattering parameters are known through single-scattering calculations, a solution of the intensity distribution in a medium composed of randomly oriented nonspherical particles in a horizontal plane may be obtained.

6.9 EQUATIONS FOR MULTIPLE SCATTERING IN THREE-DIMENSIONAL SPACE

In Section 6.1 we formulated the equation for the transfer of solar radiation in plane-parallel atmospheres. However, the general equation of transfer without any coordinate system imposed is given by Eq. (1.46). Let the extinction coefficient be $\beta_e = k_\lambda \rho$ and omit the subscript λ for simplicity, we write

$$-\frac{dI}{\beta_e\,ds} = I - J.\qquad(6.228)$$

For an inhomogeneous scattering atmosphere where the extinction property varies in space, Eq. (6.228) can be written in the form

$$-\frac{1}{\beta_e(\mathbf{s})c}\frac{\partial I(\mathbf{s},\mathbf{\Omega};t)}{\partial t}-\frac{1}{\beta(\mathbf{s})}(\mathbf{\Omega}\cdot\nabla)I(\mathbf{s},\mathbf{\Omega};t) = I(\mathbf{s},\mathbf{\Omega};t)-J(\mathbf{s},\mathbf{\Omega};t),\quad(6.229)$$

where c is the velocity of light, Ω is a unit vector specifying the direction of scattering through a position vector s, and t is the time. Under the assumption that the intensity is independent of time (steady state), Eq. (6.229) reduces to

$$-\frac{1}{\beta_e(s)}(\Omega \cdot \nabla)I(s, \Omega) = I(s, \Omega) - J(s, \Omega). \tag{6.230}$$

Analogous to Eq. (6.6), the general source function for solar radiation in any coordinate systems is given by

$$J(s, \Omega) = \frac{\tilde{\omega}(s)}{4\pi} \int_{4\pi} I(s, \Omega')P(s; \Omega, \Omega')\, d\Omega'$$

$$+ \frac{\tilde{\omega}(s)}{4\pi} P(s; \Omega, -\Omega_0)\pi F_0 \exp\left[-\int_0^s \beta_e(s')\, ds'\right]. \tag{6.231}$$

For simplicity of seeking a solution to the general equation of transfer, it is normally assumed that the medium is vertically and horizontally homogeneous in such a manner that

$$\beta_e(s) = \beta_e, \qquad \tilde{\omega}(s) = \tilde{\omega}, \qquad P(s; \Omega, \Omega') = P(\Omega, \Omega'). \tag{6.232}$$

Under this circumstance, Eqs. (6.230) and (6.231) become much simpler so that the derivation of a solution for the integro-partial-differential equation may be mathematically feasible, subject to appropriate radiation boundary conditions being imposed.

In Cartesian coordinates (x, y, z), the $\Omega \cdot \nabla$ operator is given by

$$\Omega \cdot \nabla = \Omega_x \frac{\partial}{\partial x} + \Omega_y \frac{\partial}{\partial y} + \Omega_z \frac{\partial}{\partial z}. \tag{6.233}$$

The directional cosines in this equation are simply

$$\Omega_x = \frac{\partial s}{\partial x} = \sin\theta \cos\phi = (1 - \mu^2)^{1/2} \cos\phi,$$

$$\Omega_y = \frac{\partial s}{\partial y} = \sin\theta \sin\phi = (1 - \mu^2)^{1/2} \sin\phi, \tag{6.234}$$

$$\Omega_z = \frac{\partial s}{\partial z} = \cos\theta = \mu,$$

where θ and ϕ are the zenith and azimuth angles, respectively, in polar coordinates used throughout this chapter, and $|s| = s = (x^2 + y^2 + z^2)^{1/2}$. Thus, the fundamental equation of transfer for solar radiation in Cartesian coordinates under the conditions of homogeneity may be expressed in the form

$$-\frac{1}{\beta_e}\left[(1-\mu^2)^{1/2}\cos\phi\,\frac{\partial}{\partial x}+(1-\mu^2)^{1/2}\sin\phi\,\frac{\partial}{\partial y}+\mu\,\frac{\partial}{\partial z}\right]I(x,y,z;\mu,\phi)$$

$$=I(x,y,z;\mu,\phi)-\frac{\tilde{\omega}}{4\pi}\int_0^{2\pi}\int_{-1}^{1}I(x,y,z;\mu',\phi')P(\mu,\phi;\mu',\phi')\,d\mu'\,d\phi'$$

$$-\frac{\tilde{\omega}}{4\pi}P(\mu,\phi;-\mu_0,\phi_0)\pi F_0\exp[-\beta_e(x^2+y^2+z^2)^{1/2}], \qquad (6.235)$$

where x, y, and z denote distances in X, Y, and Z coordinates, respectively. In spherical coordinates, the operator $\mathbf{\Omega}\cdot\nabla$ may be written in the form

$$\mathbf{\Omega}\cdot\nabla=\Omega_r\frac{\partial}{\partial r}+\Omega_{\theta_r}\frac{\partial}{r\,\partial\theta_r}+\Omega_{\phi_r}\frac{\partial}{r\sin\theta_r\,\partial\phi_r}, \qquad (6.236)$$

where (r,θ_r,ϕ_r) represents the coordinate system (see Fig. 6.12a), and the directional cosines may be derived by a transformation from those in Cartesian coordinates. They are given by

$$\begin{bmatrix}\Omega_r\\\Omega_{\theta_r}\\\Omega_{\phi_r}\end{bmatrix}=\begin{bmatrix}\sin\theta_r\cos\phi_r & \sin\theta_r\sin\phi_r & \cos\theta_r\\\cos\theta_r\cos\phi_r & \cos\theta_r\sin\phi_r & -\sin\theta_r\\-\sin\phi_r & \cos\phi_r & 0\end{bmatrix}\begin{bmatrix}\sin\theta\cos\phi\\\sin\theta\sin\phi\\\cos\theta\end{bmatrix}. \qquad (6.237)$$

Under the homogeneous conditions given in Eq. (6.232), the equation for the transfer of solar radiation in spherical coordinates may be written as

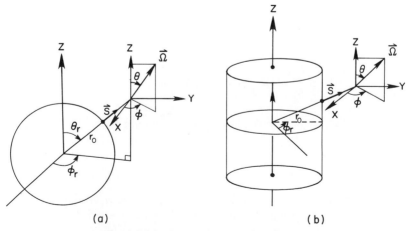

Fig. 6.12 Representations of a position vector **s** and a directional vector $\mathbf{\Omega}$ in (a) spherical coordinates and (b) cylinderical coordinates.

$$-\frac{1}{\beta_e}\Bigg\{[(1 - \mu^2)^{1/2}\sin\theta_r\cos(\phi - \phi_r) + \mu\cos\theta_r]\frac{\partial}{\partial r}$$

$$+ [(1 - \mu^2)^{1/2}\cos\theta_r\cos(\phi - \phi_r) - \mu\sin\theta_r]\frac{\partial}{r\,\partial\theta_r}$$

$$+ [(1 - \mu^2)^{1/2}\sin(\phi - \phi_r)]\frac{\partial}{r\sin\theta_r\,\partial\phi_r}\Bigg\}I(r, \theta_r, \phi_r; \mu, \phi)$$

$$= I(r, \theta_r, \phi_r; \mu, \phi) - \frac{\tilde{\omega}}{4\pi}\int_0^{2\pi}\int_{-1}^1 I(r, \theta_r, \phi_r; \mu, \phi)P(\mu, \phi; \mu', \phi')\,d\mu'\,d\phi'$$

$$- \frac{\tilde{\omega}}{4\pi}P(\mu, \phi; -\mu_0, \phi_0)\pi F_0\exp\{-\beta_e[(r_0^2 - r^2\sin\theta_r)^{1/2} - r\cos\theta_r]\},$$

$$(6.238)$$

where r_0 denotes the radius of the scattering medium.

In cylindrical coordinates, the operator $\mathbf{\Omega}\cdot\nabla$ may be written in the form

$$\mathbf{\Omega}\cdot\nabla = \Omega_r\frac{\partial}{\partial r} + \Omega_{\phi_r}\frac{\partial}{r\,\partial\phi_r} + \Omega_z\frac{\partial}{\partial z}, \qquad (6.239)$$

where (r, ϕ_r, z) represents the coordinate system (see Fig. 6.12b), and again the directional cosines may be derived by an appropriate transformation from those in Cartesian coordinates in the form

$$\begin{bmatrix}\Omega_r \\ \Omega_{\phi_r} \\ \Omega_z\end{bmatrix} = \begin{bmatrix}\cos\phi_r & \sin\phi_r & 0 \\ -\sin\phi_r & \cos\phi_r & 0 \\ 0 & 0 & 1\end{bmatrix}\begin{bmatrix}\sin\theta\cos\phi \\ \sin\theta\sin\phi \\ \cos\theta\end{bmatrix}. \qquad (6.240)$$

Subject to the homogeneous conditions specified in Eq. (6.232), the equation for the transfer of solar radiation in cylindrical coordinates is as

$$-\frac{1}{\beta_e}\Bigg[(1 - \mu^2)^{1/2}\cos(\phi - \phi_r)\frac{\partial}{\partial r} + (1 - \mu^2)^{1/2}\sin(\phi - \phi_r)\frac{\partial}{r\,\partial\phi_r}$$

$$+ \mu\frac{\partial}{\partial z}\Bigg]I(r, \phi_r, z; \mu, \phi) = I(r, \phi_r, z; \mu, \phi)$$

$$- \frac{\tilde{\omega}}{4\pi}\int_0^{2\pi}\int_{-1}^1 I(r, \phi_r, z; \mu, \phi)P(\mu, \phi; \mu', \phi')\,d\mu'\,d\phi'$$

$$- \frac{\tilde{\omega}}{4\pi}P(\mu, \phi; -\mu_0, \phi_0)\pi F_0\exp\{-\beta_e[r_0^2 - r^2\sin^2\phi_r)^{1/2} - r\cos\phi_r]\},$$

$$(6.241)$$

where r_0 is the radius of the cylindrical medium.

Equations (6.235), (6.238), and (6.241) are the basic equations describing the transfer of solar radiation in coordinate systems which may be applicable to clouds of finite dimensions. From satellite cloud pictures as well as our day-to-day experiences, we see that portions of clouds and cloud systems that cover the earth are either finite in extent or in forms of cloud bands. To what degree that the finiteness of clouds influences their reflection, transmission, and absorption properties is a question requiring further investigations. Also, the question concerning the effects of finite cumulus clouds on the heating and cooling in tropical atmospheres needs to be answered quantitatively. Perhaps, the basic equations given in this section may be of some use for an attempt to derive simplified solutions which may provide realistic flux distributions in cumulus cloudy atmospheres.

EXERCISES

6.1   Neglecting the ground reflection effect, compute and plot the reflected intensity (reflection) at the top of nonabsorbing molecular atmospheres whose optical depths are assumed to be 0.1 and 1 for $\mu_0 = 0.8$, using the single-scattering approximation. Compare the resulting calculations with those presented by Coulson *et al.* (1960, pp. 21 and 57), in whose work the the multiple-scattering effect was properly taken into account.

6.2   Derive an analytical expression for the diffuse reflection at the top of the atmosphere, utilizing the second-order scattering approximation (neglect the surface reflection). Carry out the analyses for $\mu \neq \mu'$ and $\mu = \mu'$.

6.3   Derive the two-stream solution for conservative scattering and calculate the reflection and transmission, assuming an asymmetry factor of 0.75 for optical depths of 0.25, 1, 4, and 16. Plot the results as functions of the cosine of the solar zenith angle μ_0 and compare with those computed from 16 discrete streams and the doubling method shown in Table 6.2.

6.4   Insert the expression I_1 in Eq. (6.60) into Eq. (6.59) to obtain

$$\frac{d^2 I_0}{d\tau^2} = k^2 I_0 - \eta e^{-\tau/\mu_0},$$

where $k = [3(1 - \tilde{\omega})(1 - \tilde{\omega}g)]^{1/2}$, representing the eigenvalue, and η is a certain constant associated with F_0, $\tilde{\omega}$, and g. This equation is referred to as the one-dimensional diffussion equation. Solve for I_0 in this equation and, subsequently, I_1 using the radiation boundary condition defined in Eq. (6.30).

6.5   Consider Eq. (6.15) and let

$$I(\tau, \mu) = \sum_{l=0}^{N} I_l(\tau) P_l(\mu),$$

where $P_l(\mu)$ denotes the Legendre polynomial. By utilizing the orthogonality property of the Legendre polynomials and the recurrence formula

$$\mu P_l(\mu) = \frac{l+1}{2l+1} P_{l+1} + \frac{l}{2l+1} P_{l-1},$$

show that Eq. (6.15) can be reduced to a set of first-order differential equations as

$$\frac{k}{2k-1} \frac{dI_{k-1}}{d\tau} + \frac{k+1}{2k+3} \frac{dI_{k+1}}{d\tau} = I_k \left(1 - \frac{\tilde{\omega}\tilde{\omega}_k}{2k+1}\right) - \frac{\tilde{\omega}\tilde{\omega}_k}{4} P_k(-\mu_0) F_0 e^{-\tau/\mu_0},$$

$$k = 0, 1, \ldots, N.$$

These simultaneous equations may be solved by properly setting up the boundary conditions. It is referred to as the spherical harmonics method for radiative transfer. We see that for the simple case of $k = 0, 1$, the preceding equation leads to Eqs. (6.59) and (6.60), respectively.

6.6   Formulate the transfer of infrared radiation in a scattering atmosphere having an isothermal temperature T in local thermodynamic equilibrium, assuming the intensity is azimuthally independent. By means of the discrete-ordinates method for radiative transfer, assuming isotropic scattering, show that the scattered intensity is given by

$$I(\tau, \mu_i) = \sum_{\alpha=-n}^{n} \frac{L_\alpha}{1 + \mu_i k_\alpha} e^{-k_\alpha \tau} + B_\nu(T),$$

where the L_α are unknown constants of proportionality, μ_i are the discrete streams, k_α the eigenvalues, and B_ν represents the Planck function.

6.7   A satellite radiometer measures the reflected solar radiation from a semi-infinite, isotropic-scattering atmosphere composed of particulates and gases near the vicinity of an absorption line whose line shape is given by the Lorentz profile and whose absorption coefficient can be written as

$$k_\nu = \frac{S}{\pi} \frac{\alpha}{(\nu - \nu_0)^2 + \alpha^2}.$$

Assuming that the particulates are nonabsorbing and that the scattering optical depth is equal to the gaseous absorption optical depth at the line center, calculate the reflected intensity (reflection function) as a function of wave number ν using the two-stream approximation. Do the problem by formulating (1) the single-scattering albedo as a function of ν, and (2) the reflected intensity in terms of the two-stream approximation.

6.8   For a semi-infinite, isotropic-scattering atmosphere, show that the planetary albedo

$$r(\mu_0) = 1 - H(\mu_0)\sqrt{1 - \tilde{\omega}},$$

and the spherical albedo

$$\bar{r} = 1 - 2\sqrt{1 - \tilde{\omega}} \int_0^1 H(\mu_0)\mu_0 \, d\mu_0.$$

Use the first approximation for the H function and assume single-scattering albedos of 0.4 and 0.8; compute the planetary albedo for μ_0 of 1 and 0.5 and the spherical albedo.

6.9   An optically thin layer $\Delta\tau$ is added to a finite atmosphere with an optical depth of τ_1 and all the possible transmissions of the incident beam due to the addition of the thin layer are depicted below. Formulate Eq. (6.159) using the principles of invariance discussed in Section 6.4.2. The method is also referred to as invariant imbedding. (Note that dotted lines represent direct transmission.)

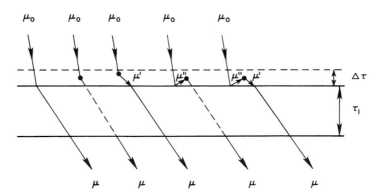

6.10   Consider a cloud layer having a total global transmission of $\bar{\gamma}$ and a global reflection (spherical albedo) of $\bar{r}$ above a Lambert surface with a surface albedo of r_s. Assuming no atmosphere between the cloud and surface, derive Eqs. (6.187)–(6.189) by means of geometrical ray-tracing for multiple reflections between the cloud and surface.

SUGGESTED REFERENCES

Ambartsumyan, V. A. (1958). *Theoretical Astrophysics*. Pergamon Press, New York. Chapters 33 and 34 contain fundamental discussions of the principles of invariance for a semi-infinite, isotropic atmosphere.

Chandrasekhar, S. (1950). *Radiative Transfer*. Dover, New York. Chapters 1, 3, 6, and 7 contain basic materials for the discrete-ordinates method for radiative transfer and the principles of invariance.

Coulson, K. L., Dave, J. V., and Sekera, Z. (1960). *Tables Related to Radiation Emerging from a Planetary Atmosphere with Rayleigh Scattering*. Univ. of California Press, Berkeley. The book contains comprehensive tables for the exact distribution and polarization of the reflected and transmitted light in plane-parallel Rayleigh scattering atmospheres.

Davison, B. (1957). *Neutron Transport Theory*. Oxford Univ. (Clarendon) Press, London and New York. Chapters 10–12 consist of advanced discussions on the sperical harmonics method for plane, spherical, and cylindrical geometries from the neutron-transport point of view.

Hansen, J. E., and Travis, L. D. (1974). Light scattering in planetary atmospheres. *Space Sci. Rev.* **16**, 527–610. This review paper gives brief summaries of various methods for radiative transfer and provides illustrative numerical examples.

Kourganoff, V. (1952). *Basic Methods in Transfer Problems*. Oxford Univ. (Clarendon) Press, London and New York. Chapter 3 contains some elementary discussions on the discrete-ordinates and spherical harmonics methods for radiative transfer.

Lenoble, J., Ed., (1977). *Standard Procedures to Compute Atmospheric Radiative Transfer in a Scattering Atmosphere*. Radiation Commission, International Association of Meteorology and Atmospheric Physics, published by National Center for Atmospheric Research, Boulder, Colorado. This monograph collects various analytical and computational methods for the solution of multiple scattering problems. It also provides comprehensive numerical comparisons for a number of computational methods.

Sobolev, V. V. (1975). *Light Scattering in Planetary Atmospheres*. Pergamon Press, New York. Chapters 2 and 3 contain discussions of the principles of invariances.

Chapter 7
APPLICATIONS OF RADIATIVE TRANSFER TO REMOTE SENSING OF THE ATMOSPHERE

7.1 INTRODUCTION

Electromagnetic waves interacting with a medium will leave a signature which may be used to identify the composition and structure of that medium. Remote sensing is contrary to in-situ measurements whereby specific observations are made within the medium. The basic principle associated with remote sensing involves the interpretation of radiometric measurements of electromagnetic radiation characterized by a specific spectral interval which is sensitive to some physical aspect of the medium. The physical principle of remote sensing is illustrated by the simple configuration in Fig. 7.1. Basically, an electromagnetic signal is recorded by a detector after it interacts with a target containing molecules and/or particulates. If T and S denote the target and signal, respectively, then we may write symbolically

$$S = F(T),$$

where F represents a function, not necessarily linear. The inverse of the above relation gives

$$T = F^{-1}(S),$$

where F^{-1} represents the inverse function of F.

The fundamental obstacle in all the inverse problems of remote sensing is the *uniqueness* of the solution. The nonuniqueness arises because the medium

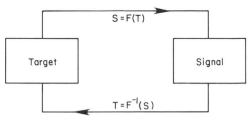

Fig. 7.1   Principle of remote sensing.

under investigation may be composed of a number of unknown parameters whose various physical combinations may lead to the same radiation signature. In addition to this physical problem, there are also mathematical problems associated with the existence and stability of the solution, and the manner in which the solution is constructed. There are two distinct principles involved in remote sensing leading to two classifications. These are the so-called *active* and *passive* remote sensing.

Active remote sensing employs a radiation source generated by artificial means such as lasers used in lidar or microwaves used in radar. The radiant energy source corresponding to a specific wavelength is sent to the atmosphere. Some of the energy is scattered back to the detector and recorded. From the recorded scattered energy, one analyzes the composition and structure of the atmosphere with which the radiant energy has interacted. Active remote sensing is normally concerned with the backscattering; i.e., the transmitter and detector are co-located.

Passive remote sensing utilizes the natural radiation sources of the sun or the earth–atmosphere system. For example, spectral solar radiation interacts with the cloud and leaves a scattered signature which may be used for its identification. Similarly, spectral thermal infrared or microwave radiation emitted from the earth–atmosphere system may be utilized to understand its thermodynamic state and composition. In reference to Fig. 1.1 regarding the electromagnetic spectrum, all wavelengths are possible from the emitting material. However, for atmospheric applications, the solar, infrared, and microwave spectra are most important.

The passive remote sensing principle leads to development of the global inference of atmospheric temperature, composition profiles, and radiative budget components from orbiting meterorological satellites. The first meteorological satellite experiment was an array of hemispheric sensors flown on the Explorer VII satellite launched in 1959 to measure the radiation balance of the earth–atmosphere system. Shortly after, a five-channel scanning radiometer was carried on board Tiros II. On the basis of the scanning radiometer, the general characteristics of sounding instrumentation were established for reserch and operational satellites during the last two decades. In April 1969, two spectrometers providing spatial infrared measurements for the determination of the vertical profiles of temperature, water vapor, and ozone were flown on the Nimbus III satellite. Also on board was another instrument for measurement of the reflected UV radiation which allowed the determination of the global ozone concentration. The launch of Nimbus 5 in December 1972, marked the first application of microwave techniques for the remote sensing of the atmospheric temperature and total water. These are the milestones of atmospheric sounding from orbiting meteorological satellites.

In this chapter, we first discuss the information content of the scattered solar radiation. Following a detailed description of the principle of temperature and gaseous profile determination from thermal infrared emission, the use of the microwave emission for atmospheric studies is introduced. These sections emphasize the remote-sensing principle for the inference of atmospheric temperature and composition from satellites. Radiation budget studies from satellites associated with radiation climatology is discussed in Chapter 8. The principle of active remote sensing utilizing the backscattered energy is then presented with emphasis on the propagation of laser energy in the atmosphere.

7.2 SCATTERED SUNLIGHT AS MEANS OF REMOTE SENSING

7.2.1 Transmitted Sunlight

7.2.1.1 *Total Ozone Determination* A classic example of utilizing the measured transmitted solar flux density as a means of inferring the composition information is the method proposed by Dobson in 1931 for the estimate of total ozone concentration from a ground-based instrument. Basically, this method uses the Beer–Bouguer–Lambert law (see Section 1.4.2) to describe the transfer of UV sunlight and neglects the effect of multiple scattering. Thus, the differential change of the incoming flux density F centered at wavelength λ at a given z in an atmosphere containing ozone, air molecules, and aerosols may be expressed by

$$dF_\lambda(z) = -F_\lambda(z)\,dz\sec\theta_0[k(\lambda)\rho_3(z) + \sigma_s^R(\lambda)N(z) + \sigma_e^M(\lambda)N_a(z)], \quad (7.1)$$

where θ_0 denotes the solar zenith angle, k the ozone mass absorption coefficient (g^{-1} cm^2), ρ_3 the ozone density, σ_s^R the Rayleigh scattering cross-section (cm^2, see Section 3.7.2), N the number of molecules (cm^{-3}), σ_e^M the Mie extinction (absorption plus scattering) cross section due to aerosols (see Section 5.4), and N_a the number of aerosols, all at height z.

To simplify Eq. (7.1), we make use of the optical depth parameter, τ^R, defined in Eq. (3.73) for Rayleigh molecules and τ^M for Mie aerosols, and define the total ozone concentration in the vertical column in the form $(g\ cm^{-2})$

$$\Omega = \int_0^\infty \rho_3(z)\,dz. \tag{7.2}$$

In addition, since ozone has a maximum concentration at about 22 km, we let the solar zenith angle in reference to this height account for the ozone slant path, and let $\theta_0 = Z$. Also for Rayleigh molecules, we use the air mass $m = \sec\theta_0$ to denote the slant path of air (see Section 2.4.1). In view of these simplifications and definitions, the solution of Eq. (7.1) may be written as

$$\ln\left[F_\lambda(0)/F_\lambda(\infty)\right] = -k(\lambda)\Omega\sec Z - \tau^R(\lambda)m - \tau^M(\lambda)m, \tag{7.3}$$

where $F_\lambda(\infty)$ and $F_\lambda(0)$ represent the incoming solar flux densities at the top and bottom of the atmosphere, respectively.

Having formulated the solar flux density attenuation, we then select a pair of wavelengths (λ_1, λ_2) in the Hartley–Huggins ozone absorption band described in Section 3.2. Assuming that the aerosol optical depths at λ_1 and λ_2 are about the same, then upon subtracting the equation for λ_2 from that for λ_1, and solving for Ω we obtain

$$\Omega = \frac{\ln[F_{\lambda1}(\infty)/F_{\lambda2}(\infty)] - \ln[F_{\lambda1}(0)/F_{\lambda2}(0)] - m[\tau^R(\lambda_1) - \tau^R(\lambda_2)]}{\sec Z[k(\lambda_1) - k(\lambda_2)]}. \tag{7.4}$$

In this equation, $[\tau^R(\lambda_1) - \tau^R(\lambda_2)]$ is computed from Rayleigh scattering theory, $[k(\lambda_1) - k(\lambda_2)]$ is determined from laboratory measurements, and $[F_{\lambda1}(\infty)/F_{\lambda2}(\infty)]$ can be obtained once and for all by means of the long method introduced in Section 2.4, by making a series of measurements over a range of zenith angles, and then by extrapolating. The zenith angles θ_0 and Z are functions of the latitude, time of year, and time of day, as discussed in Section 2.5. Thus, determination of $\ln[F_{\lambda1}(0)/F_{\lambda2}(0)]$ from the ground flux density measurements yields a total ozone value. The standard instrument for measuring the total ozone concentration is called the *Dobson spectrometer* (Dobson, 1957), which is in operational use at 80 ground stations around the world.

7.2.1.2   Turbidity and Precipitable Water Determination

The atmosphere constantly contains suspended particles ranging in size from about $10^{-3}\ \mu m$ to about 20 μm, called aerosols or referred to as pollution. These aerosols are known to be produced directly by man's activities, by natural processes having no connection with man's activities, and by natural processes which may have been intensified by man. Natural aerosols include volcanic dust occurring in the stratosphere, sea spray and its particulate products, wind-generated dust, smoke from natural forest fires, and small particles produced by chemical reactions of natural gases. Major man-made aerosols include particles directly emitted during combustion and particles formed from gases emitted during combustion.

Aerosols not only scatter, but also significantly absorb the incoming solar radiation covering the entire spectrum. Since aerosols are globally distributed, their effects on the heat balance of the earth–atmosphere system can be very significant. In recent years, there has been growing speculation and increasing concern that aerosols in general and man-made pollution in particular might be an important factor causing the worldwide disturbances of weather and climate. As a result of this concern, observations of the concentrations and physical and chemical characteristics of aerosols in the atmosphere as functions of time and space, and studies of their optical properties in the solar spectrum continuously have been carried out with considerable effort.

Observational methods to determine the dust loading of the atmosphere were developed during the 1920s by Linke and Ångström. In essence, the aerosol total optical depth, sometimes referred to as *turbidity*, is derived from spectrally dependent direct flux density observations on the ground. Wavelengths in the visible normally are employed where there is practically no water vapor absorption, and ozone absorption is at a minimum. The simple Beer–Bouguer–Lambert law is again used, and from Eq. (7.3) we obtain

$$F_\lambda(0)/F_\lambda(\infty) = \exp\{-[\tau^R(\lambda) + \tau^M(\lambda)]m\}, \tag{7.5}$$

where again m ($=\sec\theta_0$) is the air mass relative to the vertical direction, and $F_\lambda(0)$ is the observed direct solar flux density at wavelength λ. As before, $F_\lambda(\infty)$ can be determined once and for all from the long method mentioned previously. Thus, by subtracting the Rayleigh optical depth, which can be theoretically computed, the aerosol optical depth may be inferred. Correction of the small ozone absorption also may be carried out by including an additional ozone optical depth term in Eq. (7.5), which also may be theoretically estimated.

The total aerosol optical depth due to extinction may be written as

$$\tau^M(\lambda) = \int_0^\infty \beta_e(\lambda, z)\, dz. \tag{7.6}$$

Assuming that the aerosol size distribution is described by $dn(a)/da(\mathrm{cm}^{-4})$, where $dn(a)$ expresses the number of particles with radii between a and $a + da$, the extinction coefficient (cm^{-1}) is then given by Eq. (5.115) in the form

$$\beta_e(\lambda, z) = \int_{a_1}^{a_2} \sigma_e(a, \lambda) \frac{dn(a)}{da}\, da, \tag{7.7}$$

where σ_e represents the extinction cross section (cm^2) for an individual particle. The size distribution of aerosols in the atmosphere has been a subject of extensive research in the last two decades or so. To a good approximation, it may be described by the so-called *Junge distribution* in the form

$$\frac{dn(a)}{da} = C(z)a^{-(v^*+1)}, \tag{7.8}$$

where C is a scaling factor directly proportional to the aerosol concentration, and therefore is a function of height z in the atmosphere, and v^* represents a shaping constant which normally is found to lie in the range $2 \le v^* < 4$. The distribution is typically assumed to apply for sizes ranging from about $a_1 = 0.01$ μm to about $a_2 = 10$ μm.

Using the Junge size distribution, it can be shown that the aerosol optical depth is simply

$$\tau^M(\lambda) = k\lambda^{-v^*+2}, \tag{7.9}$$

where k is a certain constant (see Exercise 7.1). When $v^* = 3.3$, k is known as the *Ångström turbidity coefficient*. Clearly, if the turbidity has been measured at two wavelengths, the shaping constant may be determined, since k is a constant.

The determination of the total vertical water vapor amount using direct transmitted solar radiation observation was pioneered by Fowle in 1917. Recently, Volz (1974) proposed a multispectral instrument called a *sunphotometer* for measurements of precipitable water. Basically, a pair of wavelengths in the near infrared is selected. One wavelength is in the window of water vapor ($\lambda_1 = 0.88$ μm), while the other is in the $\rho\sigma\tau$ band of water vapor ($\lambda_2 = 0.94$ μm). It is assumed that the extinction by aerosols and molecules at these two wavelengths is about the same. Moreover, the square root approximation for the random model (see Exercise 4.5) is utilized for the spectral water vapor transmissivity.

Thus, in analogy to Eq. (7.5) the transmissivities for λ_1 and λ_2 are, respectively,

$$F_{\lambda 1}(0)/F_{\lambda 1}(\infty) = \exp\{-[\tau^M(\lambda_1) + \tau^R(\lambda_1)]m\}, \tag{7.10}$$

$$F_{\lambda 2}(0)/F_{\lambda 2}(\infty) = \exp\{-[\tau^M(\lambda_2) + \tau^R(\lambda_2)]m - K\sqrt{um}\}, \tag{7.11}$$

where $K = \sqrt{\pi S_0 \alpha}/\delta$, S_0 the mean line intensity, α the mean line half width, and δ the mean line spacing. The path length (or equivalently the precipitable water) is represented by u in units of g cm^{-2} (or cm atm.). Dividing Eq. (7.11) by Eq. (7.10), and rearranging the terms, we find

$$u(PW) \approx (K/m)[\ln(q_0/q)]^2, \tag{7.12}$$

where $q_0 = F_{\lambda 2}(0)/F_{\lambda 1}(0)$, $q = F_{\lambda 2}(\infty)/F_{\lambda 1}(\infty)$. The constant K is to be derived by comparison with sounding data, and again q_0 may be obtained by the long method previously mentioned.

7.2.2   Reflected Sunlight

7.2.2.1   Total Ozone Estimate from Reflected Intensity   The basic principle involved in the estimate of ozone concentration utilizing the reflected sunlight is to select a pair of wavelengths in the Hartley–Huggins ozone absorption band. The principle for the selection of a pair of wavelengths is the same as for the Dobson ozone spectrometer described previously. Wavelengths near the long-wavelength end of the band at which absorption is relatively weak are chosen so that most of the photons reaching the satellite instrument have passed through the ozone layer and backscattered from within the troposphere. Absorption for one of these wavelengths is stronger than for the other. The two wavelengths are separated by about 200 Å so that the scattering effect is about the same at each wavelength, whereas the relative attenuation for the pair is sensitive mostly to total ozone. A pair such as (3125, 3312 Å), for example, has been employed in the Nimbus IV satellite experiment.

The backscattering radiance in the ozone band at the point of satellite with a nadir-looking instrument depends on the attenuation of the direct solar flux through the ozone layer, the reflecting power of the atmosphere and the associated surface, and the attenuation of the diffusely reflected photons to the point of satellite. If Z denotes the solar zenith angle at the level of maximum ozone concentration (about 22 km) at the subsatellite point, then the total attenuation path of the backscattered photons through the ozone layer is proportional to $1 + \sec Z$. Let F_0 and I be the incident solar irradiance and backscattered radiance at the top of the atmosphere, respectively. We define

$$\tilde{N}(\lambda_1, \lambda_2) = \log_{10}(F_0/\tilde{I})_{\lambda 1} - \log_{10}(F_0/\tilde{I})_{\lambda 2}. \qquad (7.13)$$

The inference of total ozone concentration is made by comparing the observed $\tilde{N}$ with values precomputed for a series of different standard ozone profiles through the interpolation method.

The computational method for the transfer of solar radiation in a scattering and absorbing atmosphere usually has been developed for the atmosphere alone without considering the surface reflection effect. Assuming that the reflecting surface follows Lambert's law, then according to Eq. (6.182), the backscattered radiance at the top of the atmosphere including the surface reflection contribution may be expressed by

$$I(\Omega, \mu_0, r_s) = I(\Omega, \mu_0, 0) + \frac{r_s T(\Omega, \mu_0)}{1 - r_s \bar{r}(\Omega)}, \qquad (7.14)$$

where $T(\Omega, \mu_0) = \mu_0 F_0 \gamma(1) \gamma(\mu_0)$, $\mu_0 = \cos Z$, and Ω denotes the total ozone concentration. In this equation, all the relevant parameters in the radiative terms are included in the parentheses. Note that since the instrument is looking at the nadir direction, the azimuthal dependence may be neglected.

Figure 7.2 illustrates the computed values of N as a function of the slant path $s = 1 + 1/\mu_0$ for a pair wavelength of (3125, 3300 Å). The computations are based on the method of successive orders of scattering (Dave and Mateer, 1967). Surface albedos of 0.0 and 0.8 are used and atmospheric pressure, which is related to Rayleigh scattering, is set to be 1000 mb. For a given albedo, it is seen that values of N/S reduce as ozone concentration increases. Uncertainty in the albedo value would introduce extreme difficulty in the interpretation of the results. Consequently, from the satellite sounding point of view, it is imperative that information of the surface albedo first be determined.

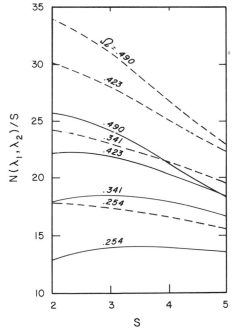

Fig. 7.2 Computed relationship between N/S and S for the various ozone distributions for $(\lambda_1, \lambda_2) = (3124, 3300$ Å$)$, $P = 1000$ mb, $r_s = 0.0$ (solid curves), and $r_s = 0.8$ (dashed curves) (after Dave and Mateer, 1967).

The three basic procedures for the estimate of total ozone concentration from observed $\tilde{N}$ are: (1) A set of tables containing the computed quantities $I(\Omega, \mu_0, 0)$, $T(\Omega, \mu_0)$, and $\bar{r}(\Omega)$ for different values of μ_0 and Ω are prepared. (2) The effective surface albedo is determined by utilizing the photometer measurement at a wavelength outside the ozone absorption band, say λ_3

(3800 Å), for example. Thus, the ozone dependence drops out of all the terms in Eq. (7.14), and a measurement of $\tilde{I}_3(\mu_0, r_s)$ permits the direct calculation of the surface albedo with the formula

$$r_s(\lambda_3) = \frac{\tilde{I}_3(\mu_0, r_s) - I_3(\mu_0, 0)}{T(\mu_0) - \bar{r}[\tilde{I}_3(\mu_0, r_s) - I_3(\mu_0, 0)]}. \tag{7.15}$$

The assumption is made that r_s is independent of wavelength so that it can be used for the pair wavelengths (λ_1, λ_2). Empirical adjustment may also be performed to correct $r_s(\lambda_3)$ to the pair wavelengths. (3) With the surface albedo known, computations are then carried out to generate $N(\lambda_1, \lambda_2)$ versus total ozone concentration Ω. Best estimates of Ω from $\tilde{N}(\lambda_1, \lambda_2)$, computed from observed radiances, may be made by an optimized search method.

The matching and search method has been used by Mateer *et al.* (1971) to estimate total ozone concentration from the Nimbus IV satellite measurements of backscattered radiances at wavelengths between 3100 and 3400 Å by means of a double monochromator, and at 3800 Å. Recently, Dave (1978) discussed the effect of atmospheric aerosols on the estimate of total ozone concentration. Influence of aerosols and clouds still appears to be the major problem in the ozone sounding study.

7.2.2.2 *Cloud Properties Inferred from Reflected Polarization* Clouds regularly cover about 50% of the planet Earth and are the most important regulators of the radiation budget of the earth–atmosphere. The transfer of radiation through cloud layers depends on the particle phase, concentration, size distribution, and the cloud thickness and shape. Information about cloud compositions and structure is of vital importance to the understanding of the radiation balance and energetics of the earth–atmosphere system.

Because of the number of variables embodied and the associated problem of multiple scattering by particles, the determination of changing cloud variables from remote sensing is very difficult. In recent years, owing to the availability of high-speed digital computers, the solution of the radiative transfer problem has been so developed that potential of estimating cloud compositions by means of reflected sunlight has been illustrated. Analogous to the total ozone concentration inference, the method involved is a direct ad hoc matching between the observed and computed intensity and/or polarization. In this section, we describe the information content of the reflected polarization of sunlight from clouds composed of particles.

In Section 5.6, some single-scattering characteristics of polydispersed spherical particles were presented and discussed; specifically, Fig. 5.11 depicts the scattering phase function and linear polarization patterns. We note that the scattering phase function is directly proportional to the scattered intensity if only single scattering is considered [see Eq. (6.31)]. Moreover, as noted in that section, the single scattered intensity and linear polarization are characterized by the strong forward peak along with rainbow and glory features. For clouds, which are normally optically thick, would the multiple scattered intensity and polarization preserve these notable features?

Shown in Fig. 7.3 are the reflected intensity and degree of linear polarization defined in Eq. (3.75a) for a plane-parallel cloud whose optical depth varies from 0.25 to 120 (essentially semi-infinite) at a wavelength of 1.2 μm when the sun is overhead. The computations (Hansen, 1971) were based on the adding method discussed in Section 6.5 in which the azimuthal dependence and four Stokes parameters associated with polarization were taken fully into account. The spherical particle size distribution used in the calculations has the form

$$n(a) = \text{const } a^{(1 - 3\beta)/\beta} e^{-a/(\alpha\beta)}, \tag{7.16}$$

where α ($= 6 \mu$m) represents the mean effective radius, and β ($= \frac{1}{9}$) is the effective dispersion of this size distribution. The abscissa in this diagram is such that $0°$ zenith angle corresponds to a $180°$ scattering angle.

In regard to the intensity pattern, the single-scattering features shown in Fig. 5.11 in the angular distribution of the scattering light are practically

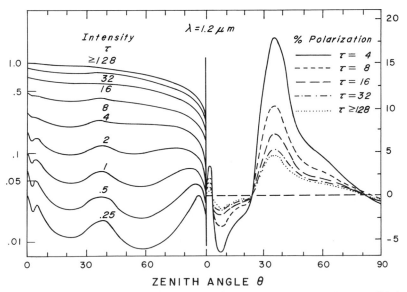

Fig. 7.3 Intensity and percent linear polarization of sunlight reflected by a plane-parallel cloud with the sun overhead ($\theta_0 = 0°$) as a function of the zenith angle. The wavelength is 1.2 μm, and results are shown for several optical thicknesses (after Hansen, 1971).

lost. The rainbow and glory features disappear completely and quickly as the optical depth increases. We note that as $\mu \to 0$ ($\theta \to 90°$), intensity decreases because of the multiple scattering effect. The decrease is referred to as *limb darkening*. The linear polarization pattern in a linear scale, however, retains the major feature, even though the polarization is largely reduced with increasing optical thickness. It should be noted that since polarization is derived from the ratio of the intensities, accuracies to within a few tenths of a percent can be obtained from observed data.

In view of the foregoing findings, it seems that the intensity measurement may yield cloud optical depth information, while polarization gives an additional dimension on the particle phase and size characteristics. The latter conclusion is based on the fact that nonspherical particles do not generate rainbow and glory as demonstrated in Section 5.7, and that the magnitude of polarization depends significantly on the sizes of particles as shown in Fig. 5.11.

The problem encountered in the direct-fitting approach is that there are several unknown parameters which are interrelated so far as the optical properties of particles are concerned. The value of each parameter must be varied in order to yield the best fit of the computed results with the observed values. Besides the problem of mathematical and physical uniqueness of the resulting computation, there is no clear-cut answer as to which values for these variables may fit the observed data best. Thus, the determination of the unknown cloud parameters would depend on intuition and ad hoc fitting and interpretation. As a result, mapping of particle characteristics over both extended space and time scales appears to be difficult.

Perhaps the most intriguing results in connection with the use of polarization data for the determination of particle sizes and optical characteristics have been found in the study of the cloud deck of Venus. Venus is the nearest, yet the most mysterious planet, as it is surrounded by a veil of clouds. Polarization observations of Venus date back to 1929 by the French astronomer Lyot using visible light. In recent years, more extensive observations have been made for wavelengths in the near infrared. Meanwhile, the advanced development of multiple scattering programs, also in recent years, has prompted the quantitative interpretation and analysis of the observed polarization data.

Hansen and Arking (1971) used the adding method for multiple scattering, and calculated the reflected polarization of sunlight from the Venus disk. Comparisons with observations revealed that the cloud particles on Venus are spherical with a refractive index of about 1.45 and a mean particle radius about 1 μm. Kattawar *et al.* (1971) made multiple scattering calculations for spherical particles employing the Monte Carlo method for a spherical atmosphere and derived a refractive index $1.45 < m_r < 1.60$. More recently,

Hansen and Hovenier (1974) reported an extensive investigation of the particle shape, size, and refractive index of the Venus cloud deck by comparing the observed linear polarization with comprehensive multiple scattering computations including Mie particles and Rayleigh molecules. They concluded that the Venus cloud layer was composed of spherical particles having a mean radius of about 1.05 μm and an effective dispersion of 0.07. The refractive index of the particles is about 1.44 at a wavelength of 0.55 μm with a normal dispersion.

Shown in Fig. 7.4 are observations and theoretical computations of the linear polarization of visible sunlight reflected by Venus. After varying α and β and m_r, the best fit to the observed data is given by the heavy curve. The maximum at the phase angle about 20° (scattering angle 160°) is the primary rainbow, the product of light rays undergoing one internal reflection, which

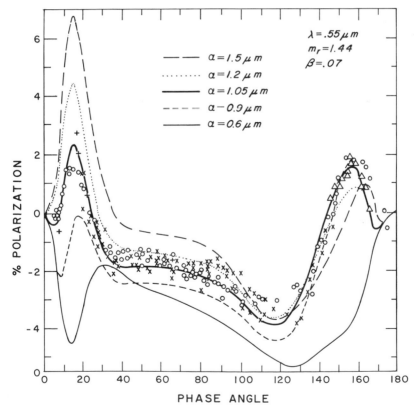

Fig. 7.4   Observations (0, ×, +, Δ) of the polarization of sunlight reflected by Venus as a function of the phase angle (180° − Θ) in the visual wavelength region and theoretical computations for $\lambda = 0.55$ μm (after Hansen and Hovenier, 1974).

nonspherical particles do not show as illustrated in Section 5.7. Consequently, the obvious conclusion is that the particles have to be largely spherical in order for the rainbow feature to be produced. Also, note that the maximum at about 155° is the feature of anomalous diffraction. This figure serves as a good illustration of the information content in the polarization data.

In view of this significant finding, photometric interpretations also may be carried out to understand the physical, optical, and chemical properties of the clouds and/or hazes that cover other planets. Recent Mariner spacecraft studies of Mars reveal that clouds of H_2O ice as well as possible CO_2 ice haze are sometimes present in the Martian atmosphere. Moreover, understanding of the likelihood of NH_3 clouds in the Jovian atmosphere and the nature of Saturn's rings would require further studies. Photometric and scattering techniques undoubtedly will provide significant data for the understanding of the physical and chemical composition of the clouds and/or haze on these planets. However, particles that occur in Mars, Jupiter, and Saturn are likely to be nonspherical. Owing to the nonsphericity and the associated orientation problem, reliable single scattering information has not been presented in a form suitable for multiple scattering investigations. Also, transfer of radiation through oriented nonspherical particles has not been theoretically formulated and understood, and it is an area that definitely requires further theoretical as well as observational studies.

7.3   INFRARED SENSING FROM SATELLITES

7.3.1   Upwelling Radiance at the Top of the Atmosphere

Assume that the satellite instrument observes in a narrow cone in the local vertical so that everywhere within the cone the cosine of the emergent angle $\mu \approx 1$, which is customarily called upwelling direction. From Eq. (4.3), the monochromatic upwelling radiance in a clear atmosphere is given by

$$I_\nu(\tau) = I_\nu(\tau_1)e^{-(\tau_1-\tau)} + \int_\tau^{\tau_1} B_\nu[T(\tau)]e^{-(\tau'-\tau)}\,d\tau'. \tag{7.17}$$

But

$$d[e^{-(\tau'-\tau)}] = -e^{-(\tau'-\tau)}\,d\tau'. \tag{7.18}$$

We have then

$$I_\nu(\tau) = I_\nu(\tau_1)e^{-(\tau_1-\tau)} - \int_\tau^{\tau_1} B_\nu[T(\tau)]\,d[e^{-(\tau'-\tau)}]. \tag{7.19}$$

The normal optical depth is defined by

$$\tau = \int_z^\infty k_\nu(z')\rho(z')\,dz', \tag{7.20}$$

where k_ν is the absorption coefficient (cm^2 g^{-1}), and ρ the density of the absorbing gases. Furthermore, the monochromatic transmission function (or transmittance) discussed in Section 4.4 may be expressed by

$$\mathscr{T}_\nu(z) = e^{-\tau} = \exp\left[-\int_z^\infty k_\nu(z')\rho(z')\,dz'\right], \qquad (7.21)$$

where the transmission function is expressed in reference to the top of the atmosphere. At the top of the atmosphere $z \to \infty$ and $\tau \to 0$. Also, when $z = 0$, $\tau = \tau_1$. Thus, the upwelling radiance at the top of the atmosphere in z coordinate may be written in the form

$$I_\nu(\infty) = I_\nu(0)\mathscr{T}_\nu(0) + \int_{z=0}^{z=\infty} B_\nu[T(z)]\frac{\partial \mathscr{T}_\nu(z)}{\partial z}\,dz, \qquad (7.22)$$

where $\partial \mathscr{T}_\nu(z)/\partial z$ is called the *weighting function* which, when multiplied by the Planck function, yields the upwelling radiance contribution from a given height z. $I_\nu(0)$ represents the upwelling radiance at the surface and is given by $\varepsilon_\nu B_\nu(T_s)$, where T_s is the surface temperature, and ε_ν the monochromatic emissivity of the surface. For all practical purposes, the emissivity ε_ν from the earth's surface can be taken as unity in the infrared region.

Sometimes it is convenient to express Eq. (7.22) in pressure coordinates. Based on the hydrostatic equation and the definition of mixing ratio $q = \rho/\rho_a$, where ρ and ρ_a are the density of gas and air, respectively, we have

$$\rho\,dz = -(q/g)\,dp. \qquad (7.23)$$

Thus, the monochromatic transmittance in pressure coordinate is

$$\mathscr{T}_\nu(p) = \exp\left[-\frac{1}{g}\int_0^p k_\nu(p')q(p')\,dp'\right], \qquad (7.24)$$

and the upwelling radiance may be expressed by

$$I_\nu(0) = B_\nu(T_s)\mathscr{T}_\nu(p_s) + \int_{p_s}^0 B_\nu[T(p)]\frac{\partial \mathscr{T}_\nu(p)}{\partial p}\,dp, \qquad (7.25)$$

where p_s is the surface pressure.

Equation (7.25) expresses the upwelling radiance at a monochromatic wave number. However, an instrument can distinguish only a finite band width $\phi(\bar{\nu}, \nu)$, where ϕ and $\bar{\nu}$ denote the instrumental response (or slit) function and the mean wave number of the band width, respectively. The measured radiance from a spectrometer over a wave number interval (ν_1, ν_2) in the normalized form is then given by

$$I_{\bar{\nu}}(0) = \int_{\nu_1}^{\nu_2} \phi(\bar{\nu}, \nu)I_\nu(0)\,d\nu \Big/ \int_{\nu_1}^{\nu_2} \phi(\bar{\nu}, \nu)\,d\nu. \qquad (7.26)$$

Upon carrying out the wave number integration of Eq. (7.26), we obtain

$$
I_{\bar{v}}(0) = \frac{1}{\int_{v_1}^{v_2} \phi(\bar{v}, v)\, dv} \left[\int_{v_1}^{v_2} \phi(\bar{v}, v) B_v(T_s) \mathcal{T}_v(p_s)\, dv \right.
$$

$$
\left. + \int_{v_1}^{v_2} \phi(\bar{v}, v) \int_{p_s}^{0} B_v[T(p)] \frac{\partial \mathcal{T}_v(p)}{\partial p}\, dp\, dv \right]. \tag{7.27}
$$

If the spectral interval (v_1, v_2) is small enough that the variation of $B_v(T)$ with respect to v is insignificant so that its value may be replaced by $B_{\bar{v}}(T)$ to a good approximation, then Eq. (7.27) becomes

$$
I_{\bar{v}}(0) = B_{\bar{v}}(T_s)\mathcal{T}_{\bar{v}}(p_s) + \int_{p_s}^{0} B_{\bar{v}}[T(p)] \frac{\partial \mathcal{T}_{\bar{v}}(p)}{\partial p}\, dp, \tag{7.28}
$$

where the spectral transmittance when the instrumental response function is taken into account is defined by

$$
\mathcal{T}_{\bar{v}}(p) = \int_{v_1}^{v_2} \phi(\bar{v}, v)\mathcal{T}_v(p)\, dv \Big/ \int_{v_1}^{v_2} \phi(\bar{v}, v)\, dv, \tag{7.29}
$$

and the spectral weighting function is given by

$$
\frac{\partial \mathcal{T}_{\bar{v}}(p)}{\partial p} = \int_{v_1}^{v_2} \phi(\bar{v}, v) \frac{\partial \mathcal{T}_v(p)}{\partial p}\, dv \Big/ \int_{v_1}^{v_2} \phi(\bar{v}, v)\, dv. \tag{7.30}
$$

Note that if $\phi(\bar{v}, v) = 1$, Eq. (7.29) reduces to Eq. (4.16).

The fundamental principle of atmospheric sounding from orbiting meteorological satellites utilizing the thermal infrared emission is based on the solution of the radiative transfer equation described by Eq. (7.28). In this equation, the upwelling radiance arises from the product of the Planck function, the spectral transmittance, and the weighting function. The Planck function consists of temperature information, while the transmittance is associated with the absorption coefficient and density profile of the relevant absorbing gases. Obviously, the observed radiance contains the temperature and gaseous profiles of the atmosphere, and therefore, the information content of the observed radiance from satellites must be physically related to the temperature field and absorbing gaseous concentration.

Perhaps at this point, it is appropriate to examine the characteristics of the infrared spectrum at the top of the atmosphere illustrated in Fig. 4.1. There are four regions over which water vapor, ozone, and carbon dioxide exhibit a significant absorption spectrum. Carbon dioxide absorbs infrared radiation in the 15 μm band from about 600 to 800 cm^{-1}. Not shown in that figure is another carbon dioxide band in the 4.3 μm region. Absorption due to ozone is largely confined in the 9.6 μm band. Water vapor exhibits absorption lines over essentially the entire infrared spectrum, but the most significant absorption lies in the vibrational–rotational band at 6.7 μm, and in the pure rotational band (< 500 cm^{-1}). From about 800 to 1200 cm^{-1} (atmospheric window), absorption of atmospheric gases shows a minimum (except the 9.6 μm ozone band), and therefore, the atmosphere is relatively transparent in this region. It should be noted that overlapping of CO_2, O_3, and H_2O absorption is relatively insignificant.

Returning to Eq. (7.28), it is clear that if observations are taken in the window region, the upwelling radiance may be approximated by

$$I_{\bar{\nu}} \approx B_{\bar{\nu}}(T_s)(1 - \alpha_1) + \alpha_2, \tag{7.31}$$

where α_1 and α_2 are small correction terms, which, if empirically determined, will give an estimate of the surface temperature from the observed radiance.

The mixing ratio of CO_2 is fairly uniform as a function of time and space in the atmosphere. Moreover, the detailed absorption characteristics of CO_2 in the infrared region are well understood, and its absorption parameters, i.e., half width, line strength, and line position, are known rather accurately. Consequently, the spectral transmittance and weighting functions for a given level may be calculated once the spectral interval and the instrumental response function have been given. To see the atmospheric temperature profile information we rewrite Eq. (7.28) in the form

$$I_{\bar{\nu}} - B_{\bar{\nu}}(T_s)\mathcal{T}_{\bar{\nu}}(p_s) = \int_{p_s}^{0} B_{\bar{\nu}}[T(p)] \frac{\partial \mathcal{T}_{\bar{\nu}}(p)}{\partial p} \, dp. \tag{7.32}$$

It is apparent that measurements of the upwelling radiance in the CO_2 absorption band contain the information of temperature values in the interval $(p_s, 0)$, once the surface temperature has been determined. However, the information content of the temperature is under the integral operator which leads to an ill-conditioned mathematical problem. In the next section, we discuss in detail this problem and a number of methods for the recovery of the temperature profile from a set of radiance observations in the CO_2 band.

Finally, to understand the information content of gaseous profile from the solution of the radiative transfer equation, we perform integration by parts on the integral term in Eq. (7.32) to yield

$$I_{\bar{\nu}} - B_{\bar{\nu}}[T(0)] = \int_{p_s}^{0} \mathcal{T}_{\bar{\nu}}(p) \frac{\partial B_{\bar{\nu}}(p)}{\partial p} \, dp. \tag{7.33}$$

Now, if measurements are made in the H_2O or O_3 spectral regions, and if temperature values are known, the transmittance profile may be inferred just as the temperature profile may be recovered when the spectral transmittance is given. To relate the gaseous concentration profile to the spectral transmittance, we refer to Eqs. (7.29) and (7.21). There we see that the density values are hidden in the exponent of an integral which is further complicated by the spectral integration over the response function. Because of these complications, retrieval of the gaseous density profile is made very difficult, and no clear-cut mathematical analyses may be followed in the inverse of the density values. Therefore, in the next section we focus our attention on the temperature inversion problem.

7.3.2 Temperature Profile Inversion Problem

Inference of atmospheric temperature profile from satellite observations of thermal infrared emission was first suggested by King (1956). In his pioneering paper, King pointed out that the angular radiance (intensity) distribution is the Laplace transform of the Planck intensity distribution as a function of the optical depth, and illustrated the feasibility of deriving the temperature profile from the satellite intensity scan measurements.

Kaplan (1959) advanced the sounding concepts by demonstrating that vertical resolution of the temperature field could be inferred from the spectral distribution of atmospheric emission. Kaplan pointed out that observations in the wings of a spectral band sense deeper into the atmosphere, whereas observations in the band center see only the very top layer of the atmosphere since the radiation mean free path is small. Thus, by properly selecting a set of different sounding wave numbers, the observed radiances could be used to make an interpretation leading to the vertical temperature distribution in the atmosphere.

In order for atmospheric temperatures to be determined by measurements of thermal emission, the source of emission must be a relatively abundant gas of known and uniform distribution. Otherwise, the uncertainty in the abundance of the gas will make ambiguous the determination of temperature from the measurements. There are two gases in the earth–atmosphere which have uniform abundance for altitudes below about 100 km, and which also show emission bands in the spectral regions that are convenient for measurement. As discussed in Section 4.2, carbon dioxide, a minor constituent with a relative volume abundance of 0.003, has infrared vibrational–rotational bands. In addition, oxygen, a major constituent with a relative volume abundance of 0.21, also satisfies the requirement of a uniform mixing ratio, and has a microwave spin-rotational band. The microwave spectrum will be discussed in the next section.

Shown in Fig. 7.5 is a spectrum of outgoing radiance in terms of the blackbody temperature at the vicinity of the 15 μm band observed by IRIS (Infrared Interferometer and Spectrometer) on the Nimbus IV satellite. The

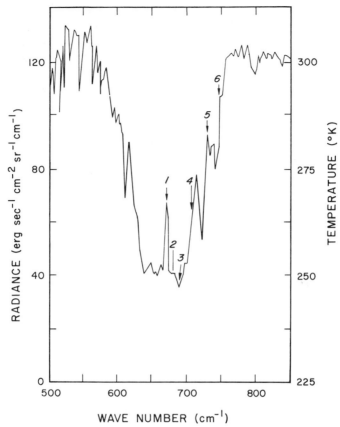

Fig. 7.5   Outgoing radiance in terms of blackbody temperature in the vicinity of 15 μm CO_2 band observed by the IRIS on Nimbus IV. The arrows denote the spectral regions sampled by the VTPR instrument (see Fig. 4.1).

equivalent blackbody temperature generally decreases as the center of the band is approached. This decrease is associated with the decrease of tropospheric temperature with altitude. Near about 690 cm^{-1}, the temperature shows a minimum which is related to the colder tropopause. Decreasing the wave number beyond 690 cm^{-1}, however, increases the temperature. This is due to the increase of the temperature in the stratosphere, since the observations near the band center see only the very top layers of the atmosphere. On the basis of the sounding principle already discussed, we could select a set of sounding wave numbers such that a temperature profile in the troposphere and lower stratosphere could be largely covered. The arrows in Fig. 7.5 indicate an example of such a selection.

Having presented the physical discussions, we now return to the basic equation (7.32). Clearly, the temperature of the underlying surface has to be determined first if the surface emission represents a significant contribution to the observed radiances, which is normally true in the wing regions. For simplicity of discussion and mathematical analyses, however, we shall drop the surface contribution term, i.e., assuming $\mathcal{T}_{\bar{v}}(p_s) = 0$, and simply write

$$I_{\bar{v}} \cong \int_{p_s}^{0} B_{\bar{v}}[T(p)] \frac{\partial \mathcal{T}_{\bar{v}}(p)}{\partial p} \, dp. \tag{7.34}$$

Upon knowing the radiances from a set of wave numbers and the associated transmittances, the fundamental problem encountered is, how to solve for the function $B_{\bar{v}}[T(p)]$.

We note that because there is a multiplicity of wave numbers at which the observations are made, the Planck function differs from one equation to another depending on the wave number. Thus, it becomes vitally important for the direct inversion problem to eliminate the wave number dependence in this function. In the vicinity of the 15 μm CO_2 band, it is sufficient to approximate the Planck function in a linear form as

$$B_{\bar{v}}[T(p)] = c_{\bar{v}} B_{\bar{v}_r}[T(p)] + d_{\bar{v}}, \tag{7.35}$$

where $\bar{v}_r$ denotes a fixed reference wave number, and $c_{\bar{v}}$ and $d_{\bar{v}}$ are empirically derived constants. Substituting Eq. (7.35) into Eq. (7.34) and assuming again that $\mathcal{T}_{\bar{v}}(p_s) = 0$, we obtain

$$g(\bar{v}) = \int_{p_s}^{0} f(p) K(\bar{v}, p) \, dp, \tag{7.36}$$

where we let

$$g(\bar{v}) = \frac{I_{\bar{v}} - d_{\bar{v}}}{c_{\bar{v}}}, \qquad f(p) = B_{\bar{v}_r}[T(p)], \qquad K(\bar{v}, p) = \frac{\partial \mathcal{T}_{\bar{v}}(p)}{\partial p}.$$

Equation (7.36) is a well-known *Fredholm equation of the first kind*. $K(\bar{v}, p)$, the weighting function, is the kernel, and $f(p)$ is the function to be recovered from a set of $g(\bar{v}_i)$, $i = 1, 2, \ldots, M$, where M is the total number of wave numbers chosen.

We shall now examine the property of the weighting function. For simplicity of discussion, we shall let the response function $\phi(\bar{v}, v) = 1$ so that the spectral transmittance may be expressed by

$$\mathcal{T}_{\bar{v}}(p) = \int_{\Delta v} \frac{dv}{\Delta v} \exp\left[-\frac{q}{g} \int_{0}^{p} k_v(p') \, dp' \right]. \tag{7.37}$$

Here we note that the mixing ratio q is a constant, and $\Delta v = v_1 - v_2$. In the lower atmosphere, collision broadening dominates the absorption process and the shape of the absorption lines is governed by the Lorentz profile

$$k_v = \frac{S}{\pi} \frac{\alpha}{(v - v_0)^2 + \alpha^2}.$$

The half width α is primarily proportional to the pressure (and to a lesser degree the temperature) according to Eq. (1.37), while the line strength S also depends on the temperature according to Eq. (4.63). Hence, the spectral transmittance may be explicitly written as

$$\mathcal{T}_{\bar{v}}(p) = \int_{\Delta v} \frac{dv}{\Delta v} \exp\left[-\frac{q}{g} \int_0^p \frac{S(p')}{\pi} \frac{\alpha(p')\,dp'}{(v - v_0)^2 + \alpha^2(p')} \right]. \tag{7.38}$$

The temperature dependence of the absorption coefficient introduces some difficulties in the sounding of the temperature profile. Nevertheless, the dependence of the transmittance on the temperature may be taken into account in the temperature inversion process by building a set of transmittances for a number of standard atmospheric profiles from which a search could be made to give the best transmittances for a given temperature profile.

As shown in Eq. (7.38), the computation of transmittances through an inhomogeneous atmosphere is rather involved, especially when the demands on accuracy are high in infrared sounding applications. Thus, accurate transmittance profiles are normally derived by means of line-by-line calculations, which involve the direct integration of monochromatic transmittance over the wave number spectral interval, weighted by an appropriate response function if so desired. Since the monochromatic transmittance is a rapidly varying function of wave number, numerical quadrature used for the integration must be carefully devised, and the required computational effort is generally enormous.

All of the earlier satellite experiments for the sounding of atmospheric temperatures of meteorological purposes have utilized the 15 μm CO_2 band. The 15 μm CO_2 band consists of a number of individual bands which contribute significantly to the absorption. The most important of these is the v_2 fundamental vibrational–rotational band mentioned in Section 4.2. In addition, there are several weak bands caused by the vibrational transitions between excited states, called the hot bands, and by molecules containing less abundant isotopes. In each of these bands there is a strong Q branch located at the center of the band with P branch and R branch lines almost equally spaced on each side of the band center.

Shown in Fig. 7.6 are the transmittance and weighting function profiles calculated for a set of Vertical Temperature Profile Radiometer (VTPR) on the NOAA 2 satellite, which is the first satellite experiment to measure atmospheric temperatures for operational meteorological use. The VTPR consists of 6 channels in the 15 μm CO_2 band with the nominal center wave numbers

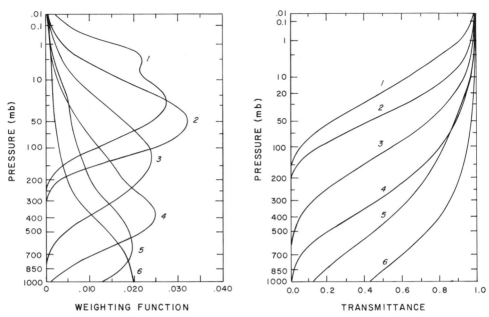

Fig. 7.6 The weighting function and transmittance for the NOAA 2 Vertical Temperature Profile Radiometer.

of 668.5, 677.5, 695.0, 708.8, 725.0, and 745.0 cm^{-1} for channels 1–6, respectively. The weighting function curves show from what part of the atmosphere the upwelling radiance arises. Each peak represents the maximum contribution to the upwelling radiance. Clearly, radiance comes mostly from progressively lower levels as the wave number moves from the center to the wing of the band. It is also apparent that the weighting functions overlap somewhat, allowing finite radiance data to define the temperature profile adequately.

7.3.3   Direct Linear Inversion Methods

Assume that N channels within the 15 μm CO_2 band have been chosen to recover the temperature profile. Let $g(\bar{v}_i) = g_i$, and $K(\bar{v}_i, p) = K_i(p)$, $i = 1$, $2, \ldots, M$. Thus, a set of M radiance observations give M integral equations as

$$g_i = \int_{p_s}^{0} f(p)K_i(p)\,dp, \qquad i = 1, 2, \ldots, M. \qquad (7.39)$$

The solution of this equation is clearly an ill-posed problem, since the unknown profile is a continuous function of pressure, and since there are only a finite number of observations. It is convenient to express $f(p)$ as a linear function of N variables in the form

$$f(p) = \sum_{j=1}^{N} f_j W_j(p),\tag{7.40}$$

where f_j are unknown coefficients, and $W_j(p)$ are the known representation functions which could be orthogonal functions, such as polynomials or Fourier series. It follows that

$$g_i = \sum_{j=1}^{N} f_j \int_{p_s}^{0} W_j(p)K_i(p)\,dp, \qquad i = 1, 2, \ldots, M.\tag{7.41}$$

Upon defining the known values in the form

$$A_{ij} = \int_{p_s}^{0} W_j(p)K_i(p)\,dp,\tag{7.42}$$

we obtain

$$g_i = \sum_{j=1}^{N} A_{ij}f_j, \qquad i = 1, 2, \ldots, M.\tag{7.43}$$

Generally, in order to find f_j $(j = 1, \ldots, N)$, one needs to have both the g_i $(i = 1, \ldots, M)$ and also $M \geq N$.

At this point, we shall define the meaning of a vector and of a matrix, and introduce a number of matrix operations required for the discussion in this section. A column vector is defined by

$$\mathbf{g} = \begin{bmatrix} g_1 \\ g_2 \\ \vdots \\ g_M \end{bmatrix}, \qquad \mathbf{f} = \begin{bmatrix} f_1 \\ f_2 \\ \vdots \\ f_N \end{bmatrix}.\tag{7.44}$$

We identify vectors with column vectors in the subsequent sections. A matrix consisting of M rows and N columns is defined by

$$\mathbf{A} = \begin{bmatrix} A_{11} & A_{12} & \cdots & A_{1N} \\ A_{21} & A_{22} & \cdots & A_{2N} \\ \vdots & \vdots & & \vdots \\ A_{M1} & A_{M2} & \cdots & A_{MN} \end{bmatrix}.\tag{7.45}$$

$\mathbf{A}$ is said to be an $(M \times N)$ matrix which is also denoted by $\|A_{ij}\|$. The vector $\mathbf{g}$ can now be considered as a $(M \times 1)$ matrix.

The transpose of a matrix $\mathbf{A}$ is defined as the interchange of the column and row elements and is given by

$$\mathbf{A}^* = \begin{bmatrix} A_{11} & A_{21} & \cdots & A_{M1} \\ A_{12} & A_{22} & \cdots & A_{M2} \\ \vdots & \vdots & & \vdots \\ A_{1N} & A_{2N} & \cdots & A_{MN} \end{bmatrix} \tag{7.46}$$

Thus, $\mathbf{A}^*$ is a $(N \times M)$ matrix, and $\mathbf{g}^*$ is then a row vector, or a $(1 \times M)$ matrix.

The product of a $(M \times N)$ matrix and a $(N \times K)$ matrix gives a $(M \times K)$ matrix. The rule is that N must be the same in the matrix product. It is clear that the existence of $\mathbf{AB}$ does not imply the existence of $\mathbf{BA}$, and vice versa, and that $\mathbf{AB} \neq \mathbf{BA}$. The product of a matrix $\mathbf{A}(M \times N)$ and a matrix $\mathbf{B}(N \times K)$ is given by

$$\mathbf{D} = \mathbf{AB} = \begin{bmatrix} D_{11} & D_{12} & \cdots & D_{1K} \\ D_{21} & D_{22} & \cdots & D_{2K} \\ \vdots & \vdots & & \vdots \\ D_{M1} & D_{M2} & \cdots & D_{MK} \end{bmatrix}, \tag{7.47}$$

where

$$D_{ik} = \sum_{j=1}^{N} A_{ij} B_{jk}.$$

Matrix products are associative, i.e.,

$$\mathbf{A(BC)} = \mathbf{(AB)C}. \tag{7.48}$$

It follows that if $\mathbf{B}$ and $\mathbf{C}$ have dimensions $(M \times N)$ and $(N \times K)$, respectively, $\mathbf{A}$ must have dimension $(L \times M)$, and the final dimension of the resulting matrix is $(L \times K)$.

On the basis of the rule of the matrix multiplication, the product of a row vector $(1 \times N)$ and a column vector $(N \times 1)$ gives a (1×1) matrix, i.e., a scalar product. Thus,

$$\mathbf{f}^*\mathbf{f} = f_1 f_1 + f_2 f_2 + \cdots + f_N f_N. \tag{7.49}$$

But the product of a column vector $(N \times 1)$ and a row vector $(1 \times N)$ gives a $(N \times N)$ matrix, i.e., a vector product. Thus

$$\mathbf{ff}^* = \begin{bmatrix} f_1 f_1 & f_1 f_2 & \cdots & f_1 f_N \\ f_2 f_1 & f_2 f_2 & \cdots & f_2 f_N \\ \vdots & \vdots & & \vdots \\ f_N f_1 & f_N f_2 & \cdots & f_N f_N \end{bmatrix}. \tag{7.50}$$

Also clear is that $A*(N \times M) \times A(M \times N)$ is a symmetric and square matrix $(N \times N)$. In addition, it is straightforward to prove that $(\mathbf{AB})* = \mathbf{B}*\mathbf{A}*$.

The inverse of a matrix is denoted by $\mathbf{A}^{-1}$. Generally speaking, the inverse exists only when $\mathbf{A}$ is a square $(N \times N)$ and the determinant of the array $\det(\mathbf{A})$ is not zero (nonsingular). The procedures for inverting a matrix are quite involved, and normally a computer is needed to perform the inversion. A number of useful rules governing the product of matrices involving an inverse matrix are

$$(\mathbf{AB})^{-1} = \mathbf{B}^{-1}\mathbf{A}^{-1}, \qquad \mathbf{AA}^{-1} = \mathbf{A}^{-1}\mathbf{A} = \mathbf{1}, \qquad (7.51)$$

where $\mathbf{1}$ denotes an identity matrix.

With the preceding introduction of matrix definitions and useful operations, Eq. (7.43) can be expressed in matrix form

$$\mathbf{g} = \mathbf{Af}. \qquad (7.52)$$

Thus, using Eq. (7.51), we have

$$\mathbf{f} = \mathbf{A}^{-1}\mathbf{g} = (\mathbf{A}*\mathbf{A})^{-1}\mathbf{A}*\mathbf{g}. \qquad (7.53)$$

To find the solution $\mathbf{f}$, one requires the inverse of a symmetric and square matrix.

It has been pointed out by many studies that the solution derived from Eq. (7.53) is unstable because the equation is underconstrained. Furthermore, the instability of this solution may also be traced to the following sources of error: (1) the errors arising from the numerical quadrature used for the calculation of A_{ij} in Eq. (7.42); (2) the approximation to the Planck function; and (3) the numerical round-off errors. In addition, sounding radiometers possess inherent instrumental noise, and thus the observed radiances generate errors probably in a random fashion. All of these errors make the direct inversion from the solution of transfer equation impractical. In the following section, we discuss a number of methods which can be utilized to stablize the solution, and for certain instances give reasonable results.

7.3.3.1 *Constrained Linear Inversion* Consider the ill-posed problem

$$g_i = \sum_{j=1}^{N} A_{ij}f_j, \qquad i = 1, \ldots, M. \qquad (7.54)$$

Since, in practice, the true g_i are never known because they always contain certain measurement errors ε_i, the measured data may be expressed by

$$\hat{g}_i = g_i + \varepsilon_i. \qquad (7.55)$$

Thus, to within the measurement error, solution f_j is not unique, and the ambiguity can be removed only by imposing an additional condition which would enable one of the possible sets f_j to be chosen.

Next, consider the function which utilizes a least square method with quadratic constraints in the form

$$\sum_i \varepsilon_i^2 + \gamma \sum_{j=1}^{N} (f_j - \bar{f})^2, \tag{7.56}$$

where γ is an arbitrary smoothing coefficient which determines how strongly the solution f_j is constrained to be near the mean $\bar{f}$; i.e., the constraint is given by the variance of f_j.

A simple criterion for a solution is that the measurement error is minimized while the solution is constrained to be close to the mean $\bar{f}$. Thus, we set

$$\frac{\partial}{\partial f_k}\left[\sum_i \left(\sum_{j=1}^{N} A_{ij}f_j - \hat{g}_i\right)^2 + \gamma \sum_{j=1}^{N}(f_j - \bar{f})^2\right] = 0, \tag{7.57}$$

where $k = 1, \ldots, j, \ldots, N$. This leads to

$$\sum_i \left(\sum_{j=1}^{N} A_{ij}f_j - \hat{g}_i\right)A_{ik} + \gamma(f_k - \bar{f}) = 0. \tag{7.58}$$

But the mean value is

$$\bar{f} = N^{-1}\sum_{k=1}^{N} f_k, \tag{7.59}$$

so that

$$f_k - \bar{f} = -N^{-1}f_1 - \cdots + (1 - N^{-1})f_k - \cdots - N^{-1}f_N.$$

In terms of matrix forms, Eq. (7.58) may be written as

$$\mathbf{A}^*\mathbf{A}\mathbf{f} - \mathbf{A}^*\mathbf{g} + \gamma\mathbf{H}\mathbf{f} = 0, \tag{7.60}$$

where $\mathbf{H}$ is a $(N \times N)$ matrix given by

$$\mathbf{H} = \begin{bmatrix} 1 - N^{-1} & -N^{-1} & \cdots & -N^{-1} \\ -N^{-1} & 1 - N^{-1} & \cdots & -N^{-1} \\ \vdots & \vdots & & \vdots \\ -N^{-1} & -N^{-1} & \cdots & 1 - N^{-1} \end{bmatrix}. \tag{7.61}$$

It follows that

$$\mathbf{f} = (\mathbf{A}^*\mathbf{A} + \gamma\mathbf{H})^{-1}\mathbf{A}^*\mathbf{g}. \tag{7.62}$$

This is the equation for constrained linear inversion derived by Phillips (1962) and Twomey (1963). The quadrature constraint for smoothing can also be imposed on the first differences, i.e., $\sum(f_{j-1} - f_j)^2$, or the second differences, i.e., $\sum(f_{j-1} - 2f_j - f_{j+1})^2$, and so on.

In some inversion problems, there may be a considerable amount of background data available by means of direct methods. It is sometimes desirable to construct from the past data an appropriate set of base function for approximating the unknown f. This may be accomplished by deriving the mean $\bar{f}$ of all past data, and finding a constrained solution which would minimize the mean square departure from this mean. If $\bar{\mathbf{f}}$ is the known mean value vector, then Eq. (7.58) may be written in the matrix forms

$$A^*A\mathbf{f} - A^*\hat{\mathbf{g}} + \gamma(\mathbf{f} - \bar{\mathbf{f}}) = 0. \tag{7.63}$$

It follows that

$$\mathbf{f} = (A^*A + \gamma\mathbf{1})^{-1}(A\hat{\mathbf{g}} + \gamma\bar{\mathbf{f}}), \tag{7.64}$$

where $\mathbf{1}$ is an $(N \times N)$ identity matrix. This formula will give an improvement if there is a reasonable base for selecting $\bar{\mathbf{f}}$.

7.3.3.2 *Statistical Method* In many sounding problems, errors encountered are statistical in nature, and it is desirable to consider the inversion problem by taking into account the statistical nature of the measurement errors and other relevant information. In the statistical approach, it is generally assumed that the deviation of the predicted parameter f (in the present case it is the temperature) from the climatological mean may be expressed as a linear combination of the deviation of the measured data (in this case the radiances). Thus, we write

$$\hat{f}_j - \bar{f}_j = \sum_{i=1}^{M} D_{ji}(\hat{g}_i - \bar{g}_i), \qquad j = 1, 2, \ldots, N, \tag{7.65}$$

where $\hat{f}_j$ is the predicted values of f_j, which represent the true temperature values, $\bar{f}_j$ the climatological mean of f_j, $\hat{g}_i$ the observed data, $\bar{g}_i$ the mean of the observed data, and $\|D_{ji}\|$ a certain predictor matrix.

We wish to find a linear predictor which will give the minimum mean square deviation of the predicted (or estimated) profile from the true profile when Eq. (7.65) is applied to a statistical ensemble of temperature profiles; that is, to find the minimum value for

$$\sum_{x=1}^{X} (f_{jx} - \hat{f}_{jx})^2 = \sum_{x=1}^{X} [(f_{jx} - \bar{f}_j) - (\hat{f}_{jx} - \bar{f}_j)]^2, \tag{7.66}$$

where the subscript x denotes the membership of the statistical sample whose size is X. We note that $\bar{f}_j = X^{-1}\sum_{x=1}^{X} f_{jx}$. Utilizing the assumption postulated in Eq. (7.65), we may write

$$G(D_{ji}) = \sum_{x=1}^{X} [(f_{jx} - \bar{f}_j) - \sum_{i=1}^{M} D_{ji}(\hat{g}_{ix} - \bar{g}_i)]^2. \tag{7.67}$$

To find the minimum with respect to the linear predictor, we set

$$\frac{\partial G(D_{ji})}{\partial D_{jk}} = 0 = -2 \sum_{x=1}^{X} \left[(f_{jx} - \bar{f_j}) - \sum_{i=1}^{M} D_{ji}(\hat{g}_{ix} - \bar{g}_i) \right] (\hat{g}_{kx} - \bar{g}_k),$$

$$k = 1, 2, \ldots, M. \quad (7.68)$$

This leads to

$$\sum_{x=1}^{X} (f_{jx} - \bar{f_j})(\hat{g}_{kx} - \bar{g}_k) - \sum_{i=1}^{M} D_{ji} \left[\sum_{x=1}^{X} (\hat{g}_{ix} - \bar{g}_i)(\hat{g}_{kx} - \bar{g}_k) \right] = 0. \quad (7.69)$$

In terms of matrix operations, we write

$$\sum_{x=1}^{X} (\mathbf{f}_x - \bar{\mathbf{f}})(\hat{\mathbf{g}}_x - \bar{\mathbf{g}})^* - \mathbf{D} \sum_{x=1}^{X} (\hat{\mathbf{g}}_x - \bar{g})(\hat{\mathbf{g}}_x - \bar{\mathbf{g}})^* = 0, \quad (7.70)$$

where $\mathbf{f}$ is a $N \times 1$ matrix, $\mathbf{g}$ is a $M \times 1$ matrix, and $\mathbf{D}$ is a $N \times M$ matrix. The covariance matrix for any two variables is defined by

$$\mathbf{C}(\mathbf{f}_x, \hat{\mathbf{g}}_x) = \frac{1}{X} \sum_{x=1}^{X} \mathbf{f}_x \hat{\mathbf{g}}_x^*. \quad (7.71)$$

Consequently, the predictor matrix may be expressed in terms of the covariance matrix in the form

$$\mathbf{D} = \mathbf{C}(\mathbf{f}_x - \bar{\mathbf{f}}, \hat{\mathbf{g}}_x - \bar{\mathbf{g}})\mathbf{C}^{-1}(\hat{\mathbf{g}}_x - \bar{\mathbf{g}}, \hat{\mathbf{g}}_x - \bar{\mathbf{g}}). \quad (7.72)$$

Infrared and microwave radiometers possess inherent instrumental noise which may be accounted for in the analyses. Let the experimental random error vector due to system noise be ε. Thus, the data vector is

$$\hat{\mathbf{g}} = \mathbf{g} + \varepsilon, \quad (7.73)$$

where $\mathbf{g}$ is the exact value. Then the covariance matrices are given by

$$\mathbf{C}(\mathbf{f}_x - \bar{\mathbf{f}}, \hat{\mathbf{g}}_x - \bar{\mathbf{g}}) = \frac{1}{X} \sum_{x=1}^{X} (\mathbf{f}_x - \bar{\mathbf{f}})(\hat{\mathbf{g}}_x - \bar{\mathbf{g}})^* = \mathbf{C}(\mathbf{f}_x - \bar{\mathbf{f}}, \mathbf{g}_x - \bar{\mathbf{g}}), \quad (7.74)$$

where we note that $\sum_x (\mathbf{f}_x - \bar{\mathbf{f}})\varepsilon = 0$, and

$$\mathbf{C}(\hat{\mathbf{g}}_x - \bar{\mathbf{g}}, \hat{\mathbf{g}}_x - \bar{\mathbf{g}}) = \frac{1}{X} \sum_{x=1}^{X} (\mathbf{g}_x - \bar{\mathbf{g}})(\mathbf{g}_x - \bar{\mathbf{g}})^* + \frac{1}{X} \sum_{x=1}^{X} \varepsilon\varepsilon^*$$

$$= \mathbf{C}(\mathbf{g}_x - \bar{\mathbf{g}}, \mathbf{g}_x - \bar{\mathbf{g}}) + \mathbf{C}(\varepsilon, \varepsilon), \quad (7.75)$$

where $\mathbf{C}(\varepsilon, \varepsilon)$ represents the noise covariance matrix.

The covariance matrix can be constructed experimentally by collecting coincidences of radiances derived from remote sounders with temperature

values obtained from direct radiosonde or rocket sounding. Normally, the
D matrix is built to give a dimension $N \times M$ such that $N > M$ so that more
temperature values may be inferred from a limited set of radiance observations. The **D** matrix can be estimated entirely from experiment, and the
information of the weighting functions is not needed.

7.3.3.3 *Backus–Gilbert Inversion Method*   If there are not enough measurements and constraints to make the inverse problem well imposed, and
if the available measurements and constraints do not reduce the solution
error covariance sufficiently, then the inverse problem is not solvable, and
additional considerations need to be given. This is the basic concept that
Backus and Gilbert (1970) adopted in their approach to the sounding of the
solid earth using seismic waves.

In reference to Eq. (7.39), we wish to find a solution profile seen at finite
resolution, which is given by a linear function of the measurements in the
form

$$\hat{f}(x) = \sum_{i=1}^{M} D_i(x)\hat{g}_i, \tag{7.76}$$

where M denotes the total number of channels for the spectral interval, and
x a reference level. Inserting Eq. (7.39) into Eq. (7.76) and defining the scanning
function

$$A(x, p) = \sum_i D_i(x)K_i(p), \tag{7.77}$$

we obtain

$$\hat{f}(x) = \int_{p_s}^{0} A(x, p)\hat{f}(p) \, dp. \tag{7.78}$$

It is clear that the only way in which $\hat{f}(x)$ can exactly equal $\hat{f}(p)$ is for the
scanning function to be a Dirac delta function $\delta(p - x)$ centered at $p = x$.
However, for a finite number of terms involved in the scanning function, it is
not possible, and $A(x, p)$ will have some finite spread about each level x. A
useful measure of this spread given by Backus and Gilbert has the form

$$s(x) = 12 \int_{p_s}^{0} (x - p)^2 A^2(x, p) \, dp. \tag{7.79}$$

The normalization factor 12 is chosen such that when $A(x, p)$ is a rectangular
function of width l centered on x, $s(x)$ will have a spread equal to its width l.
$A(x, p)$ must also have a unit area, i.e.,

$$\int_{p_s}^{0} A(x, p) \, dp = 1. \tag{7.80}$$

To take the instrumental noise into account in the analysis, we insert unknown random errors in the measured data to obtain

$$f(x) = \sum_i D_i(x)(\hat{g}_i + \varepsilon_i) = \sum_i D_i(x)\hat{g}_i + \sum_i D_i(x)\varepsilon_i. \qquad (7.81)$$

The variance of the solution incurred at level x due to random errors is then given by

$$\sigma^2(x) = \left[\sum_i D_i(x)\varepsilon_i\right]^2. \qquad (7.82)$$

Theoretically, we would like to be able to select $D_i(x)$ such that both the spread $s(x)$ and the error variance $\sigma^2(x)$ are minimized. Unfortunately, this cannot be done, but it is possible to minimize a linear combination of $s(x)$ and $\sigma^2(x)$ in the form

$$Q(x) = s(x) + \gamma\sigma^2(x), \qquad (7.83)$$

where γ is a parameter between zero and infinity depending on whether the minimization of the spread or the error should be emphasized. In carrying out the minimization, it is convenient to normalize the weighting function in order that

$$\int_{p_s}^0 K_i(p)\,dp = u_i = 1, \qquad (7.84)$$

where we let $u_i = 1$ for mathematical convenience. Thus, from Eqs. (7.77) and (7.80), we obtain

$$\sum_i D_i(x) = 1, \qquad \text{or} \qquad \mathbf{D}^*(x)\mathbf{u} = \mathbf{1}. \qquad (7.85)$$

We now minimize $Q(x)$ with respect to $D_k(x)$ subject to the unit area constraint denoted in Eqs. (7.84) and (7.85)

$$\frac{\partial Q(x)}{\partial D_k(x)} = 0 = \frac{\partial}{\partial D_k(x)} \left\{ 12\int_{p_s}^0 (x-p)^2 \left[\sum_i D_i(x)K_i(p)\right]^2 dp \right.$$

$$\left. + \gamma \left[\sum_i D_i(x)\varepsilon_i\right]^2 + \eta\sum_i D_i(x)u_i \right\}, \qquad (7.86)$$

where η is an arbitrary constant. This leads to

$$\sum_i D_i(x)[S_{ik}(x) + \gamma E_{ik}] + (\eta/2)u_k = 0, \qquad (7.87)$$

where

$$E_{ik} = \varepsilon_i\varepsilon_k, \qquad S_{ik}(x) = 12\int_{p_s}^0 (x-p)^2 K_i(p)K_k(p)\,dp. \qquad (7.88)$$

In terms of matrix notation (note that summation is over index i), we write

$$\mathbf{D}^*(x)[\mathbf{S}(x) + \gamma\mathbf{E}] + (\eta/2)\mathbf{u}^* = 0, \qquad (7.89)$$

where $\mathbf{S}$ and $\mathbf{E}$ are ($M \times M$) matrices. Thus,

$$\mathbf{D}^*(x) = -(\eta/2)\mathbf{u}^*[\mathbf{S}(x) + \gamma\mathbf{E}]^{-1}. \qquad (7.90)$$

But from Eq. (7.85), we have

$$\mathbf{D}^*(x)\mathbf{u} = 1 = -(\eta/2)\mathbf{u}^*[\mathbf{S}(x) + \gamma\mathbf{E}]^{-1}\mathbf{u}. \qquad (7.91)$$

Eliminating η from Eqs. (7.90) and (7.91), we obtain

$$\mathbf{D}^*(x) = \mathbf{u}^*[\mathbf{S}(x) + \gamma\mathbf{E}]^{-1}/\{\mathbf{u}^*[\mathbf{S}(x) + \gamma\mathbf{E}]^{-1}\mathbf{u}\}. \qquad (7.92)$$

Here $\mathbf{S}$ and $\mathbf{E}$ are to be obtained from Eq. (7.88). We also note that as $\gamma \to 0$, we get the best resolution but poor noise. On the other hand, as $\gamma \to \infty$, we get poor resolution but best noise consideration. This trade-off parameter between resolution and noise must be chosen according to applications.

7.3.4   Numerical Iteration Methods

7.3.4.1   *Chahine's Relaxation Method*   The difficulty in reconstructing the temperature profile from the radiance is due to the fact that the Fredholm equation with fixed limits may not always have a solution for an arbitrary function. Since the radiances are obtained from measurements which are only approximate, the reduction of this problem to a linear system is mathematically improper, and a nonlinear approach to the solution of the full radiative transfer equations appears to become necessary.

The basic radiance equation is

$$I_i = B_i(T_s)\mathscr{T}_i(p_s) + \int_{p_s}^{0} B_i[T(p)]\frac{\partial\mathscr{T}_i(p)}{\partial \ln p}\,d\ln p, \qquad i = 1,2,\ldots,M, \quad (7.93)$$

where i denotes the number of the spectral channel, and the weighting function is expressed in logarithmic scale. The Planck function in the wave number domain may be written in the form

$$B_i(T) = av_i^3/(e^{bv_i/T} - 1) \qquad (7.94)$$

with $a = 2hc^2$ and $b = hc/K$. In reference to the weighting function depicted in Fig. 7.6, we note that for a given wave number range, the integrand reaches a strong maximum at different pressure levels. From the mean-value theorem, the observed upwelling radiance $\tilde{I}_i$ may be approximated by

$$\tilde{I}_i - B_i(T_s)\mathscr{T}_i(p_s) \approx B_i[T(p_i)]\left[\frac{\partial\mathscr{T}_i(p)}{\partial \ln p}\right]_{p_i}\Delta_i\ln p, \qquad (7.95)$$

where p_i denotes the pressure level at which the maximum weighting function is located, and $\Delta_i \ln p$ is the difference of the pressure at the ith level and is defined as the effective width of the weighting functions. Let the guessed temperature at p_i level be $T'(p_i)$. Thus, the expected upwelling radiance I'_i is given by

$$I'_i - B_i(T_s)\mathcal{T}'_i(p_s) = B_i[T'(p_i)]\left[\frac{\partial \mathcal{T}'_i(p)}{\partial \ln p}\right]_{p_i}\Delta_i \ln p. \qquad (7.96)$$

Upon dividing Eq. (7.95) by Eq. (7.96), and noting that the dependence of the Planck function on temperature variations is much stronger than that of the weighting function, we obtain

$$\frac{\tilde{I}_i - B_i(T_s)\mathcal{T}_i(p_s)}{I'_i - B_i(T_s)\mathcal{T}'_i(p_s)} \approx \frac{B_i[T(p_i)]}{B_i[T'(p_i)]}. \qquad (7.97)$$

When the surface contribution to the upwelling radiance is negligible or dominant, the equation may be approximated by

$$\frac{\tilde{I}_i}{I'_i} \approx \frac{B_i[T(p_i)]}{B_i[T'(p_i)]}. \qquad (7.98)$$

This is the relaxation equation developed by Chahine (1970).

Since most of the upwelling radiance at the strong absorption bands arises from the upper parts of the region, whereas the radiance from the less attenuating bands comes from progressively lower levels, it is possible to select a set of wave numbers to recover the atmospheric temperature at different pressure levels. The size of a set of sounding wave numbers is defined by the degree of the vertical resolution required and is obviously limited by the capacity of the sounding instrument.

Assuming now that the upwelling radiance is measured at a discrete set of M spectral channels, and that the composition of carbon dioxide, the instrumental slit function $\phi(\bar{v}, v)$, and the level of the weighting function peak p_i are all known, the following iteration procedures are utilized to recover the temperature profile $T^{(n)}(p_i)$ at level p_i, where n is the order of iterations:

(a)   Make an initial guess for $T^{(n)}(p_i)$, $n = 0$.

(b)   Substitute $T^{(n)}(p_i)$ into Eq. (7.93) and use an accurate quadrature formula to evaluate the expected upwelling radiance $I_i^{(n)}$ for each sounding channel.

(c)   Compare the computed radiance values $I_i^{(n)}$ with the measured data $\tilde{I}_i$. If the residuals $R_i^{(n)} = |\tilde{I}_i - I_i^{(n)}|/\tilde{I}_i$ are less than a preset small value (say, 10^{-4}) for each sounding channel, then $T^{(n)}(p_i)$ is a solution.

(d)   If the residuals are greater than the preset criterion, we apply the relaxation equation (7.98) M times to generate a new guess for the temperature values $T^{(n+1)}(p_i)$ at the selected i pressure levels. From Eqs. (7.94) and (7.98), we have

$$T^{(n+1)}(p_i) = bv_i/\ln\{1 - [1 - \exp(bv_i/(T^{(n)}(p_i)))]I_i^{(n)}/\tilde{I}_i\}, \qquad i = 1, 2, \ldots, M. \tag{7.99}$$

In this calculation, each sounding channel acts at one specific pressure level p_i to relax $T^{(n)}(p_i)$ to $T^{(n+1)}(p_i)$.

(e) Carry out the interpolation between the temperature value at each given level p_i to obtain the desirable profile (it is sufficient to use linear interpolation).

(f) Finally, with this new temperature profile, go back to step (b) and repeat until the residuals are less than the preset criterion.

7.3.4.2 *Smith's Iteration Method* Smith (1970) developed an iterative solution for the temperature profile retrieval, which differs somewhat from that of the relaxation method introduced by Chahine. As before, let $\tilde{I}_i$ denote the observed radiance and $I_i^{(n)}$ the computed radiance in the nth iteration. Then the upwelling radiance expression in Eq. (7.93) may be written as

$$I_i^{(n)} = B_i^{(n)}(T_s)\mathcal{T}_i(p_s) + \int_{p_s}^0 B_i^{(n)}[T(p)] \frac{\partial \mathcal{T}_i(p)}{\partial \ln p} \, d\ln p. \tag{7.100}$$

Further, for the $(n+1)$ step we set

$$\tilde{I}_i = I_i^{(n+1)} = B_i^{(n+1)}(T_s)\mathcal{T}_i(p_s) + \int_{p_s}^0 B_i^{(n+1)}[T(p)] \frac{\partial \mathcal{T}_i(p)}{\partial \ln p} \, d\ln p. \tag{7.101}$$

Upon subtracting Eq. (7.100) from Eq. (7.101), we obtain

$$\tilde{I}_i - I_i^{(n)} = [B_i^{(n+1)}(T_s) - B_i^{(n)}(T_s)]\mathcal{T}_i(p_s)$$

$$+ \int_{p_s}^0 \{B_i^{(n+1)}[T(p)] - B_i^{(n)}[T(p)]\} \frac{\partial \mathcal{T}_i(p)}{\partial \ln p} \, d\ln p. \tag{7.102}$$

An assumption is made at this point that for each sounding wave number the Planck function difference for the sensed atmospheric layer is independent of the pressure coordinate. Thus, Eq. (7.102) may be simplified to give

$$\tilde{I}_i - I_i^{(n)} = B_i^{(n+1)}[T(p)] - B_i^{(n)}[T(p)]. \tag{7.103a}$$

That is,

$$B_i^{(n+1)}[T(p)] = B_i^{(n)}[T(p)] + [\tilde{I}_i - I_i^{(n)}]. \tag{7.103b}$$

This is the iteration equation developed by Smith. Moreover, from Eq. (7.94) we have

$$T^{(n+1)}(p, v_i) = bv_i/\ln\{1 + av_i^3/B_i^{(n+1)}[T(p)]\}. \tag{7.104}$$

Since the temperature inversion problem now depends on the sounding wave number v_i, the best approximation of the true temperature at any level p would be given by a weighted mean of independent estimates so that

$$T^{(n+1)}(p) = \sum_{i=1}^{M} T^{(n+1)}(p, v_i)W_i(p) \bigg/ \sum_{i=1}^{M} W_i(p), \qquad (7.105)$$

where the proper weights based on Eq. (7.102) should be approximately

$$W_i(p) = \begin{cases} d\mathcal{T}_i(p), & p < p_s \\ \mathcal{T}_i(p), & p = p_s \end{cases}.$$

It should be noted that the numerical technique presented above makes no assumption about the analytical form of the profile imposed by the number of radiance observations available. The following iteration schemes for the temperature retrieval may now be employed:

(a) Make an initial guess for $T^{(n)}(p)$, $n = 0$.

(b) Compute $B_i^{(n)}[T(p)]$ from Eq. (7.94) and $I_i^{(n)}$ from Eq. (7.100).

(c) Compute $B_i^{(n+1)}[T(p)]$ and $T^{(n+1)}(p, v_i)$ from Eqs. (7.103) and (7.104), subsequently, for the desirable levels.

(d) Make a new estimate of $T^{(n+1)}(p)$ from Eq. (7.105).

(e) Compare the computed radiance values $I_i^{(n)}$ with the measured data $\tilde{I}_i$. If the residuals $R_i^{(n)} = [\tilde{I}_i - I_i^{(n)}]/\tilde{I}_i$ are less than a preset small value, then $T^{(n+1)}(p)$ would be the solution. If not, repeat steps (b)–(d) until convergence is achieved.

Figure 7.7 illustrates a retrieval exercise using both Chahine's and Smith's methods. The transmittances used are those depicted in Fig. 7.6, and the true temperature profile is shown in the figure. An isothermal profile of $300°K$ was used as an initial guess, and the surface temperature was fixed at $279.5°K$. The observed radiances utilized were obtained by direct computations for six VTPR channels at 669.0, 676.7, 694.7, 708.7, 723.6, and 746.7 cm^{-1} using a forward difference scheme. Numerical procedures already outlined were followed, and a linear interpolation with respect to $\ln p$ was used in the relaxation method to get the new profile. With the residual set at 1%, the relaxation method converged after only four iterations, and results are given by the solid line with black dots. Since the top level at which the temperature was calculated was about 20 mb, extrapolation to the level of 1 mb was carried out in which the true temperature at that level was used. Recovered results using Smith's method are displayed by the dashed line. No interpolation is necessary since this method gives temperature values at desirable levels. It took about 20 iterations to converge the solution to within 2%. Reducing the residual did not improve the solution, however. Both methods

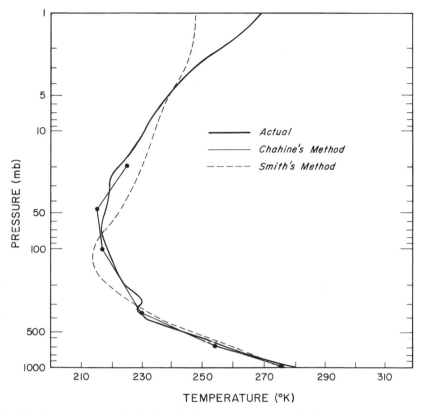

Fig. 7.7   Temperature retrieval using Chahine's relaxation and Smith's iterative methods for VTPR channels.

do not adequately recover the temperature at upper levels due to the fact that the highest weighting function peak is at about 30 mb. It should be noted that the retrieval exercise presented here does not account for random errors and therefore, it is a hypothetical one.

7.3.5   Limb Scanning Method

In Sections 7.3.1 and 7.3.2, we introduced the concept of downward viewing infrared spectral methods for the sounding of temperature profiles. There is another important technique for probing the atmosphere called the limb scanning method. An instrument having a very small optical field of view scans the limb of the earth and receives radiation from the atmosphere over a relatively narrow layer in height. The geometry of limb viewing is illustrated in Fig. 7.8. A radiometer receives radiation emitted by the atmosphere along a ray path which may be identified by the tangent height closest to the surface. The atmosphere may be scanned by sweeping the view direction vertically or horizontally. The advantageous features of limb scanning for atmospheric probes are: (1) Emission originates in the few kilometers immediately above the tangent point because of the rapid decrease of atmospheric density and pressure. Thus, a high inherent vertical resolution may be obtained. (2) All radiation received comes from the atmosphere only. Variation of a changing underlying surface, which occurs when a nadir viewing instrument is utilized, is absent. (3) There is large opacity involved along a horizontal path. Hence, it is particularly useful for the determination of minor gases in the upper atmosphere. (4) The viewing direction from the satellite can be oriented in any azimuthal direction relative to the satellite motion and covers a large area. A disadvantage associated with these features is the interference of clouds along the ray path acting as bodies of infinite opacity, which may cause a considerable alteration in the emerging radiation. Also, the sharp vertical weighting function, related to a horizontal region stretching 200 km or more along the ray path, leads to problems of interpretation of large changes in the atmospheric state over this distance. For these reasons, the limb scanning technique may be most useful for the exploration of composition and structure of the stratosphere and mesosphere.

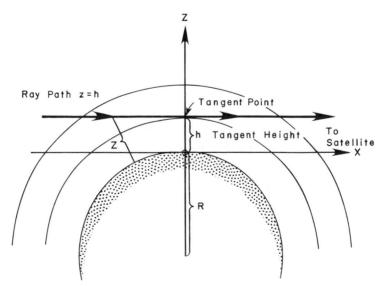

Fig. 7.8   The geometry of limb viewing.

In view of the limb viewing geometry, the solution of the fundamental radiative transfer equation for a nonscattering atmosphere in local thermodynamic equilibrium may be expressed by

$$I_{\tilde{v}}(h) = \int_{-\infty}^{\infty} B_{\tilde{v}}[T(x)] \frac{\partial \mathcal{T}_{\tilde{v}}(h, x)}{\partial x} dx, \tag{7.106}$$

where x is the distance coordinate along the ray path with the origin at the tangent point. The principle of the temperature and gaseous profile inversion problems is similar to that discussed in the previous sections. However, because of the spherical geometry, it is necessary to change from variable x to z in order to obtain the temperature and gaseous profile as functions of height in the atmosphere. Thus, the limb viewing radiance may be written in the form

$$I_{\tilde{v}}(h) = \int_{0}^{\infty} B_{\tilde{v}}[T(z)] K_{\tilde{v}}(h, z) \, dz, \tag{7.107}$$

where $K_{\tilde{v}}$ represents the weighting function which is a function of the geometrical factors and the band model used with respect to height z. Figure 7.9 shows the limb viewing weighting function for a hypothetical instrument with an infinitesimal vertical field of view for a wide spectral band 585–705 cm^{-1} covering most of the 15 μm CO_2 band. It is apparent that for tangent heights above 25 km, the major part of the contribution comes from within about 3 km of the tangent height. Below 25 km, the weighting function takes on the broader shape of the nadir viewing weighting functions, although a spike still remains at the tangent point. Inversion of limb radiance measurements, in principle, may be carried out utilizing the same techniques as in the nadir looking radiance observations.

As pointed out previously, the most significant application of the limb scanning radiometer is to derive the temperature structure and minor gaseous concentrations in the upper atmosphere, when the nadir looking radiometer is incapable of deriving sufficient information for their recovery. In the stratosphere, to a good approximation, the geostrophic approximation may be applicable, and the geostrophic wind is related to the horizontal gradient of the thickness. The horizontal equation of motion is given by

$$\left(\frac{d\mathbf{v}}{dt} \right)_{\mathrm{H}} = -g \nabla_{\mathrm{H}} p - f \mathbf{k} \times \mathbf{v},$$

where $\nabla_{\mathrm{H}} p = \nabla_p z$ is the pressure gradient force, $\mathbf{k}$ the unit vector in the vertical direction, and f the Coriolis parameter. The balance between the pressure gradient force and the Coriolis force gives the geostrophic wind

$$\mathbf{v}_{\mathrm{g}} = -(g/f) \nabla_p z \times \mathbf{k}. \tag{7.108}$$

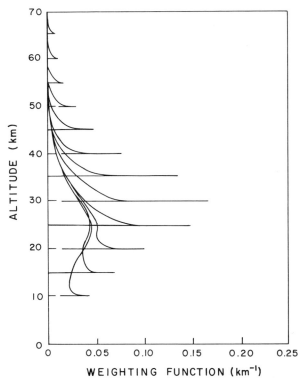

Fig. 7.9   Limb viewing weighting function for the ideal case of an instrument with an infini-tesimal vertical field of view for the spectral band 585–705 cm^{-1} covering most of the 15 μm band of CO_2 (after Gille and House, 1971).

Let v_{g_1} and v_{g_2} correspond to heights z_1 and z_2, such that $z_2 > z_1$, respectively. Then the variation of the geostrophic wind is

$$\Delta v_g = v_{g_2} - v_{g_1} = -(g/f)\nabla_p(\Delta z) \times k. \qquad (7.109)$$

But from the hydrostatic equation we have

$$dp = -\rho_a g \, dz,$$

where ρ_a is the air density. And upon using the equation of state

$$P = \rho_a RT,$$

we obtain

$$\Delta z = \int_{z_1}^{z_2} dz = z_2 - z_1 = -\frac{R}{g}\int_{p_1}^{p_2} T(p)\frac{dp}{p}, \qquad (7.110)$$

where R is the gas constant. Consequently, if the temperatures are known between P_1 and P_2, then from Eqs. (7.109) and (7.110), the vertical change in the geostrophic wind between two pressure surfaces can be estimated. If the geostrophic wind at pressure level p_1 is known, an estimate of the geostrophic wind at pressure level p_2 can then be derived.

Winds are basic to the study of the dynamics of the upper atmosphere. Balloon soundings of winds normally are made up to about 10 mb and are included in conventional map analyses. The limb scanning radiometer is capable of determining the vertical temperature from about 10 mb to 1 mb. The temperature data from satellites would then provide the wind data to the level of about 1 mb, which is essential to the understanding of the dynamics of the stratosphere.

7.3.6 Infrared Sensing of Cirrus Clouds

Owing to their semitransparent appearance in the infrared region, cirrus clouds have been noted to introduce serious difficulties in infrared sensing of atmospheric temperature and gaseous profiles, and surface conditions. Moreover, the radiative properties of cirrus, which consist of nonspherical ice crystals of various sizes and shapes possibly oriented in a horizontal plane, are largely unknown. Thus, from the infrared satellite sounding point of view, it is extremely important to derive reliable and accurate radiative properties of cirrus from an independent set of sounding wave numbers in conjunction with the recovery of atmospheric temperatures and gaseous concentrations. Further, determination of the vertical ice content over the global atmosphere is equally important from the point of view of climatology and weather prediction.

To demonstrate cirrus clouds are global in nature, a global infrared photograph from the sun synchronous GOES satellite for February 18, 1976, is shown in Fig. 7.10. In northern hemisphere midlatitudes, a very well-defined cyclone, depicted by an extensive comma-shaped cloud is centered near 50°N and 175°W. North of this cloud system is the bright cirriform canopy. Also a thin band of anticyclonically curved cirrus is located on the southeast edge of the comma tail and extends eastward. In the north-central Pacific Ocean, an elongated area of high clouds points east-northeast from a tropical disturbance in the western Pacific. Moreover, another cirriform cloud area is present over the east-central United States. In the southern hemisphere, the very long band of mostly high cloudiness extends from beyond the southern tip of South America to the western Pacific Ocean.

Because of the complexity of the cloud interaction with the radiation field of the atmosphere, study on the quantitative estimate of cloud properties by means of passive infrared sensing in the past has been limited.

Fig. 7.10   A global infrared photograph from the sun synchronous GOES satellite.

Houghton and Hunt (1971) explored the feasibility of passive remote sensing of ice clouds by means of two wavelengths in the far infrared where absorption due to ice and water is significantly different. Recently, Liou (1977) proposed a theoretical retrieval technique for the determination of surface temperature, cloud thickness, cloud transmissivity, and the fraction of cloud cover by means of four radiance measurements in the 10 μm window region. The proposed technique utilizes the principle of radiative transfer and presents an analytical approach to the cloud sounding problem. We shall introduce the analyses involved in the method through which some fundamental problems concerning clouds will also become evident.

Assume that within the field of view of the satellite radiometer the atmosphere contains η portion of cloudiness, and let the cloud base height and thickness be z_b and Δz, respectively. The spectral upwelling radiance measured by the satellite radiometer at the top of the atmosphere in partly cloudy condition is given by

$$\tilde{I}^{pc}_{\bar{v}} = \eta I^{c}_{\bar{v}} + (1 - \eta)I^{nc}_{\bar{v}}, \tag{7.111}$$

where η denotes the fraction of cloud cover in the satellite field of view. The clear column radiance in height coordinates is simply [see Eq. (7.22)]

$$I^{nc}_{\bar{v}} = B_{\bar{v}}(T_s)\mathcal{T}_{\bar{v}}(\infty, 0) + \int_{z=0}^{z=\infty} B_{\bar{v}}[T(z)]\, d\mathcal{T}_{\bar{v}}(\infty, z), \tag{7.112}$$

and the cloudy radiance may be expressed by

$$I^{c}_{\bar{v}} = I_{\bar{v}}(z_b)\mathcal{T}^{c}_{\bar{v}}(\Delta z)\mathcal{T}_{\bar{v}}(\infty, z_b + \Delta z) + \int_{z=z_b+\Delta z}^{z=\infty} B_{\bar{v}}[T(z)]\, d\mathcal{T}_{\bar{v}}(\infty, z). \tag{7.113}$$

In Eq. (7.113) the cloud transmissivity $\mathcal{T}^{c}_{\bar{v}}$ is defined as the ratio of the up-welling radiance at the cloud top to that at the cloud base, and it can be written as

$$\mathcal{T}^{c}_{\bar{v}} = I_{\bar{v}}(z_b + \Delta z)/I_{\bar{v}}(z_b). \tag{7.114}$$

The upwelling radiance reaching the cloud base may be expressed analogously to Eq. (7.112) in the form

$$I_{\bar{v}}(z_b) = B_{\bar{v}}(T_s)\mathcal{T}_{\bar{v}}(z_b, 0) + \int_{z=0}^{z=z_b} B_{\bar{v}}[T(z)]\, d\mathcal{T}_{\bar{v}}(z_b, z). \tag{7.115}$$

Since thin cirrus are normally fairly high in the atmosphere with their top heights on the order of 10 km or higher, it is ideal to select spectral wave numbers in the 10 μm window where the effect of water vapor absorption above the cirrus is so small that for all practical purposes it can be neglected. Thus, we find

$$\int_{z=z_b+\Delta z}^{z=\infty} B_{\bar{v}}[T(z)]\, d\mathcal{T}_{\bar{v}}(\infty, z) \approx 0, \qquad \mathcal{T}_{\bar{v}}(\infty, z_b + \Delta z) \approx 1. \tag{7.116}$$

Using Eq. (7.116), Eq. (7.111) may now be simplified to give

$$\tilde{I}^{pc}_{i} \approx (1 - \eta + \eta\mathcal{T}^{c}_{i})\left\{ B_{i}(T_s)\mathcal{T}_{i}(z_b, 0) + \int_{z=0}^{z=z_b} B_{i}[T(z)]\, d\mathcal{T}_{i}(z_b, z) \right\}, \tag{7.117}$$

where we change $\bar{v}$ to i. The unknown parameters in this equation are the fraction of cloud cover, cloud base height, surface temperature, cloud transmissivity which is wave number dependent, and the temperature and water vapor transmittance profiles. Even with the simplification given in Eq. (7.116), there are several unknown variables involved associated with the presence of clouds. The cloud base height is a difficult parameter to determine, and its contribution to the upwelling radiance enters in the transmittance term $\mathcal{T}_{\bar{v}}(z_b, 0)$ and the integral term. Moreover, the temperature and water vapor profiles in a partly cloudy atmosphere are also unknown. It is obvious that physical approximations must be made for these two terms. We let

$$\mathscr{I}_i^* = \int_{z=0}^{z=z_b} B_i[T(z)] \, d\mathscr{T}_i(z_b, z), \qquad \mathscr{T}_i^* = \mathscr{T}_i(z_b, 0), \qquad (7.118)$$

where * is used to denote the values to be estimated from prior knowledge of the cloud base height, and temperature and water vapor profiles below the cloud height. Such estimates can be done using climatological data.

The cloud transmissivity defined in Eq. (7.114) depends generally on the wave number. Examination of the transmissivity calculations using 900, 950, 1100, and 1150 cm^{-1} reveals that a linear relationship exists between the transmissivities with respect to the cloud thickness in the form

$$\mathscr{T}_i^c = (a_i \Delta z + b_i)\mathscr{T}_1^c, \qquad i = 1, 2, 3, 4, \qquad (7.119)$$

where the transmissivities of 950 (2), 1100 (3), and 1150 (4) cm^{-1} are expressed in terms of the transmissivity of 900 cm^{-1} (1). The coefficients $a_1 - a_4$ are 0, -0.02, -0.25, and -0.04, respectively, and for $b_1 - b_4$, they are 1, 0.99, 1.23, and 1.01, respectively.

Upon substituting Eqs. (7.118) and (7.119) into Eq. (7.117), we obtain

$$Q_i(T_s) = \frac{\tilde{I}_i^{pc}}{B_i(T_s)\mathscr{T}_i^* + \mathscr{I}_i^*} = 1 - \eta + \eta(a_i \Delta z + b_i)\mathscr{T}_1^c, \quad i = 1, 2, 3, 4. \quad (7.120)$$

This is a nonlinear equation and contains only four unknown parameters, namely T_s, η, Δz, and $\mathscr{T}_1^c$, and consequently, four radiance measurements are required to determine these values. At this point, we close the complicated transfer equation based on a number of physical assumptions and postulations.

Successive eliminations of η, $\mathscr{T}_1^c$ and Δz give the equations

$$(Q_1 - 1)[(a_i \Delta z + b_i)\mathscr{T}_1^c - 1] - (Q_i - 1)(\mathscr{T}_1^c - 1) = 0, \qquad i = 2, 3, 4, \quad (7.121)$$

$$\frac{(Q_1 - Q_2)[(Q_1 - 1)(a_i \Delta z + b_i) - (Q_i - 1)]}{(Q_1 - Q_i)[(Q_1 - 1)(a_2 \Delta z + b_2) - (Q_2 - 1)]} - 1 = 0, \qquad i = 3, 4, \quad (7.122)$$

$$\frac{[Q_1(b_3 - b_2) + Q_2(1 - b_3) + Q_3(b_2 - 1)][Q_1(a_2 - a_4) + Q_2 a_4 - Q_4 a_2]}{[Q_1(b_4 - b_2) + Q_2(1 - b_4) + Q_4(b_2 - 1)][Q_1(a_2 - a_3) + Q_2 a_3 - Q_3 a_2]} - 1 = 0. \quad (7.123)$$

Equation (7.123) represents a complicated nonlinear equation consisting of an unknown parameter T_s, which can be determined from a set of four measured radiances $\tilde{I}_i^{pc}$ in a partly cloudy atmosphere. On deriving the surface temperature, the cloud thickness, the cloud transmissivity, and the fractional cloud cover subsequently can be evaluated. Numerical experiments have been carried out to investigate the determination of these four parameters simultaneously using synthetic atmospheric data, and the feasibility of this method of inferring the cloud parameters has been illustrated.

7.4 MICROWAVE SENSING FROM SATELLITES

7.4.1 Microwave Spectrum
and Microwave Radiation Transfer

Water vapor and molecular oxygen exhibit absorption lines in the microwave region. Below 40 GHz (1 GHz = 10^9 cycles/sec; note 1 cm = 30 GHz) only the weakly absorbing pressure broadened 22.235 GHz water vapor line is dominant. This resonance absorption line arises from transitions between the rotational states 5_{23} and 6_{16}. At about 31.4 GHz, air is relatively transparent. It is the window between the reasonance water vapor line and the strongly absorbing oxygen complex of lines centered around 60 GHz. The oxygen molecule has a magnetic dipole moment arising from the combined spins of two impaired electrons in its $^3\sum_g^-$ electronic ground state. Changes in the orientation of the electronic spin relative to the orientation of the molecular rotation produce a band of magnetic dipole transitions near 60 GHz and a single transition at 118.75 GHz. For frequencies greater than 120 GHz, water vapor absorption again becomes dominant due to the strongly absorbing line at 183 GHz. Figure 7.11 show the vertical atmospheric transmittance as a function of frequency for a standard atmosphere.

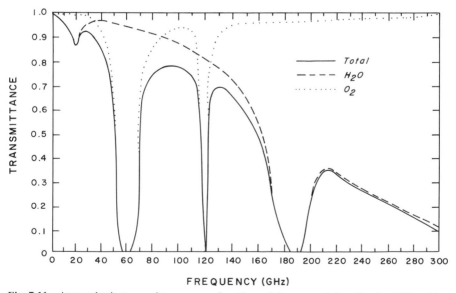

Fig. 7.11 Atmospheric transmittance as a function of frequency (after Grody, 1976, with modification).

On the basis of the discussion in Section 7.3.1, the solution of the transfer equation for a nonscattering atmosphere in local thermodynamic equilibrium is given by

$$I_{\tilde{v}}(0) = I_{\tilde{v}}(p_s)\mathscr{T}_{\tilde{v}}(p_s,0) + \int_{p_s}^{0} B_{\tilde{v}}[T(p)]\frac{\partial \mathscr{T}_{\tilde{v}}(p,0)}{\partial p}\,dp, \qquad (7.124)$$

where $\tilde{v}$ denotes the frequency, $I_{\tilde{v}}(p_s)$ represents the radiance contribution from the surface, and the transmittance $\mathscr{T}_{\tilde{v}}(p,0)$ is expressed with respect to the top of the atmosphere. The emissivity in the microwave region is normally less than unity. Thus, there is a reflection contribution from the surface. The radiance emitted from the surface would therefore be given by

$$I_{\tilde{v}}(p_s) = \varepsilon_{\tilde{v}}B_{\tilde{v}}(T_s)\mathscr{T}_{\tilde{v}}(p_s,0) + (1 - \varepsilon_{\tilde{v}})\int_{0}^{p_s} B_{\tilde{v}}[T(p)]\frac{\partial \mathscr{T}_{\tilde{v}}(p_s,p)}{\partial p}\,dp. \qquad (7.125)$$

The first term in the right-hand side of Eq. (7.125) denotes the surface emission contribution, whereas the second term represents the emission contribution from the entire atmosphere to the surface, which is reflected back to the atmosphere at the same frequency. The transmittance $\mathscr{T}_{\tilde{v}}(p_s,p)$ is now expressed with respect to the surface.

Inserting the lower boundary condition into Eq. (7.124), the upwelling radiance can now be expressed as

$$I_{\tilde{v}} = \varepsilon_{\tilde{v}}B_{\tilde{v}}(T_s)\mathscr{T}_{\tilde{v}}(p_s,0) + (1 - \varepsilon_{\tilde{v}})\mathscr{T}_{\tilde{v}}(p_s,0)\int_{0}^{p_s} B_{\tilde{v}}[T(p)]\frac{\partial \mathscr{T}_{\tilde{v}}(p_s,p)}{\partial p}\,dp$$

$$+ \int_{p_s}^{0} B_{\tilde{v}}[T(p)]\frac{\partial \mathscr{T}_{\tilde{v}}(p,0)}{\partial p}\,dp. \qquad (7.126)$$

In the frequency domain, the Planck function is given by

$$B_{\tilde{v}}(T) = 2h\tilde{v}^3/[c^2(e^{h\tilde{v}/KT} - 1)]. \qquad (7.127)$$

In the microwave region $h\tilde{v}/KT \ll 1$, the Planck function may be approximated by

$$B_{\tilde{v}}(T) \approx (2K\tilde{v}^2/c^2)\,T. \qquad (7.128)$$

Thus, the Planck radiance is linearly proportional to the temperature, which is referred to as the *Rayleigh–Jeans law*. Analogous to the above approximation, we may define an equivalent brightness temperature T_B such that

$$I_{\tilde{v}} = (2K\tilde{v}^2/c^2)\,T_B(\tilde{v}). \qquad (7.129)$$

Substituting Eqs. (7.128) and (7.129) into Eq. (7.126), the solution of microwave radiative transfer may now be written in terms of temperature as

$$T_B(\tilde{v}) = \varepsilon_{\tilde{v}} T_s \mathcal{T}_{\tilde{v}}(p_s, 0) + (1 - \varepsilon_{\tilde{v}}) \mathcal{T}_{\tilde{v}}(p_s, 0) \int_0^{p_s} T(p) \frac{\partial \mathcal{T}_{\tilde{v}}(p_s, p)}{\partial p} dp$$

$$+ \int_{p_s}^{0} T(p) \frac{\partial \mathcal{T}_{\tilde{v}}(p, 0)}{\partial p} dp. \tag{7.130}$$

Contribution of each term to the brightness temperature at the top of the atmosphere is illustrated in Fig. 7.12.

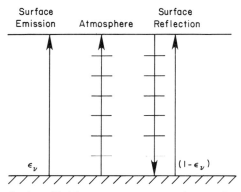

Fig. 7.12 Contribution of brightness temperature at the top of a clear atmosphere.

The transmittance is generally available with respect to the top of the atmosphere; i.e., $\mathcal{T}_{\tilde{v}}(p) = \mathcal{T}_{\tilde{v}}(p, 0)$. Thus, for computational purposes, it is desirable to express $\mathcal{T}_{\tilde{v}}(p_s, p)$ in terms of $\mathcal{T}_{\tilde{v}}(p, 0)$. For monochromatic frequencies, the transmittance is an exponential function of the optical depth [see Eq. (7.24)]. Hence, we may express

$$\mathcal{T}_{\tilde{v}}(p_s, p) = \exp\left[-\frac{1}{g} \int_p^{p_s} k_{\tilde{v}}(p')q(p') dp' \right]$$

$$= \exp\left[-\frac{1}{g} \int_0^{p_s} k_{\tilde{v}}(p')q(p') dp' + \frac{1}{g} \int_0^{p} k_{\tilde{v}}(p')q(p') dp' \right]$$

$$= \mathcal{T}_{\tilde{v}}(p_s, 0) / \mathcal{T}_{\tilde{v}}(p, 0), \tag{7.131}$$

where $\mathcal{T}_{\tilde{v}}(p_s, 0)$, the transmittance of the entire atmosphere, is a constant value. Thus,

$$\frac{\partial \mathcal{T}_{\tilde{v}}(p_s, p)}{\partial p} = -\frac{\mathcal{T}_{\tilde{v}}(p_s, 0)}{[\mathcal{T}_{\tilde{v}}(p, 0)]^2} \frac{\partial \mathcal{T}_{\tilde{v}}(p, 0)}{\partial p}. \tag{7.132}$$

Substituting Eq. (7.132) into Eq. (7.130), rearranging terms and letting $\mathcal{T}_{\tilde{\nu}}(p, 0) = \mathcal{T}_{\tilde{\nu}}(p)$, we find

$$T_B(\tilde{\nu}) = \varepsilon_{\tilde{\nu}} T_s \mathcal{T}_{\tilde{\nu}}(p_s) + \int_{p_s}^{0} J_{\tilde{\nu}}(p) \frac{\partial \mathcal{T}_{\tilde{\nu}}(p)}{\partial p} \, dp, \tag{7.133}$$

where the atmospheric source term is given by

$$J_{\tilde{\nu}}(p) = \{1 + (1 - \varepsilon_{\tilde{\nu}})[\mathcal{T}_{\tilde{\nu}}(p_s)/\mathcal{T}_{\tilde{\nu}}(p)]^2\} T(p). \tag{7.134}$$

In microwave sounding, the transmittances are computed by including the antenna gain characteristics.

A special problem area in the use of microwave for atmospheric sounding from a satellite platform is surface emissivity. In the microwave spectrum, emissivity values of the earth's surface vary over a considerable range, from about 0.4 to 1.0. The emissivity of the sea surface typically ranges between 0.4 and 0.5, depending upon such variables as salinity, sea ice, surface roughness, and sea foam. In addition, there is a frequency dependence with higher frequencies displaying higher emissivity values. Over land, the emissivity depends on the moisture content of the soil. Wetting of a soil surface results in a rapid decrease in emissivity. The emissivity of dry soil is on the order of 0.95 to 0.97, while for wet bare soil it is about 0.80 to 0.90, depending on the frequency. The surface emissivity appearing in the first term has a significant effect on the brightness temperature value.

7.4.2 Atmospheric Water Information from Microwave Sensing

One of the important applications of microwave sounding has been the determination of atmospheric liquid water and water vapor amount since microwaves see through heavy clouds and precipitation, which, however, have large opacity in the infrared wavelengths. In order to derive the liquid water and water vapor amount, it is necessary to develop an empirical equation in which the liquid water content and water vapor amount are explicitly given.

We return to Eq. (7.130) and perform integration by parts on the two integral terms to obtain

$$T_B(\tilde{\nu}) = \left[T(0) - \int_{p_s}^{0} \mathcal{T}_{\tilde{\nu}}(p) \frac{\partial T(p)}{\partial p} \, dp \right] - (1 - \varepsilon_{\tilde{\nu}}) \mathcal{T}_{\tilde{\nu}}^2(p_s)$$

$$\times \left[T(0) + \frac{1}{\mathcal{T}_{\tilde{\nu}}(p_s)} \int_{0}^{p_s} \mathcal{T}_{\tilde{\nu}}(p_s) \frac{\partial T(p)}{\partial p} \, dp \right], \tag{7.135}$$

where for simplicity we use one pressure variable in the argument of the transmittance. Moreover, we define

$$x_{\tilde{v}} = \frac{T(0)}{T_s} - \frac{1}{T_s} \int_{p_s}^{0} \mathcal{T}_{\tilde{v}}(p) \frac{\partial T(p)}{\partial p} dp$$

$$= 1 + \frac{1}{T_s} \int_{p_s}^{0} [1 - \mathcal{T}_{\tilde{v}}(p)] \frac{\partial T(p)}{\partial p} dp,$$

$$\tag{7.136}$$

$$y_{\tilde{v}} = \frac{T(0)}{T_s} + \frac{1}{\mathcal{T}_{\tilde{v}}(p_s)T_s} \int_{0}^{p_s} \mathcal{T}_{\tilde{v}}(p) \frac{\partial T(p)}{\partial p} dp$$

$$= \frac{1}{\mathcal{T}_{\tilde{v}}(p_s)} + \frac{T(0)}{T_s}\left[1 - \frac{1}{\mathcal{T}_{\tilde{v}}(p_s)}\right] + \frac{1}{\mathcal{T}_{\tilde{v}}(p_s)T_s} \int_{0}^{p_s} [1 - \mathcal{T}_{\tilde{v}}(p)] \frac{\partial T(p)}{\partial p} dp.$$

Hence,

$$T_B(\tilde{v}) = T_s[x_{\tilde{v}} - y_{\tilde{v}}\mathcal{T}_{\tilde{v}}^2(p_s)(1 - \varepsilon_{\tilde{v}})]. \tag{7.137}$$

On inspection of the microwave spectrum, we find that below about 40 GHz, the transmittance $\mathcal{T}_{\tilde{v}}(p_s) \approx 1$ so that $x_{\tilde{v}} \approx y_{\tilde{v}} \approx 1$. As a result of this simplification, the brightness temperature may be approximated by

$$T_B(\tilde{v}) \approx T_s[1 - \mathcal{T}_{\tilde{v}}^2(p_s)(1 - \varepsilon_{\tilde{v}})]. \tag{7.138}$$

The transmittance for frequencies lower than 40 GHz is mainly due to the absorption of water vapor and liquid water. It may be expressed by

$$\mathcal{T}_{\tilde{v}}(p_s) = \mathcal{T}_{\tilde{v}}(\text{vapor})\mathcal{T}_{\tilde{v}}(\text{liquid}). \tag{7.139}$$

For frequencies lower than about 40 GHz, the transmittance of liquid water may be approximated by

$$\mathcal{T}_{\tilde{v}}(\text{liquid}) \approx \exp(-Q/Q_0) \approx 1 - Q/Q_0(\tilde{v}), \tag{7.140}$$

where $Q_0(\tilde{v})$ is a constant which depends on the frequency and cloud temperature. In a similar manner, if we select a frequency at about 22 GHz, the water vapor transmittance may be approximated by

$$\mathcal{T}_{\tilde{v}}(\text{vapor}) \approx 1 - W/W_0(\tilde{v}), \tag{7.141}$$

where again $W_0(\tilde{v})$ is a constant. Upon substituting these two expressions into Eq. (7.138), and neglecting second-order terms involving Q and W, we obtain

$$T_B(\tilde{v}) \approx \varepsilon_{\tilde{v}}T_s + 2(1 - \varepsilon_{\tilde{v}})T_s(Q/Q_0 + W/W_0). \tag{7.142}$$

Assuming that the surface temperature T_s and surface emissivity $\varepsilon_{\tilde{v}}$ are known parameters, it is evident that two brightness temperature observations at about 40 and 22 GHz can be used to determine Q and W. It is straightforward to show that

$$Q = q_0 + q_1 T_B(\tilde{v}_1) + q_2 T_B(\tilde{v}_2),$$
$$W = w_0 + w_1 T_B(\tilde{v}_1) + w_2 T_B(\tilde{v}_2),$$

(7.143)

where w and q are quantities related to the frequencies chosen, the surface temperature, emissivity, and empirical parameters Q_0 and W_0. Normally, q_i and w_i are determined statistically from a sample of known brightness temperatures and liquid water and water vapor amount in known atmospheric profiles.

Staelin *et al.* (1976), and Grody (1976) used these equations to infer water content and total water vapor amount from Nimbus 5 NEMS 22.235 and 31.4 GHz channels over the ocean. The coefficients were obtained by a multidimensional regression analysis based upon computed brightness temperatures with known atmospheric temperature and water profiles from radiosondes. Computations involving the contribution of clouds and precipitation in the brightness temperature did not include the scattering contribution due to cloud and rain drops. These authors have demonstrated the feasibility of mapping the large scale features of liquid water and total water vapor patterns from the water vapor (22.235 GHz) and window (31.4 GHz) channel data. Recently, Liou and Duff (1979) derived the liquid water content from Nimbus 6 SCAMS (Scanning Microwave Spectrometer) data utilizing Eqs. (7.143) over land. In their simulation study of microwave radiative transfer in nonprecipitating and precipitating cloudy atmospheres, they have taken into account effects of scattering and absorption properties of hydrometeors in an inhomogeneous absorbing gaseous atmosphere. As pointed out previously, the emissivity over land varies with wet or dry condition of the surface. Moreover, satellite observations over the land surface do not give information of the wet or dry conditions. Thus, they developed an empirical means using observed brightness temperatures for SCAMS channels 2 and 3 to determine the surface characteristics, and showed some success in the inference of meso-scale liquid water content for a data base during a two-week period in August 1975. The problem of mapping atmospheric moisture over land, however, is an area still requiring further investigation and verification.

7.4.3   Temperature Retrieval from Microwave Sounders

The basic concept of inferring atmospheric temperatures from satellite observations of thermal microwave emission in the oxygen spectrum was developed by Meeks and Lilley (1963) in whose work the microwave weighting functions were first calculated. The prime advantage of microwave over infrared temperature sounders is that the longer wavelength microwaves are much less influenced by clouds and precipitation. Consequently, microwave sounders can be effectively utilized to infer atmospheric temperatures in all-weather conditions.

The first application of microwave techniques for the temperature profile determination from an orbiting satellite was the Nimbus 5 Microwave Spectrometer experiment (NEMS). The experiment was designed to evaluate passive microwave techniques for use on operational meteorological satellites. NEMS consisted of three channels centered at frequencies 53.65, 54.90, and 58.80 GHz in the oxygen band. The 53.65 GHz channel whose peak of the weighting function is near the surface is effected by the surface emissivity. The approach to recover the temperature profile utilizing the microwave sounder has been by means of the statistical method discussed in Section 7.3.3.2. Basically, the predictor **D** matrix is derived from a priori atmospheric data from radiosonde observations. Waters *et al.* (1975) presented the recovered temperature profiles from NEMS data in which the data vector of the brightness temperature was composed of seven values based on three measured data. Influence of land, water, or a combination of the two on the temperature retrieval was taken into account by generating different **D** matrices for these three conditions.

The first microwave sounder intended for operational use was flown aboard the Air Force DMSP (Defense Meteorological Satellite Program) Block 5D satellite system lauched in June 1979. This microwave sensor (SSM/T) contains seven channels at 50.5, 53.2, 54.35, 54.9, 58.4, 58.825, and 59.4 GHz. Because of the surface reflectivity effect, the weighting function $K_{\tilde{v}}(p)$ is defined by

$$T_{B}(\tilde{v}) = \varepsilon_{\tilde{v}} T_s \mathscr{T}_{\tilde{v}}(p_s) + \int_{p_s}^{0} T(p) K_{\tilde{v}}(p)\, dp. \tag{7.144}$$

From Eq. (7.134) we have

$$K_{\tilde{v}}(p) = 1 + (1 - \varepsilon_{\tilde{v}}) \left[\frac{\mathscr{T}_{\tilde{v}}(p_s)}{\mathscr{T}_{\tilde{v}}(p)}\right]^2 \frac{\partial \mathscr{T}_{\tilde{v}}(p)}{\partial p}. \tag{7.145}$$

Displayed in Fig. 7.13 are weighting functions of seven SSM/T channels for an incident (emergent) angle of 0° and a surface emissivity of 0.97 using a nominal spacecraft orbit altitude of 834 km. Absorption due to molecular oxygen and water vapor along with antenna gain characteristics are included in the transmittance calculations. Channel 1 is a window channel responding strongly to the earth's surface characteristics, dense clouds, and rain. It is used to correct the other channels for these background effects. The weighting

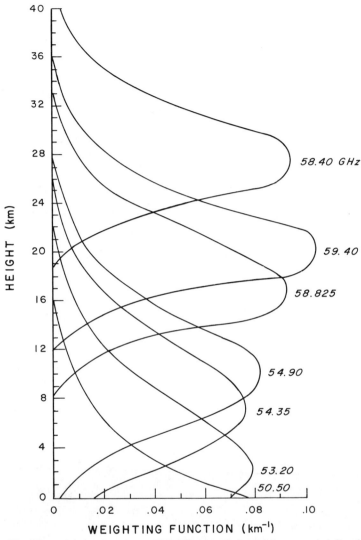

Fig. 13   The weighting functions of SSM/T channels (nadir) for an emissivity of 0.97.

function peaks of channels 1 to 4 are below about 10 km, and therefore, there would be some effects of dense clouds and precipitation on the temperature retrieval.

Because of the surface emissivity effect, it would be desirable to remove it in the statistical method of temperature retrieval so that the predictor matrix **D** could be constructed over all surface conditions (Rigone and Strogryn,

1977). For simplicity of analyses, we define

$$T_d(\tilde{v}) = \int_0^{p_s} T(p) \frac{\partial \mathscr{T}_{\tilde{v}}(p)}{\partial p} \, dp$$

$$T_u(\tilde{v}) = \int_{p_s}^0 T(p) \frac{\partial \mathscr{T}_{\tilde{v}}(p)}{\partial p} \, dp, \qquad (7.146)$$

so that Eq. (7.130) is written in the form

$$T_B(\tilde{v}) = \varepsilon_{\tilde{v}} T_s \mathscr{T}_{\tilde{v}}(p_s)\left[1 - T_d(\tilde{v})/T_s\right] + T_a(\tilde{v}), \qquad (7.147a)$$

$$T_a(\tilde{v}) = T_u(\tilde{v}) + T_d(\tilde{v})\mathscr{T}_{\tilde{v}}(p_s). \qquad (7.147b)$$

The second term in the right-hand side of Eq. (7.147a) denotes the contribution to the brightness temperature caused by the atmosphere only, and the surface effects are contained in the first term. Since channel 1 centered at 50.5 GHz has a weighting function peak at the surface, it is utilized in the context of removing the surface contribution for other channels. Based on Eq. (7.147a), we may define the contribution to the brightness temperature caused by the atmosphere only for channels 2 to 7 in the form

$$T_a(\tilde{v}_j) = T_B(\tilde{v}_j) - \left[T_B(\tilde{v}_1) - T_a(\tilde{v}_1)\right]a(\tilde{v}_j), \qquad j = 2, 3, \ldots, 7, \quad (7.148)$$

where

$$a(v_i) = \frac{\varepsilon_{\tilde{v}_i} T_s \mathscr{T}_{\tilde{v}_i}(p_s)\left[1 - T_d(\tilde{v}_i)/T_s\right]}{\varepsilon_{\tilde{v}_1} T_s \mathscr{T}_{\tilde{v}_1}(p_s)\left[1 - T_d(\tilde{v}_1)/T_s\right]}, \qquad \text{and} \qquad a(\tilde{v}_1) = 1.$$

The statistical method described in Section 7.3.3.2 assumes a correlation between the atmospheric temperature and the measured data, which in the present case is T_a given by Eq. (7.148). Thus,

$$
\begin{aligned}
(\hat{T}_i - \bar{T}_i) &= \sum_j D_{ij}(\hat{T}_{aj} - \bar{T}_{aj}) \\
&= \sum_j D_{ij}\left[\hat{T}_{Bj} - (\hat{T}_{B1} - \hat{T}_{a1})a_j - \bar{T}_{aj}\right] \\
&= \sum_{j \neq 1} D_{ij}\hat{T}_{Bj} - \hat{T}_{B1}\sum_j D_{ij}a_j + \sum_j D_{ij}(\hat{T}_{a1}a_j - \bar{T}_{aj}), \quad (7.149)
\end{aligned}
$$

where we note that T_{B1} is not defined in Eq. (7.148), and so the first term contains $j = 2, \ldots, 7$. In matrix notations, we find

$$\hat{\mathbf{T}} = \mathbf{D}'\hat{\mathbf{T}}_B + \mathbf{R}, \qquad (7.150)$$

where

$$\mathbf{R} = \bar{\mathbf{T}} + \hat{\mathbf{T}}_{a1}\mathbf{Da} - \mathbf{D}\bar{\mathbf{T}}_a,$$

and $\mathbf{D}'$ is a matrix whose first column is $-\mathbf{Da}$ and whose remaining columns are the columns of $\mathbf{D}$. It is clear that the retrieval technique contains elements depending mainly on the atmosphere but not on the surface, and so it should be valid over land, water, or mixed surface conditions. The $\mathbf{D}$ [see Eq. (7.72)] and $\mathbf{R}$ could be determined from a large number of upper air soundings for a wide range of meteorological conditions which have been achieved over the years and the brightness temperatures calculated for a given atmosphere.

Shown in Fig. 7.14 is an exercise of temperature retrieval using the statistical covariance method. The midlatitude Spring/Fall profile of a standard atmosphere (solid curve) is used and the observed brightness temperatures used for the seven SSM/T channels are values theoretically calculated. The exercise has been carried out for cases over ocean and land. It is apparent that the procedure outlined above has very successfully removed surface effects from the temperature retrieval. Also shown are the temperature retrievals when a 2-km-thick precipitation layer with a base height set at 1 km, having various rainfall rates, have been added to the atmosphere. It is seen that the surface temperature suffers increased degradation as the

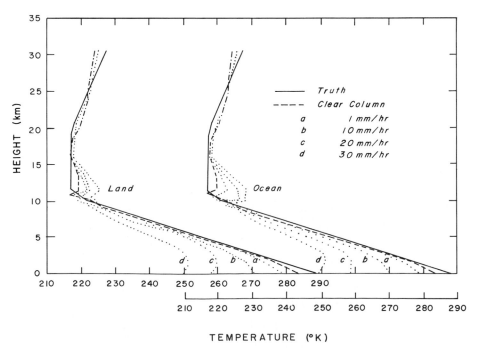

Fig. 7.14 Hypothetical temperature retrieval exercise over land and ocean using the statistical covariance method.

rainfall rate increases. Large errors in the recovered temperature profile would be anticipated, even with microwave sounders, when the atmosphere within the satellite field of view contains precipitation and heavy clouds.

7.5   LIDAR (OR RADAR) BACKSCATTERING

During the sixties, the advent of the laser as a source of energy opened up a number of possibilities for new remote sensing techniques of probing the atmosphere. The laser energy at optical frequencies is highly mono-chromatic and coherent. And with the development of Q-switching tech-niques, very short pulses of high power could be generated. The recognition of the applicability of the high power pulsed laser by a number of atmospheric scientists (e.g., Schotland, 1969; Collis, 1969) has prompted the development of backscattering techniques for the detection of the composition and struc-ture of clouds, aerosols, and minor gases in the atmosphere. The word *lidar*, which originally stands for *LI*ght *D*etection *A*nd *R*anging, is an acronym analogous to radar which utilizes energy source in the microwave region. Perhaps, it may be appropriate to speak of lidar as a laser radar. Since the advance of laser radar, which employs the same backscattering principle as microwave radar, techniques have been developed to map and track con-centrations of particulate matter, to study the density profile of the atmo-sphere, and cloud compositions. More recently, advanced developments of lidar techniques for atmospheric probes have proved fruitful. These include the use of multiple wavelength lidars for determining the minor gaseous composition by means of differential absorption techniques, the use of Doppler techniques for determining the motion of particulates and molecules, the use of depolarization techniques for inferring the water and ice in clouds, and the use of the Raman scattering technique in which a weak scattering occurs at a shifted wavelength for water vapor measurements. In this section, in consistency with the theory of light scattering and radiative transfer introduced previously, we present the basic lidar (or radar) equation which is fundamental to all backscattering techniques, and discuss the possible difficulties encountered in the interpretation of backscattering returns.

From the scattering theory developed in Chapter 5, the scattered flux density for a single particle or molecule can be expressed as [see Eqs. (5.84) and (5.111a,b), also Eq. (3.65)]

$$F_{l,r}^{s} = F_{l,r}^{i} \, \frac{\sigma_s}{r^2} \frac{P_{2,1}(\Theta)}{4\pi}, \tag{7.151}$$

where the subscripts $l(2)$ and r(1) denote the radiation component parallel and perpendicular to the scattering plane, respectively. Also, r is the distance at which scattering takes place, σ_s the scattering cross section, and $P(\Theta)$ the phase function. Neglect the subscripts on the flux density and phase function for the convenience of the following discussions; the backscattered flux density due to a single particle may be written in the form

$$F^s(\pi) = F^i \frac{\sigma_s}{r^2} \frac{P(\pi)}{4\pi}. \tag{7.152}$$

At this point, it may be convenient to define the *backscattering cross section* σ_π as the area which, when multiplied by the incident flux density, gives the total power radiated by an isotropic source such that it radiates the same power in the backward direction as the scatterer. Thus,

$$F^i \sigma_\pi = F^s(\pi) 4\pi r^2, \tag{7.153}$$

where $4\pi r^2$ represents the surface area of a sphere. Consequently, the backscattering cross section is given by

$$\sigma_\pi = \sigma P(\pi). \tag{7.154}$$

Let P_t denote the transmitted power so that the incident flux density may be expressed by

$$F^i = P_t/A_t, \tag{7.155}$$

where A_t is the cross sectional area at distance r. Let A_r be the collecting aperture; it follows that the backscattered power received is given by

$$P_{r0} = F^s(\pi) A_r = \frac{P_t}{A_t} \frac{\sigma P(\pi)}{4\pi r^2} A_r. \tag{7.156}$$

In view of the volume scattering cross section given by Eq. (5.116), we may define an averaged scattering cross section as

$$\bar{\sigma}_s = \beta_s/N,$$

where N denotes the particle number density. Hence, after performing the particle-size distribution integration, the averaged backscattered power is given by

$$\bar{P}_{r0} = P_t \frac{A_r}{r^2} \frac{P(\pi)}{4\pi} \frac{\bar{\sigma}_s}{A_t}. \tag{7.157}$$

In reference to Fig. 7.15, let the pulse length transmitted by a lidar (or radar) system be Δh. For a given instance of time, the lidar receiver collects the scattered energy from half of the pulse length. That is to say, the scattering signal of the bottom portion returns simultaneously as the top portion which undergoes round trip backscattering event. The total number of particles

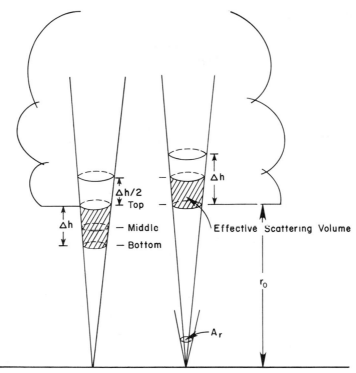

Fig. 7.15   The backscattering geometry of a pulsed lidar system where r_0 denotes a reference range.

within the effective scattering volume would therefore be $N A_t \, \Delta h/2$. Upon utilizing the definition of the volume scattering coefficient, the backscattered power is now given by

$$\bar{P}_{r0} = P_t \frac{A_r}{r^2} \frac{P(\pi)}{4\pi} \beta_s \frac{\Delta h}{2}. \tag{7.158}$$

During the backscattering event, the energy pulse also undergoes attenuation. On the basis of the Beer–Bouguer–Lambert law developed in Section 1.4.2, the actual backscattered power at the point of the receiver is

$$\bar{P}_r = \bar{P}_{r0} \exp\left\{ -2 \int_0^r \beta_e(r') \, dr' \right\}, \tag{7.159}$$

where 2 represents the round trip attenuation effect, $r = 0$ is the position corresponding to $\bar{P}_{r0}$, and β_e is the volume extinction coefficient including both scattering and absorption effects. Inserting Eq. (7.158) into Eq. (7.159), the backscattered power observed by the receiver now may be written

$$\bar{P}_r(r) = P_t \frac{A_r}{r^2} \frac{P(\pi)}{4\pi} \beta_s \frac{\Delta h}{2} \exp\left\{-2\int_0^r \beta_e(r')\,dr'\right\}. \qquad (7.160a)$$

This is the basic lidar (or radar) equation. Note here that r is the range. In this development, we have neglected the energy gains corresponding to the transmitter and receiver. For lidar or radar applications, it is customary to use the volume backscattering coefficient similar to that given in Eq. (7.154), i.e., $\beta_\pi = P(\pi)\beta_s$. Thus, we rewrite Eq. (7.160a) to obtain

$$\bar{P}_r(r) = \frac{P_t A_r \beta_\pi(r)\,\Delta h}{8\pi r^2} \exp\left\{-2\int_0^r \beta_e(r')\,dr'\right\}, \qquad (7.160b)$$

where the collecting aperture A_r, the transmitted power P_t, and the pulse length Δh are all known parameters. The range r is a function of time t. But there are two unknown parameters β_π and β_e which relate to the optical properties and concentration of particles and/or molecules. It is not possible to evaluate the information content of the return power in absolute terms unless the volume backscattering coefficient β_π, and the volume extinction coefficient β_e are uniquely related.

For Rayleigh scattering, the scattering phase function for unpolarized light at the backscattering direction from Eq. (3.64) is

$$P^R(\pi) = \tfrac{3}{4}(1 + \cos^2 \pi) = 1.5.$$

Thus,

$$\beta_\pi^R = 1.5\beta_s^R.$$

Clearly, the ratio of backscattering to the extinction coefficient in this case is a constant of 1.5 and is not subject to fluctuations. For a single Mie spherical particle, however, the scattering phase function depends on the size parameter $2\pi a/\lambda$ and is characterized by strong forward scattering (see Fig. 5.4). $P^M(\pi)$ as a function of the size parameter fluctuates greatly and is normally less than unity. But for a sample of polydispersed spheres, the fluctuations tend to average out [see Figs. (5.11) and (5.12)], and useful approximate values can be determined for $P^M(\pi)$. For water clouds it has been found that for many cases

$$\beta_\pi^M \approx 0.625\beta_s^M.$$

This relation is also found to be a reasonable approximation for spherical aerosols.

Rayleigh scattering parameters are known and are constants. Thus, assuming that a unique relation between the volume backscattering coefficient and volume extinction coefficient for particles could be derived, then from

Eq. (7.160b) a profile of volume extinction coefficients for particles can be inferred from the continuous backscattering power returned to the detector. This can be done by means of successive determination of the volume extinction coefficients from the edge to the center of the target area. The volume extinction coefficients are in turn related to the sizes and concentrations of aerosols and cloud particles.

So far, we have neglected polarization considerations in the development of lidar equation. In reference to Eq. (7.156), the transmitted power P_t can be generated to be vertically or horizontally polarized, while the detector can be devised so that both polarization components are measured. This provides additional information in regard to the particle characteristics. Because of the geometrical symmetry, backscattering from spheres retains the same polarization state as the incident energy. On the contrary, however, backscattering from nonspherical particles produces a crosspolarized component in addition to the polarization state of the incident energy. The ratio of the cross-polarized component to the component which retains the same polarization as the incident beam is defined as the backscattering *depolarization ratio*. The depolarization ratio of the backscattering return, if only single scattering is considered, is associated with the deviation of the particle shape from the symmetrical sphere. Thus, discrimination between spherical water drops and irregular ice crystals has been shown to be feasible through the interpretation of depolarization values. The ice and water information is of vital importance in the investigation of cloud microstructures and in weather modification experiments. One complication in the backscattering and depolarization experiments has been the effects of multiple scattering. Through multiple scattering events, the incident electric vector is transferred from the initial reference plane to the plane of scattering, and therefore, partial depolarization is produced. The relative significance of multiple scattering in the backscattering experiment when cloud particles are involved in general is associated with the field of view of the detector. Although progress has been made in the last 10 years or so, the definitive and quantitative assessment of multiple scattering effects in the backscattering experiment is still an area requiring further investigation.

EXERCISES

7.1 (a) Derive Eq. (7.9) from Eqs. (7.7) and (7.8) and show that

$$k = \pi(2\pi)^{v^*-2} \int_0^\infty C(h)\,dh \int_{x_1}^{x_2} \frac{Q_e(x)\,dx}{x^{v^*-1}},$$

where the extinction efficiency Q_e [see Eq. (5.94)] is expressed in terms of the size parameter x.

(b)   Direct solar radiation measurements are made with a multiple wavelength radiometer. The aerosol optical depth derived from the observation are 0.17 and 0.1 at 0.5 and 0.8 μm, respectively. Assuming a Junge size distribution and a constant k, what would be the shaping factor v ?

7.2   Prove Eq. (7.60) from Eq. (7.58) by assuming $M = N = 2$. Using the first, second, and third differences, derive the H matrices similar to that shown in Eq. (7.61).

7.3   Consider the following Fredholm equation of the first kind

$$g(k) = \int_0^1 e^{-kx} f(x)\,dx,$$

where the kernel is given by a simple exponential function. Let the unknown function be given in the form

$$f(x) = x + 4x(x - \tfrac{1}{2})^2.$$

(a)   Derive an analytical expression for $g(k)$ and compute $g(k_i)$ for k_i in the interval $(0, 10)$ using $\Delta k_i = 0.5$ $(i = 1, 2, \ldots, 20)$.
(b)*   Write the integral equation in a summation form as

$$g(k_i) = \sum_{j=1}^{20} f(x_j) e^{-k_i x_j} \Delta x_j, \qquad i = 1, 2, \ldots, 20.$$

Let $\Delta x_i = 0.05$, compute $g(k_i)$ again, and compare with those computed from the exact integration.
(c)*   Let

$$A_{ij} = e^{-k_i x_j} \Delta x_j.$$

Compute $\|A_{ij}\|$ which is a 20×20 matrix. Use the direct linear inversion method to recover $f(x_j)$. Compare the retrieved results with the exact values.
(d)*   Perform the inversion again utilizing the linear constrained method [Eq. (7.62)], and let the smoothing parameter γ be 1 and 10^{-7}. Compare the retrieved results with exact values.

7.4   Given the temperature profile and transmittances for the six VTPR channels (669.0, 676.7, 694.7, 708.7, 723.6, and 746.7 cm^{-1}) in the accompanying table:
(a)   Compute and plot the weighting functions $\Delta \mathcal{T}_i(p)/\Delta \ln p$ as functions of the pressure on a logarithmic scale. What is the physical meaning of the weighting function?
(b)*   Assume that the calculated radiances are the values observed from the NOAA 4 VTPR instrument, recover the temperature profile from

* Simple computer programming is required.

Atmospheric Profiles and Transmission Functions for VTPR *Channels*

Pressure (mb)	Temp (°K)	Transmittances					
		1	2	3	4	5	6
0.8	270.7	.9198	.9817	.9890	.9922	.9931	.9968
1.4	265.0	.8846	.9733	.9837	.9891	.9906	.9953
2.1	256.4	.8429	.9637	.9777	.9861	.9879	.9940
3.1	248.3	.7979	.9508	.9704	.9817	.9848	.9925
4.4	241.9	.7520	.9344	.9612	.9783	.9810	.9906
5.9	238.2	.7061	.9139	.9497	.9732	.9763	.9885
7.9	235.0	.6591	.8890	.9358	.9670	.9709	.9859
10.3	232.7	.6094	.8591	.9188	.9597	.9645	.9828
13.1	228.8	.5562	.8239	.8980	.9506	.9570	.9790
16.5	226.3	.5001	.7831	.8740	.9403	.9485	.9747
20.4	222.5	.4423	.7369	.8470	.9290	.9392	.9701
24.9	221.1	.3840	.6853	.8168	.9167	.9290	.9652
30.2	219.5	.3266	.6291	.7831	.9033	.9180	.9600
36.1	219.0	.2716	.5691	.7458	.8887	.9065	.9546
42.9	218.5	.2203	.5064	.7051	.8728	.8945	.9489
50.5	217.9	.1738	.4424	.6609	.8555	.8821	.9431
59.1	217.4	.1329	.3785	.6136	.8366	.8695	.9370
68.6	216.8	.0980	.3160	.5638	.8162	.8567	.9307
79.2	217.3	.0693	.2563	.5119	.7941	.8437	.9241
90.9	218.0	.0468	.2008	.4584	.7699	.8304	.9173
103.8	218.8	.0299	.1510	.4043	.7436	.8163	.9101
117.9	219.7	.0179	.1080	.3508	.7152	.8029	.9026
133.3	220.7	.0100	.0727	.2988	.6847	.7884	.8946
150.2	221.7	.0052	.0456	.2496	.6520	.7731	.8861
168.5	222.6	.0024	.0264	.2042	.6175	.7570	.8771
188.4	223.6	.0010	.0139	.1634	.5812	.7397	.8674
209.9	225.2	.0004	.0066	.1275	.5431	.7212	.8569
233.1	227.5	.0001	.0028	.0968	.5033	.7011	.8454
258.0	229.7	.0000	.0011	.0711	.4615	.6792	.8325
284.8	230.2	.0000	.0004	.0508	.4195	.6561	.8187
313.6	231.8	.0000	.0001	.0354	.3782	.6321	.8043
344.3	232.8	.0000	.0000	.0237	.3365	.6064	.7883
377.2	234.2	.0000	.0000	.0151	.2940	.5782	.7701
412.2	235.5	.0000	.0000	.0090	.2514	.5475	.7493
449.5	236.9	.0000	.0000	.0050	.2099	.5142	.7253
489.2	241.6	.0000	.0000	.0026	.1706	.4785	.6992
531.2	245.4	.0000	.0000	.0012	.1343	.4402	.6687
575.8	249.0	.0000	.0000	.0005	.1017	.3993	.6326
623.1	252.8	.0000	.0000	.0002	.0740	.3565	.5917
673.0	256.8	.0000	.0000	.0000	.0516	.3127	.5467
725.7	260.5	.0000	.0000	.0000	.0346	.2689	.4983
781.3	263.9	.0000	.0000	.0000	.0221	.2261	.4476
839.9	267.5	.0000	.0000	.0000	.0134	.1852	.3952
901.5	272.1	.0000	.0000	.0000	.0076	.1456	.3371
966.3	277.0	.0000	.0000	.0000	.0039	.1064	.2682
1019.8	279.5	.0000	.0000	.0000	.0019	.0770	.2099

these radiances utilizing the relaxation method outlined in Section 7.3.4.1 Use a linear interpolation between the recovered temperatures and use the true temperatures for the surface and the top layer. Plot the retrieved temperature profile in a logarithmic scale and compare with the true temperature profile.

SUGGESTED REFERENCES

Derr, V. E., Ed. (1972). *Remote Sensing of the Troposphere.* Supt. of Documents, US Govt. Printing Office, Washington, D.C. This book collects for the first time organized review papers on the diverse subjects of passive and active remote sensing.

Fymat, A. L., and Zuev V. E., Eds. (1978). *Remote Sensing of the Atmosphere: Inversion Methods and Applications.* Elsevier, New York. This book collects papers representing current research on remote sensing problems associated with the temperature and composition profiles and the aerosol size distribution.

Hinkley, E. D., Ed. (1976). *Laser Monitoring of the Atmosphere.* Springer-Verlag, New York. This book contains updated state-of-the-art review papers in the field of active remote sensing by means of lidar.

Rodgers, C. D. (1976). Retrieval of atmospheric temperature and composition from remote measurements of thermal radiation. *Rev. Geophys. Space Phys.* **14**, 609–624. This paper provides a comprehensive review on various inversion methods.

Twomey, S. (1977). *Introduction to the Mathematics of Inversion in Remote Sensing and Indirect Measurements.* Elsevier, New York. This book presents the fundamental theory and background materials on the inversion process in remote sensing.

Chapter 8
RADIATION CLIMATOLOGY

8.1 GENERAL SURVEY OF RADIATION BUDGET STUDIES OF THE EARTH-ATMOSPHERE

Knowledge about the radiation budget of the earth–atmosphere dates back to the work of Simpson in 1928, which was based on a great many simplifying assumptions and very sketchy radiative transfer data. It was not until 1954 that Houghton made comprehensive calculations of the annual heat balance of the northern hemisphere. Houghton utilized the Elsasser radiation chart described in Chapter 4 to compute the infrared flux at the top of the atmosphere, and used observations from a pyrheliometric network to derive the solar radiation reaching the surface. Rayleigh scattering and absorption by water vapor and carbon dioxide were considered through somewhat crudely parameterized methods. Absorption by ozone, however, was neglected. Houghton performed a useful computation of zonally average surface albedos, which had not been done previously. Upon using the available information about cloud albedos, he computed reasonable values for the mean annual albedo of the hemisphere and zonally averaged latitude belts. The global albedo was determined to be 0.34.

London (1957) developed a radiation balance model for the northern hemisphere, which included for the first time results for the vertical, latitudinal, and seasonal distributions of radiative heating and cooling, and vertical fluxes of solar and thermal infrared radiation. The Elsasser radiation chart again was used to calculate thermal infrared flux, while empirical expressions were employed for the absorption and scattering of solar radiation by water vapor and aerosols. Ozone and carbon dioxide effects were not considered in the calculation. A cloud distribution consisting of heights and thicknesses of six cloud types was used along with their climatological values of fractional cloudiness at 10-degree latitude belts. This cloud climatology is the only one of its kind still in existence. London derived the global albedo to be 0.35.

Davis (1963) presented the atmospheric heat budget along with computations of net flux of latent heat and sensible heat from the earth's surface to the atmosphere for the latitude belt 20°N and 70°N. Infrared cooling rates were computed with the use of several approximations for flux transmittances. Empirical expressions for absorption by water vapor, carbon dioxide, and ozone were used to compute the solar heating rates. Scattering by clouds and aerosols, however, was not considered in the study.

The radiation budget model of Katayama (1966) for the northern hemisphere troposphere is extremely detailed and complete, including seasonal, latitudinal, zonal, and hemispheric distributions of the radiative heating, and a comprehensive discussion of energy balances. Graphical methods and the Yamamoto radiation chart were used to compute thermal infrared fluxes. Clouds were treated as blackbodies in the infrared, with the exception of cirrus, which was considered to be gray. In the solar spectrum, Katayama relied upon empirical equations integrated over the entire spectral range for absorption by water vapor, Rayleigh scattering, reflection by clouds, and attenuation by dust. Effects of cloud and aerosol scattering were accounted for only by simplified approximations. Katayama first obtained a global albedo of 0.374; from this he subtracted 2.8% of the incident solar flux to account for absorption by stratospheric ozone which was not considered, and obtained a corrected value of 0.346. This value is very close to the values of 0.34 and 0.35 obtained by Houghton and London, respectively. Moreover, Katayama also derived useful values for the latitudinal distribution of zonally averaged albedos for January and July in the northern hemisphere.

Rodgers (1967) calculated the radiative energy budget for the region 0–70° north, and from 1000 to 10 mb for January, April, July, and October. In this work, parameterized equations were used in both the solar and infrared spectra. The Goody random model was used with the Curtis–Godson adjustment for water vapor and carbon dioxide absorption (see Chapter 4). Rodgers assumed all clouds to be black in the infrared region, except cirrus, which he took to be 50% black. In the solar spectrum, Rodgers used a simply modeled ray-tracing technique, following individual rays through the atmosphere from the top to the final destination in the atmosphere, on the surface, or back in space. For clouds, a single absorption coefficient was employed throughout the entire solar spectrum. Aerosols, however, were not considered in the analyses.

The radiation budget of the southern hemisphere had been largely ignored until Sasamori *et al.* (1972) performed their comprehensive calculations. This study followed the general techniques of Houghton, London, and Katayama. Calculations of radiation fluxes in the vertical were reduced to integration of formulas for upward flux at the top of the atmosphere and downward flux at the bottom. Thus, the study does not present vertical distributions of the radiative parameters. Distributions of the fractional cloud amount were obtained from recent observed data. London's values for heights of cloud tops and bases again were used since there is little data available for the southern hemisphere. Sasamori *et al.* compared their results with computations of other researchers and with the satellite observations. The global albedo derived for the southern hemisphere by Sasamori *et al.* is 0.347.

Dopplick (1972) reported on radiative heating of the global atmosphere, and in this study provided monthly and annual zonal mean global heating rates in the form of latitudinal cross sections. Dopplick also presented seasonal profiles of the contribution of each atmospheric constituent. For infrared transfer calculations, in addition to the use of the Goody random model with the Curtis–Godson approximation, continuum absorption also was utilized to represent water vapor transmission. Empirical fits for measured band absorption were used to compute the solar heating rate. Three classes of the cloud distribution, determined largely from satellite observations, were considered. However, cloud scattering, and scattering and absorption by aerosols were not considered.

In a recent paper, Hunt (1977) addressed the sensitivity of the various components of the radiation budget to changes in cloud properties. A zonally averaged model atmosphere for both northern and southern hemispheres and a simple radiative transfer model of the sort to be used in a general circulation model were employed to calculate the sensitivity of heating and cooling to changes of cloud conditions and cloud radiative properties.

More recently, in an effort to investigate the climatic effects of increasing cirrus cloudiness in the northern hemisphere, Freeman and Liou (1979) constructed a comprehensive atmospheric radiation budget model based on rigorous transfer methods for solar and thermal infrared radiation. Particular emphasis was focused on the scattering effects of clouds and aerosols. Their analyses and some of their resulting calculations form the basis for discussions presented in Section 8.4.

Satellite observations of the global radiation budget were reported by Vonder Haar and Suomi (1971). Three very important radiation components were given in this study. These included the reflected solar flux, absorbed solar flux, and the outgoing thermal infrared flux at the top of the atmosphere. Seasonal and annual latitudinal variations as well as global horizontal distributions of these components were presented. In addition, significant radiation budget studies from satellite observations also were carried out by Winston (1969), Raschke and Bandeen (1970), and more recently by Raschke *et al.* (1973) and Smith *et al.* (1977).

In this chapter we introduce broadband radiation observations, and radiation balance studies from satellites. These discussions include radiation budget studies of latitudinal zones and of the globe, as well as the importance of the radiation budget component in the global energy balance. Following these discussions, we present some significant radiation budget studies, including heating and cooling within the atmosphere, and the total radiation balance at the top of the atmosphere and the surface. Finally, simple radiation and climatic models are introduced.

8.2 BROADBAND RADIATION OBSERVATIONS FROM SATELLITES

8.2.1 Low Resolution Wisconsin Sensors

The first generation (TIROS-type) satellites carried the so-called Wisconsin sensors designed by Suomi and described in detail by Suomi *et al.* (1967). The instrument consists of a matched pair of spherical black and white sensors utilizing thermistor detectors to measure the sensor temperature. In a short time after exposure to various radiative components involving the direct solar flux, solar flux reflected by the earth and atmosphere (short-wave), and thermal infrared flux emitted by the earth and atmosphere (long-wave), each sensor achieves radiative equilibrium. It is assumed that the absorptivity of the black sensor A_b is the same for short-wave and long-wave radiation. However, the absorptivity of the white sensor for short-wave and long-wave radiation are given by A_w^s and A_w^l, respectively.

Let the temperatures measured by the black sensors and white sensors be T_b and T_w, respectively. On the basis of the Stefan–Boltzmann and Kirchhoff laws, introduced in Chapter 1, radiative equilibrium equations for both sensors may be expressed by

$$4\pi r^2 A_b \sigma T_b^4 = \pi r^2 A_b (F_0 + F_S' + F_{IR}'), \tag{8.1}$$

and

$$4\pi r^2 A_w^l \sigma T_w^4 = \pi r^2 [A_w^s (F_0 + F_S') + A_w^l F_{IR}']. \tag{8.2}$$

These two equations show that the emitted energy per unit time is equal to the absorbed energy per unit time, where $4\pi r^2$ and πr^2 represent the emission and absorption areas, respectively, for the two spherical sensors each with radius r. The flux densities of the reflected short-wave and long-wave radiation for spherical sensors are defined by

$$F_S' = \int_0^\Omega I_S \, d\Omega, \qquad F_{IR}' = \int_0^\Omega I_{IR} \, d\Omega, \tag{8.3}$$

where Ω is the solid angle by which the sensor sees the earth, I_S and I_{IR} are the radiant intensities reflected and emitted from the earth, respectively. Note that F_0 denotes the direct solar flux density.

Upon solving the sum of the short-wave flux densities and the long-wave flux density, we obtain

$$F_0 + F'_S = [4\sigma A^l_w/(A^l_w - A^s_w)](T^4_b - T^4_w), \tag{8.4}$$

and

$$F'_{IR} = [4\sigma/(A^l_w - A^s_w)](A^l_w T^4_b - A^s_w T^4_w). \tag{8.5}$$

The direct solar flux density F_0 can be evaluated from the solar constant, which is specified prior to the experiment.

In order to derive the reflected solar flux density and the emitted thermal infrared flux density in terms of the measured values expressed in Eq. (8.3), the following evaluation procedures are made. According to the definition of the flux density and the isotropic radiation assumption discussed in Chapter 1, the reflected solar flux density is given by

$$F_S = \int_0^{2\pi} \int_0^{\pi/2} I_S(\theta, \phi) \cos\theta \sin\theta \, d\theta \, d\phi = \pi I_S.$$

Thus, the planetary albedo r can be expressed by

$$r = \pi I_S/(F_0 \cos\theta_0), \tag{8.6}$$

where the denominator represents the solar flux density available at the top of the atmosphere normal to the plane-parallel stratification. On the basis of Eqs. (8.3) and (8.6), we have

$$F'_S = \frac{F_0}{\pi} \int_0^{\Omega} r \cos\theta_0 \, d\Omega. \tag{8.7}$$

Since $\cos\theta_0$ does not vary greatly over the viewing area of the satellite, it can be removed from the integral. Moreover, we define the average planetary albedo of the viewing area as

$$\bar{r} = \frac{1}{\Omega} \int_0^{\Omega} r \, d\Omega. \tag{8.8}$$

It follows that

$$\bar{r} = \pi F'_S/(F_0 \cos\theta_0 \Omega). \tag{8.9}$$

In a similar manner, under the assumption of isotropic radiation, the emitted thermal infrared flux density is given by

$$F_{IR} = \pi I_{IR} = \sigma T^4_e, \tag{8.10}$$

where T_e denotes the equivalent blackbody temperature of the earth–atmosphere. Upon defining the average equivalent blackbody temperature of the viewing area in the form

$$\overline{T_e^4} = \frac{1}{\Omega} \int_0^\Omega T_e^4 \, d\Omega, \tag{8.11}$$

we obtain

$$\overline{T_e^4} = \pi F_{IR}'/(\sigma\Omega). \tag{8.12}$$

The solid angle through which the sensor sees the earth is shown in Fig. 8.1 (see Exercise 1.2), and it is given by

$$\Omega = 2\pi \left[1 - \frac{(2a_e h + h^2)^{1/2}}{a_e - h} \right]. \tag{8.13}$$

Hence, the average planetary albedo $\bar{r}$ and equivalent blackbody temperature T_e can be evaluated from the black and white sensors through F_S' and F_{IR}'. The average reflected solar flux density and emitted thermal infrared flux density as functions of location and time now may be expressed, respectively, by

$$\overline{F}_S = \bar{r} F_0 \cos\theta_0, \tag{8.14}$$

and

$$\overline{F}_{IR} = \sigma \overline{T_e^4}. \tag{8.15}$$

Averaging process with respect to time and space will be discussed in the next section.

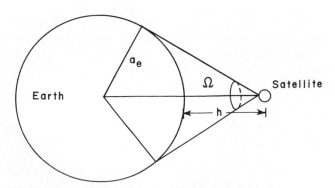

Fig. 8.1   The definition of the solid angle for the earth and satellite system.

8.2.2 Medium Resolution Scanning Radiometer

Radiation budget determination from the first-generation satellites employs self-integrating radiometers. The area coverage is normally large, and the assumption of isotropic radiation is required to derive the reflected solar flux and emitted infrared flux. The medium resolution scanning radiometer on board the Nimbus satellite series is an improved instrument for radiation budget studies. The scanning radiometer permits radiation budget analyses with a higher spatial resolution over areas on the order of 250 km by 250 km (e.g., see Raschke and Bandeen, 1970). In addition, some aspects of the anisotropy of the atmospheric radiation field are incorporated into the data reduction process. Below is a description of the evaluation procedures by which the reflected solar and emitted infrared flux densities are determined.

The daily average reflected flux density of solar radiation is defined by

$$F_S(\lambda, \psi) = \int_{day} F_S[\lambda, \psi; \theta_0(t)] \, dt$$

$$= \int_{day} dt \int_0^{2\pi} \int_0^{\pi/2} I_S[\lambda, \psi; \theta, \phi; \theta_0(t)] \cos\theta \sin\theta \, d\theta \, d\phi, \quad (8.16)$$

where λ and ψ denote the latitude and longitude, respectively. For a given location (λ, ψ), the broadband scanning radiometer measures the scattered radiance or intensity (energy/area/time/solid angle), which depends on the zenithal and azimuthal angles of the outgoing radiation as well as the position of the sun in terms of the solar zenith angle θ_0. The configuration of the scattering geometry has been shown in Fig. 6.2.

Since the scanning radiometer detects the reflected solar radiance only at a given scan angle, certain empirical adjustments are required in order to evaluate the daily reflected flux. We define the empirical anisotropic scattering function as

$$X(\theta, \phi; \theta_0) = F_S(\theta_0)/[\pi I_S(\theta, \phi; \theta_0)]. \quad (8.17)$$

Prior to satellite experiments, the X function may be determined based on aircraft and balloon observations for selected localities. Raschke and Bandeen (1970) derived the empirical X function shown in Fig. 8.2 for solar zenith angles between about 35 and 60°. Once the X values have been determined, and the assumption is made that they are independent of the area, the daily averaged reflected solar flux can be evaluated by

$$F_S(\lambda, \psi) = \int_{day} X[\theta, \phi; \theta_0(t)] \pi I_S[\lambda, \psi; \theta, \phi; \theta_0(t)] \, dt. \quad (8.18)$$

It follows that the daily planetary albedo now can be defined by

$$r(\lambda, \psi) = F_S(\lambda, \psi)/Q(\lambda), \quad (8.19)$$

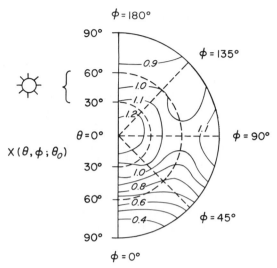

Fig. 8.2 The dependence of the empirical anisotropic scattering function X on the zenith angle θ and azimuthal angle ϕ. The data here pertain to the solar zenith angle interval $35° < \theta_0 < 60°$ (after Raschke and Bandeen, 1970, with modification).

where the daily insolation discussed in Chapter 2 is given by

$$Q(\lambda) = \int_{\text{day}} F_0 \cos \theta_0(t)\, dt. \tag{8.20}$$

In a similar manner, we may define an empirical function for the emitted infrared radiation. Since the outgoing infrared radiation is generally assumed to be azimuthally independent, and there is no dependence on the sun's position, the function may be written in the form

$$X(\theta) = F_{\text{IR}}/[\pi I_{\text{IR}}(\theta)]. \tag{8.21}$$

This function again is to be obtained prior to satellite experiments. Thus, the daily average thermal infrared flux density can be evaluated by

$$
\begin{aligned}
F_{\text{IR}}(\lambda, \psi) &= \int_{\text{day}} F_{\text{IR}}(\lambda, \psi; t)\, dt \\
&= 2\pi \int_{\text{day}} dt \int_0^{\pi/2} I_{\text{IR}}(\lambda, \psi; \theta; t) \cos \theta \sin \theta\, d\theta \\
&= \pi \int_{\text{day}} X(\theta) I_{\text{IR}}(\lambda, \psi; \theta; t)\, dt. \tag{8.22}
\end{aligned}
$$

Note here that radiances measured by the scanning radiometer normally are corrected to the nadir angle ($\theta = 0°$).

We may now define the radiation balance equation using Eqs. (8.19), (8.20), and (8.22). For a given locality with latitude λ and longitude ψ, the net daily flux density may be expressed by

$$R(\lambda, \psi) = Q(\lambda)[1 - r(\lambda, \psi)] - F_{IR}(\lambda, \psi). \tag{8.23}$$

To derive the zonally averaged quantities, we perform the integration over the longitudinal direction to give

$$R(\lambda) = Q(\lambda)[1 - r(\lambda)] - F_{IR}(\lambda), \tag{8.24}$$

where the operator

$$R(\lambda) = \int R(\lambda, \psi) \, d\psi / \Delta\psi.$$

Moreover, the global value may be evaluated by carrying out the integration over the latitudinal direction as

$$\bar{R} = \bar{Q}(1 - \bar{r}) - \bar{F}_{IR}, \tag{8.25}$$

where the operator

$$\bar{(\,)} = \int_{\lambda} \int_{\psi} (\,) \, d\psi \, d\lambda / (\Delta\psi \, \Delta\lambda).$$

Finally, time averaging also can be carried out to obtain the monthly and annual radiation budget values.

8.3   RADIATION BUDGET STUDIES FROM SATELLITE OBSERVATIONS

8.3.1   Radiation Budget of the Globe

Figures 8.3–8.5 (Vonder Haar and Suomi, 1971) illustrate satellite observed mean annual maps of planetary albedo, infrared radiation, and net radiation, respectively, from 1962 to 1965. Figure 8.3 depicts the map of planetary albedo. The planetary albedos of middle and high latitude in northern and southern hemispheres, poleward from 30°N and S, show an interesting contrast. A simple zonal pattern of albedo isopaths is seen in the southern hemisphere owing to the less complex distribution of land and sea. The highest albedos in the tropics are associated with deserts, such as the Sahara, and with continental convective cloudiness in central Africa and convergent ocean areas in the case of the equatorial eastern Pacific. Mean cloudiness and ice and snow fields appear to be the dominant factors for the high albedos.

The mean annual long-wave radiation loss from the earth–atmosphere system is shown in Fig. 8.4. The regional long-wave radiation from the earth–atmosphere is primarily related to the effective mean temperature. Similar patterns in northern and southern hemispheres are observed. The poles reflect a large amount of incoming solar radiation and consequently depress

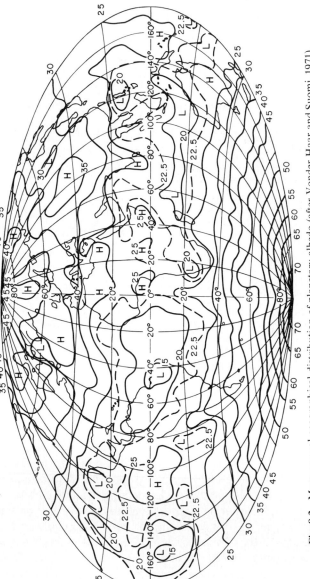

Fig. 8.3   Mean annual geographical distribution of planetary albedo (after Vonder Haar and Suomi, 1971).

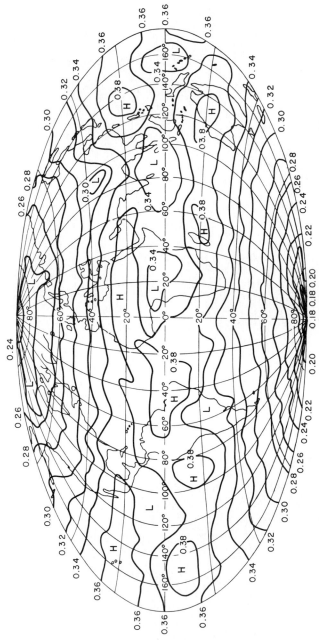

Fig. 8.4   Mean annual geographical distributions of long-wave radiation (after Vonder Haar and Suomi, 1971).

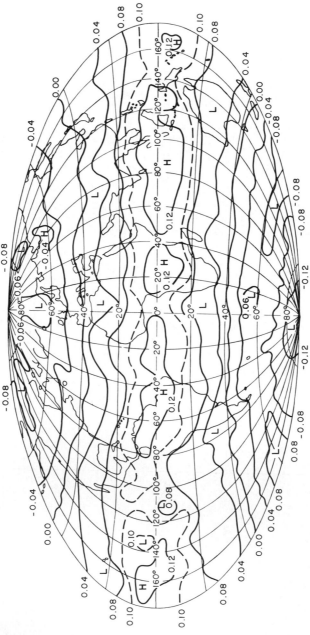

Fig. 8.5 Mean annual geographical distribution of net radiation using a solar constant of 2 cal cm^{-2} min^{-1} (after Vonder Haar and Suomi, 1971).

the local rate of outgoing long-wave radiation emission. In low latitudes an inverse relationship between albedo and long-wave radiation also is evident, because strong cloud covers trap emission from the warmer underlying surface to some extent. A notable exception to this inverse relationship principle is in North Africa, where high albedo is associated with brightly reflective desert surfaces from which high rates of long-wave radiation emission are permitted by the relatively dry and cloud-free atmosphere over the Sahara.

In Fig. 8.5, the net radiation gain and loss, using a solar constant of 2 cal cm^{-2} min^{-1}, is shown. The net radiation gain is evident through almost the entire zone from 40°N to 40°S. It is flanked by radiation sinks which generally deepen toward the poles. In the tropical zones the greatest minima are found over the oceanic deserts west of South America and Africa. Here the low, bright, warm clouds strongly reflect the solar energy and also emit well in the infrared. Maxima of net radiation sources occur over clear oceanic regions. Note that the radiation budget map is for the mean annual case, and thus all daily and even seasonal anomalies are smoothed a great deal. Even so, a definite distribution of relative energy gain and loss is seen within a zone. In higher latitudes, more zonal patterns are shown, whereas in the tropics, variations of radiation sources with longitude appear significant.

8.3.2 Radiation Budgets of Latitudinal Zones

Based on an extended series of measurements of the earth's radiation budget from the first (TIROS-type) and second (Nimbus and ESSA) generation United States meteorological satellites, Vonder Haar and Suomi (1971) derived values of planetary albedo, infrared radiant emittance, and the resulting net radiation budget for 39 months during the period 1962–1966. Eighty percent of all observations considered in their study have been acquired by the lower resolution Wisconsin-type sensors described in the previous section, while the remainder has been obtained from medium resolution scanning radiometers.

Figure 8.6 shows the satellite measured meridional profiles of the planetary albedo, absorbed solar energy, and infrared loss. Latitudinally averaged budgets are useful for indicating the effect of changing surface and atmospheric conditions associated with the large-scale radiation pattern. It is clear that the annual receipt of solar energy along the equator is some 2.5 times higher than that at the poles. On the contrary, equatorial regions show a larger thermal infrared loss as compared to polar regions, but with variations much less from the equator to the poles. Without horizontal net

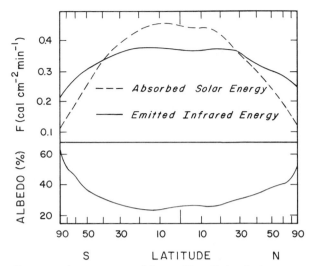

Fig. 8.6 Zonally averaged components of the earth's radiation budget measured during the period 1962–1966. Note that the abscissa is scaled by the cosine of latitude (after Vonder Haar and Suomi, 1971, with modification).

energy flows from the radiation sources in low latitudes to the sinks in higher latitudes, the tropical regions would become increasingly hotter, while the polar belts would grow steadily colder. It has been estimated that about four fifths of the total energy transferred poleward is carried by the atmosphere with the remainder by the ocean. We also notice that there is a lack of equatorial symmetry with more energy being retained in the southern subtropics than at the same latitudes in the north. On the other hand, the Arctic is observed to gain more solar energy than the Antarctic. With respect to the planetary albedo, a value as low as 23% is found in the subtropics. The planetary albedo ranges from 50% near the North Pole and 70% at the South Pole. The higher values of the planetary albedo in middle and high latitudes are associated with the strong cloudiness of the extratropical baroclinic zone belts and high-latitude ice and snow. The global albedo is 30% for the whole earth, and for both northern and southern hemispheres separately. Vonder Haar and Suomi also indicated significant seasonal and annual variations of the mean meridional profiles of the radiation budget.

    Figure 8.7 shows the radiation balance as a function of latitude presented by Raschke and Bandeen (1970) for the summer months, based on the Nimbus II Medium Resolution Scanning Radiometer measurements. Their results indicate higher values of planetary albedo and long-wave radiation loss as compared with those obtained by Vonder Haar and Suomi. The global values for the planetary albedo and long-wave radiation loss derived

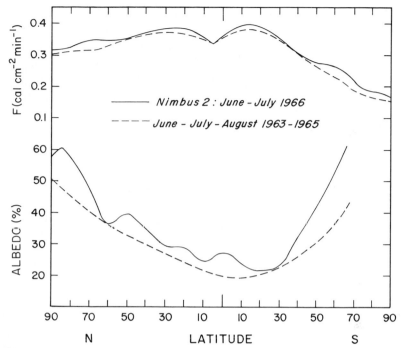

Fig. 8.7 Comparison of zonally averaged albedo and long-wave radiation derived from Nimbus II measurements during June and July 1966 with those derived by Vonder Haar and Suomi (after Raschke and Bandeen, 1970, with modification).

from their study are 30% and 0.345 cal cm^{-2} min^{-1}, respectively, as compared with 25% and 0.330 cal cm^{-2} min^{-1} obtained by Vonder Haar and Suomi for the same summer months. The differences between these two investigations simply might indicate actual variation of the earth's radiation field in the different years. However, they might be caused by the different measuring techniques and evaluation methods used.

8.3.3 Radiation Budget Component in the Global Energy Balance

In the previous section, we noted that in the mean, there is a radiation excess in the tropical region and a radiation deficit in the middle and high latitudes. Thus, there must be a poleward energy transport in order to balance the radiation surplus and deficit. Clearly, the energy exchange within the earth–atmosphere system involves a number of mechanisms of which radiative transfer represents only one component of the total energy budget.

The heating mechanism for an atmospheric column may be associated with (1) the absorption of solar radiation denoted previously; (2) the condensation of water vapor LC_{v1}, where L is the latent heat of vaporization (~ 590 cal g^{-1}) and C_{v1} is the gram mass of water vapor condensed per time per area; (3) the horizontal flux of sensible heat carried by the atmospheric motion into the column C_1; and (4) the horizontal flux of sensible heat carried by ocean currents into the column F_1. On the other hand, the cooling mechanism for the column may be related to (1) the emission of thermal infrared radiation to space denoted previously; (2) the evaporation of water LC_{v2}, where C_{v2} denotes the gram mass of water evaporated per time per area; (3) the horizontal flux of sensible heat carried by the atmospheric motion out of the column C_2; and (4) the horizontal flux of sensible heat carried by ocean currents out of the column F_2.

Let the net heating or cooling rate be Q_N, then the energy balance for an atmospheric column may be expressed by

$$Q_N = Q(1 - r) + LC_{v1} + C_1 + F_1 - F_{IR} - LC_{v2} - C_2 - F_2. \quad (8.26)$$

But from Eq. (8.24) we have

$$R = Q(1 - r) - F_{IR}. \quad (8.27)$$

Upon defining $\Delta C_v = C_{v2} - C_{v1}$, $\Delta C = C_2 - C_1$, and $\Delta F = F_2 - F_1$, and noting that over the period of a year or so, the net heating or cooling of the column will be relatively small compared with the remaining terms, then the annual energy balance equation of the earth–atmosphere system simply is given by

$$R = L\,\Delta C_v + \Delta C + \Delta F, \quad (8.28)$$

where the last three terms represent, respectively, the net flux out of the column of latent heat, sensible heat by atmospheric currents, and sensible heat by ocean currents.

We note that the net flux of sensible heat by atmospheric currents, i.e., the rate of diabatic (nonadiabatic) heating or cooling of the column air, may be expressed in terms of the rate of change of the enthalpy C_pT and geopotential Φ through the first law of thermodynamics and hydrostatic equilibrium. The quantity $(C_pT + \Phi)$ represents the sum of the internal and gravitational potential energy per unit mass.

Figure 8.8a shows the latitudinal values of the components of the energy balance equation. The net fluxes of latent heat and sensible heat by ocean currents are estimated by Sellers (1965), and the radiation balance term is calculated from Fig. 8.6. The atmospheric flux then is evaluated from Eq. (8.28) based on these components.

In Fig. 8.8a net flux of sensible heat due to ocean currents is seen to have a maximum between about 20°N and 20°S. Net flux of latent heat, on the

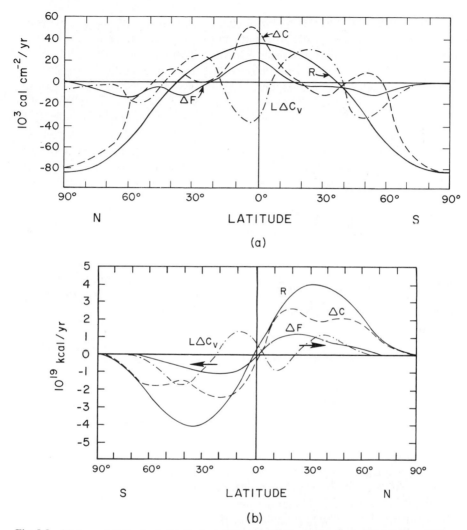

Fig. 8.8 (a) Annual latitudinal distribution of energy balance components (after Sellers, 1965, with modification). (b) Total poleward energy transport required by the radiation budget and other major components (after Vonder Haar and Suomi, 1971).

other hand, has a minimum between about 10°N and 10°S owing to excess precipitation in the tropical convective zone. There are also minima between about 40 and 60°N and S, the baroclinic zones where frequent cylonic storm activities produce surplus precipitation. Since evaporation rates are highest in the dry subtropics, net flux of latent heat shows a peak between about 20 and 30° in both hemispheres. The peak is much more pronounced in the southern hemisphere than in the northern hemisphere where subtropical desert regions are incapable of generating enough moisture for the atmosphere. Moreover, radiation excess is located between about 40°S and 40°N, whereas radiation deficits are found poleward of 40°S and N. From these components, net flux of atmospheric sensible heat shows three peaks at the equator and at about 40°N and S.

On the basis of the preceding values, the poleward energy transport, i.e., the differential net fluxes, taking into account the latitudinal cross section area, may be evaluated and is shown in Fig. 8.8b. Oceanic transport of energy, which occurs in low latitudes, is found to account for about 25% and 20% of the total transport in northern and southern hemispheres, respectively. Latent heat is transferred both toward the equator and poleward from about 25°N and S where, as mentioned previously, the evaporation maxima are located. The latent heat flux accounts for about 20 and 25% of the total transport in northern and southern hemispheres, respectively. Atmospheric potential energy transport shows double maxima in both hemispheres. One peak is located in the subtropics between about 15 and 25°, and the other in midlatitudes between about 50 and 60°. These poleward energy transports are required to balance the radiative transfer component.

8.4 THEORETICAL RADIATION BUDGET STUDIES

8.4.1 Model Description

In order to model the atmospheric radiation balance, radiative transfer calculations must be accurate and efficient so that the spectral dependence of radiation can be covered, the absorption, scattering, and emission of the molecules and particulates can be treated adequately, and the inhomogeneity of the atmosphere can be taken into account. Many radiative transfer methods as discussed in Chapter 6 undoubtedly are available for the comprehensive radiation budget calculations. However, because of the requirement of computational effort, few of these techniques meet the criteria of accuracy and efficiency for the purpose of performing the extensive calculations involved in the radiative budget model. In this section we describe some significant results of the recent comprehensive radiative budget studies reported by Freeman and Liou (1979), in whose work the rigorous discrete-ordinates method for radiative transfer was utilized in the radiative budget computations.

The spectral distribution of electromagnetic energy of importance to the global radiation budget divides into two major segments as discussed in Chapters 3 and 4. Radiation from the sun is essentially restricted to a wavelength range of about 0.1 to 5 μm, with the peak energy located at about 0.47 μm, and with negligible solar flux contained outside this range. On the other hand, the radiation from the earth–atmosphere is contained in the wavelength range from 5 to 100 μm with peak energy occurring at about 10 μm. Owing to this natural characteristic of electromagnetic radiation, the earth's radiation budget can be treated separately in the solar and thermal infrared radiation bands.

The theoretical computational results reported below are based on a radiation model utilizing the discrete-ordinates method for monochromatic radiative transfer with applications to inhomogeneous atmospheres by matching the intensity components at predivided homogeneous layers. The solar spectrum is divided into nine spectral intervals according to the position of the absorption bands (see Fig. 2.6). Within the solar spectrum, the radiative transfer algorithm includes the absorption contribution by water vapor, oxygen, ozone, and carbon dioxide, and absorption and scattering contributions by clouds and aerosols. As for the thermal infrared radiation, band-by-band calculations for the 6.3 μm, continuum, and rotational water vapor bands, the 15 μm carbon dioxide band, and the 9.6 μm ozone band (see Fig. 4.1) are performed to obtain the distribution of infrared radiation. The bulk of the data required for the global radiation computations are the atmospheric profile, the geometrical and physical properties of clouds, the global fractional cloudiness for each cloud type, the earth's surface albedo, the duration of sunlight, and the zenith angle of the sun. The atmospheric profile includes the vertical profiles of pressure, molecular and aerosol densities, water vapor, and ozone.

The atmospheric profiles used in the radiative budget calculations were based on comprehensive compilations reported by McClatchey et al. (1971) in which the water vapor, ozone, pressure, density, and temperature profiles for tropical (0–30°), midlatitude (30–60°), and arctic (60–90°) atmospheres were given for both winter and summer seasons. The concentrations of the uniformly mixed gases of interest, CO_2 and O_2 were taken to be 5.11×10^{-4} and 0.236 gm cm^{-2}/mb, respectively, constant with season and latitude.

Clouds were divided into six types which include (1) high clouds (Ci, Cs, Cc), (2) middle clouds (As, Ac,) (3) low clouds (St, Sc), (4) cumulus (Cu), (5) cumulonimbus (Cb), and (6) nimbostratus (Ns). The fractional cloud cover and the cloud top and base heights for each cloud type as a function of the latitude and season were taken from London (1957) and Katayama (1966) for the northern hemisphere. Cloud data for the southern hemisphere were obtained from values provided by Sasamori et al. (1972). The atmospheric aerosol model used was a light backgroud concentration providing about 23 km surface visibility. The size distribution assumed was a modification of the bimodal log-normal distribution, and the particular aerosol utilized was a water-soluble particle. For scattering calculations, all cloud (except cirrus) and aerosol particles were assumed to be spherical. The high cirrus were considered to be composed exclusively of ice cylinders, randomly oriented in a horizontal plane, with a mean length of 200 μm and a mean radius of 30 μm. The drop-size distributions for the low, middle, and stratus clouds were based on observed data. Within the infrared spectrum all clouds except cirrus were considered to be blackbodies.

The surface albedo of the earth is also an important parameter which determines the amount of the transmitted solar radiation reaching the surface and reflecting back into the atmosphere to be absorbed or scattered, or to escape back into space as a component of the earth's global albedo. Values of surface albedo for the northern and southern hemispheres were taken from the work of Katayama and Sasamori *et al.*, respectively.

The duration of sunlight and the solar zenith angle are important parameters in determining the radiation balance of the earth–atmosphere system. The solar zenith angle can be computed from the angles associated with the latitude, the declination of the sun, and the hour angle of the sun [see Eq. (2.16)]. In general, the solar zenith angle varies significantly each hour during the day, except in the arctic summer. Sunlight duration varies with season and latitude.

For the computations of the interaction of solar radiation with the atmosphere and surface, simultaneous absorption and scattering by gas molecules, clouds, aerosols and the underlying surfaces for every spectral band, and every layer in the model atmosphere must be taken into account. In the solar band, 15 layers were used with their thicknesses varied to better resolve the clouds and lower layers of the troposphere. For thermal infrared radiation, the computational problem is somewhat simplified because surface reflection and scattering by gas molecules and clouds (except cirrus) can be neglected, and there is no diurnal, seasonal, or latitudinal zenith angle dependence to be dealt with. Thus, only flux calculations need be performed in the clear atmosphere. Also, effects of a light aerosol concentration are important only in the window region. In the terrestrial infrared band, 100 layers were utilized for the flux calculations. We shall now discuss some physical significance of radiation features derived from the climatological data.

8.4.2   Solar Heating and Thermal Cooling Within the Atmosphere

The broad-scale features of the planetary climate are determined by the distribution of solar radiation over the globe. In addition to providing the ultimate energy source for the earth's general circulation, the differential heating of the equatorial and polar regions is also responsible for causing the climatic extremes between the tropical and polar latitudes. Every portion of the earth in sunlit sky receives energy from the sun and is warmed to a greater or lesser degree. The primary factors which determine the degree of solar warming received by a particular region on the average are the cloud cover and the latitude. The latitude is related to the range of solar zenith angles experienced by the area. Additional factors are the annual range of surface albedos, the presence of aerosols in greater or lesser concentrations, and the water vapor and ozone contents of the atmosphere.

The zonally averaged meridional profiles of the solar heating rate are illustrated in Figs. 8.9 and 8.10 for January and July. Maximum solar heating of about 2.2°C/day is observed at an altitude of about 3 to 4 km in the tropical and subtropical regions of the summer hemisphere. Second maxima occur in the troposphere in the summer polar regions. These are of about 1.5 to 2.0°C/day and are caused by the duration of daylight in these regions as well as by the occurrence of a maximum of cloudiness in the subarctic summer. Broad, flat minima occur in the upper troposphere and lower stratosphere in both months. Heating rates of only about 0.02 to 0.04°C/day are observed, owing to the lack of both clouds and absorbing gases at these altitudes. Maxima again occur in the stratosphere at about 25 km, almost exclusively due to the presence of ozone. The heating produced is on the order of 1.5 to 1.8°C/day, and is an important source of heating for the upper atmosphere. Effects of aerosols appear to increase the heating rate in the atmosphere. Latitudinal cross sections of zonally averaged solar heating were constructed using the climatological cloud and surface conditions applied at 10° latitude intervals and using the three atmospheric profiles described previously.

While the radiation from the sun warms the earth's atmosphere everywhere, the role of terrestrial infrared radiation is more complex. In the main, the thermal radiation serves to cool the atmosphere, radiating away to space an amount of energy equivalent to the solar input, maintaining the radiative balance. Under certain conditions, however, thermal radiation adds to the warming of the atmosphere at particular levels and locations.

The zonally averaged meridional cooling profiles for January and July are shown in Figs. 8.11 and 8.12. The maximum cooling takes place in the summer stratosphere, due exclusively to ozone and carbon dioxide. Indeed, almost all cooling above the tropopause is due to these two gases since above about 10 km the water vapor concentration decreases drastically to a negligible amount. Ozone is also responsible for the obvious region of thermal heating found above the tropopause in tropical and subtropical latitudes for both seasons. This heating is associated with the increase of ozone concentration with height to about 23 km, resulting in a convergence of flux into

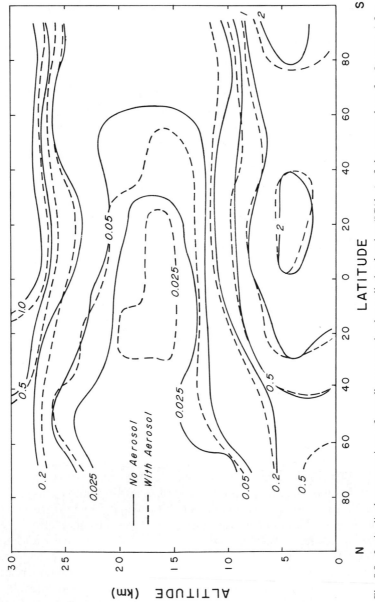

Fig. 8.9 Latitudinal cross sections of zonally averaged solar radiative heating (°C/day) of the atmosphere for January (after Freeman and Liou, 1979).

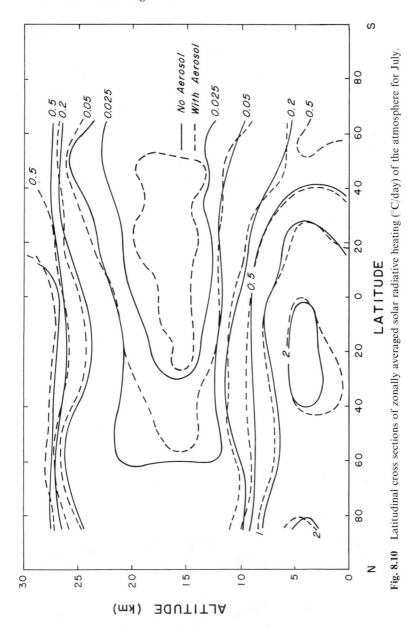

Fig. 8.10 Latitudinal cross sections of zonally averaged solar radiative heating (°C/day) of the atmosphere for July.

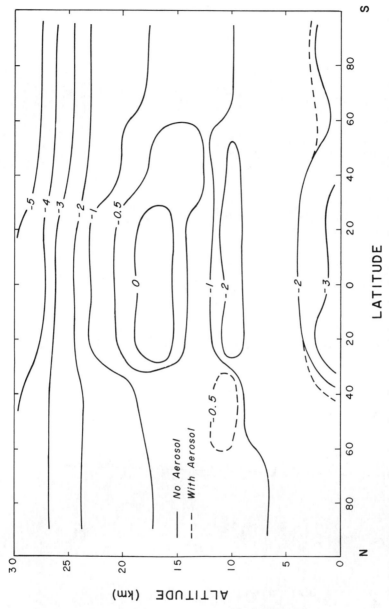

Fig. 8.11 Latitudinal cross sections of zonally averaged thermal radiative cooling (°C/day) of the atmosphere for January (after Freeman and Liou, 1979).

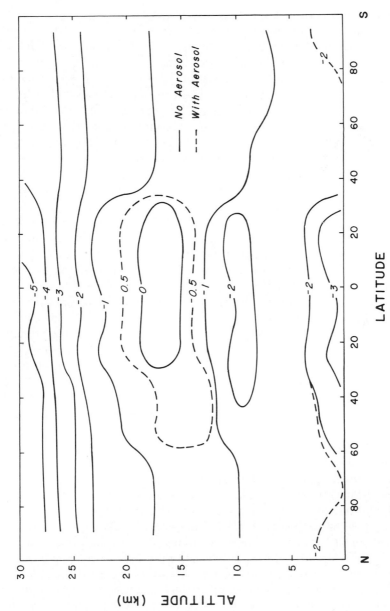

Fig. 8.12 Latitudinal cross sections of zonally averaged thermal radiative cooling (°C/day) of the atmosphere for July.

the region. The heating in this region is supplemented by a smilar region of heating due to carbon dioxide at the tropical tropopause and due to the higher temperatures found both above and below the tropopause. The effect of water vapor is to cool the clear atmosphere everywhere since there is an increase of flux with height as the water vapor concentration decreases. A secondary maximum of cooling occurs in tropical latitudes within the troposphere, associated with the large vertical gradients of water vapor and temperature. The effects of clouds, which tend to increase the cooling above their tops and decrease the cooling below their bases, also are included in this region. In the vicinity of the tropopause, water vapor exhibits a minimum of cooling in the tropics, again due to the warmer temperatures above and below the tropopause and the resulting convergence of thermal flux into the area. Above the tropopause, because of the low concentration, very small cooling results from water vapor; only on the order of $-0.2°C/day$ or less. Near the surface, below 4 km, owing to large water vapor density and temperature gradients, another maximum of cooling occurs in the tropics and summer midlatitudes. This cooling is offset somewhat by the increase in warming below the cloud bases.

In summary, the net thermal cooling is dominated by water vapor below the tropopause with maximum cooling on the order of $-2.0°C/day$ occurring in low latitudes near 10 km altitude. Another maximum at the surface, of about -2.0 to $-3.0°C/day$, and within the same latitude belts, is also due to water vapor. At the tropical tropopause there exists a relatively uniform level of heating on the order of $0.3°C/day$, resulting from the interactions of carbon dioxide, water vapor, and clouds. Above the tropopause, the thermal cooling effects are due to ozone and carbon dioxide. Here steadily increasing cooling to space is found toward the summer hemisphere in the upper stratosphere, where cooling on the order of $-5.6°C/day$ is found.

The net heat budget was computed by summing the heating and cooling rates presented earlier for each month at each latitude and atmospheric layer. The net heating cross sections for the two months are presented in Figs. 8.13 and 8.14. Radiative cooling dominates the solar heating almost everywhere. In the upper stratosphere, above 25 km, intense cooling due to ozone and carbon dioxide is found. The thermal cooling of -4 to $-5°C/day$ completely overshadows the solar heating by ozone to produce a net cooling of -4 to $-4.5°C/day$. The large cooling is due, in part, to the effect of colder cloud tops. At the tropical tropopause, near 18 km, there is a maximum longwave heating of about $0.35°C/day$. This occurs in the region of minimum solar heating, about $0.025°C/day$ to produce net heating. Below this region of heating is a region of maximum cooling with values near $-2.0°C/day$ and it is associated with large vertical gradients of water vapor and temperature.

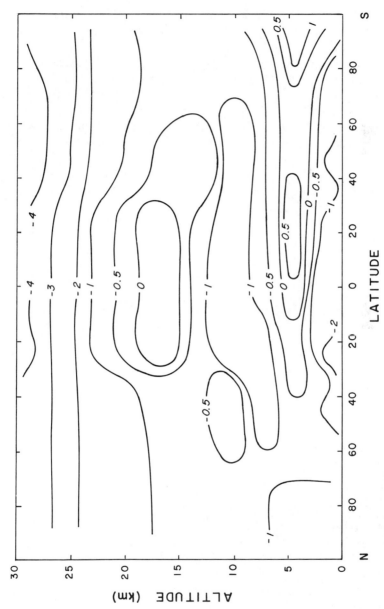

Fig. 8.13 Latitudinal cross sections of zonally averaged net radiative heating (°C/day) of the atmosphere for January (after Freeman and Liou, 1979).

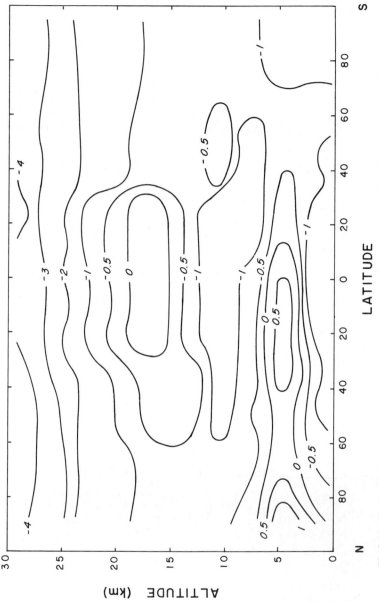

Fig 8.14 Latitudinal cross sections of zonally averaged net radiative heating (°C/day) of the atmosphere for July.

It is apparent from the cross section of radiative heating presented in Figs. 8.13 and 8.14 that cooling by thermal infrared radiation outweighs solar heating at every latitude for both seasons. The cooling is due primarily to water vapor, and thus has a maximum in the tropics. The presence of clouds tends to moderate the cooling in the lower levels of the atmosphere by reducing the cooling below their bases and by producing strong solar heating at their tops. Their effect varies with latitude and season, as the cloud distribution varies. The moderately strong feature of net heating, extending from the summer pole into the tropical latitudes of the winter hemisphere, at a level of about 5 km, is due to strong water vapor absorption in the near solar infrared region augmented by solar geometry. Heating by clouds also contributes significantly to this feature. The solar heating at the cloud tops is partially offset by the increased thermal cooling above the clouds, whereas below the cloud bases the small solar heating is supplemented by the reduced thermal cooling. The maximum heating is found in this region near the summertime pole where the length of the period of solar heating offsets the low solar zenith angle. Because of the low temperatures and the cloud effects, thermal infrared cooling in this region at a height of 4 to 5 km is relatively small. In both hemispheres cooling maxima of about $-2.0°C/day$ are found in the surface layer of the winter tropics. It is due to the maximum water vapor concentration near the surface layer and also to a relative minimum of cloudiness in the wintertime tropics as compared with the summer hemisphere tropics.

8.4.3 Zonally Averaged Total Radiation Budgets

In Fig. 8.15 the zonally averaged total absorption of solar radiation is shown for January and July, and the annual mean obtained from the average of the two months. Solid lines are results obtained by Freeman and Liou (1979) for both northern and southern hemispheres, while dashed and dotted lines represent those of London (1957) and Sasmori *et al.* (1972), respectively, for northern and southern hemispheres. Generally, the absorption computed in the Freeman and Liou exceeds the absorption computed by the London and Sasamori *et al.* This can be accounted for by the increased absorption due to aerosols and the effects of scattering by aerosols and clouds, which increase the amount of energy available for absorption, in effect, by increasing the optical path lengths through the atmosphere. Also note that London and Sasamori *et al.* used empirical parameterized expressions for the scattering by clouds and aerosols, but not the direct computations from Mie and multiple scattering. On the whole, very good agreement for the annual mean of absorption is obtained between the calculations of Freeman and

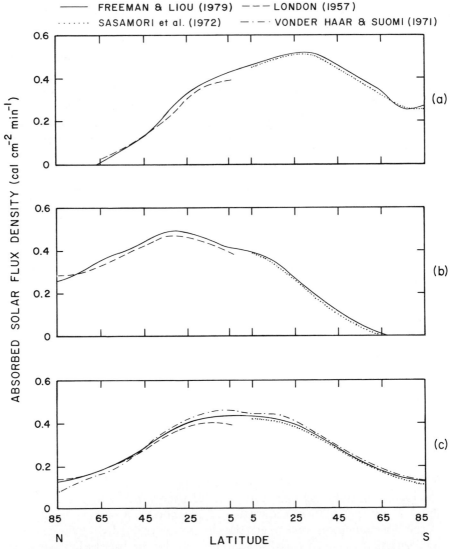

Fig. 8.15 Absorption of solar radiation by the earth–atmosphere system: (a) January, (b) July, (c) annual (after Freeman and Liou, 1979).

Liou and the satellite observations presented by Vonder Haar and Suomi (1971) in which five years of satellite data were analyzed as described in the previous section. The largest differences occur in the polar regions, particularly in the northern hemisphere. Perhaps most of the difference may be accounted for by departures from reality of the cloud and aerosol distributions used in the model, and in part, by the radiative transfer method of accounting for scattering and absorption by aerosols and clouds. Some portion of the differences also may be due to different ozone models used, which would change the total absorption in the 0.3 and 0.5 μm solar bands.

Shown in Fig. 8.16 are the latitudinal distributions of upwelling long-wave flux at the top of the atmosphere. Differences between the calculations performed by various authors appear to be due primarily to the different water vapor and temperature distributions used. On the whole, the theoretical results do an adequate job of reproducing the satellite-observed upwelling

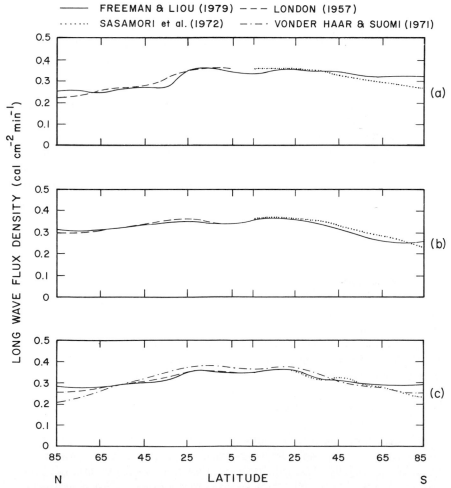

Fig. 8.16   Total upwelling long-wave radiation from the atmospheric top: (a) January, (b) July, (c) annual (after Freeman and Liou, 1979).

long-wave flux, except at the poles, where overestimation of the upward flux occurs. Again, the distribution of clouds used in the models may play a significant role in determining the differences between calculated and observed values.

The values of planetary albedos are presented in Fig. 8.17. The theoretical values include the work of Freeman and Liou (1979), London (1957), Sasamori *et al.* (1972), and Katayama (1967). Freeman and Liou's results give values of the planetary albedos with and without the effect of aerosols. Along with the computed values, satellite-observed planetary albedos as reported by Vonder Haar and Suomi (1971) and Raschke and Bandeen (1970) are also depicted. On the average, all the computed values, both present and past, are larger than the satellite observed values. In some cases the computed results may exceed the satellite values by as much as 15%. Probably the most important reason for the differences is the overestimation of cloudiness, particularly in the tropics. Another factor is the underestimation of absorption in the atmosphere, due to uncertainties in the treatment of aerosols.

The zonally averaged net radiation budgets at the top of the atmosphere, the surface, and the atmosphere as a whole are depicted in Fig. 8.18 for January, July, and the annual mean. The radiation budget for the top of the atmosphere is determined by subtracting the outgoing infrared flux at the top of the atmosphere from the incoming solar radiation absorbed by the earth and atmosphere. The result indicates a net gain of energy at the top of the atmosphere in the tropics and summer hemisphere midlatitudes for both January and July. In the annual case, the gain occurs between $40°N$ and $40°S$ with losses poleward of that region. The annual global average is a net loss of 0.023 cal cm^{-2} min^{-1}.

At the earth's surface, the radiation budget is calculated by subtracting the net upward terrestrial infrared radiation from the solar radiation absorbed by the surface. This quantity is positive in all of the summer hemisphere and through about $35°$ latitude in the winter hemisphere in both months. Annually, there is a net gain in the tropics and midlatitudes, and a net loss in the subarctic regions of the northern hemisphere. In the southern hemisphere, the net gain in the antarctic region is very small and may be within the expected error of these calculations. Globally, the annual mean shows a net gain of 0.094 cal cm^{-2} min^{-1}. The negative values in the polar regions are caused by the high surface albedos and low water vapor contents in the subarctic atmospheres.

By combining the solar radiation absorbed by the atmosphere with the divergence of terrestrial infrared radiation, the net radiation loss for the atmosphere may be obtained. In the annual case, a net global loss of 0.120 cal cm^{-2} min^{-1} is observed. This quantity represents a radiative deficit which must be balanced by the transfer of latent and sensible heat to the atmosphere

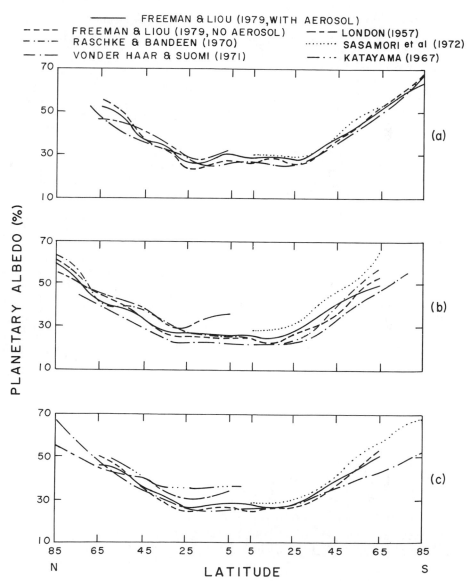

Fig. 8.17 Zonally averaged planetary albedo: (a) January, (b) July, (c) annual (after Freeman and Liou, 1979).

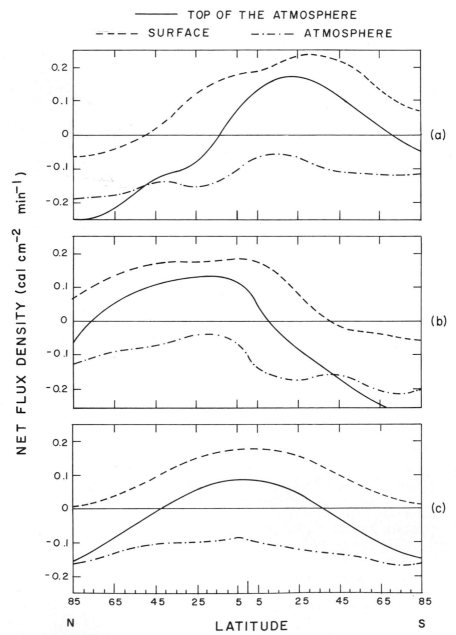

Fig. 8.18 Zonally averaged net radiation budgets at the top of the atmosphere, the surface, and the atmosphere as a whole: (a) January, (b) July, (c) annual.

from the earth's surface; if the atmosphere as a whole is considered to be in a steady state energetically and if no heat transfer across the equator is considered. Sasamori *et al.* (1972) suggest that about 77% of this deficit is made up by the release of latent heat of condensation in the southern hemisphere and about 70% in the northern hemisphere. The remainder of the deficit then must be compensated for by the transport of sensible heat into the lower layers of the atmosphere from the surface.

8.4.4 Global Radiation Budget

The global radiation budget of the earth–atmosphere system presented here is based on the recent computations by Wittman (1978), who used the radiation program described earlier. The major factors considered in the determination of the radiation balance are the atmospheric profile, the geometrical and physical properties of clouds, the global fractional cloudiness for each cloud type, the earth's surface albedo, the duration of sunlight, and the zenith angle of the sun. The atmospheric profile used in this study was an average of the five model atmospheres. Each atmosphere includes vertical profiles of pressure, temperature, molecular and aerosol densities, water vapor, and ozone. The clear atmospheric aerosol loading has a ground visibility of 23 km. The cloud geometrical properties considered are height in the atmosphere, thickness, and horizontal extent. Mean values were obtained for each of four cloud types (low cloud, middle cloud, high cloud, and stratus) by averaging the cloud height distributions and the distributions of fractional cloudiness.

The surface albedo of the earth is also an important parameter since it determines the amount of transmitted solar radiation reaching the surface, which is reflected back into the atmosphere to be absorbed or scattered, or to escape back into space as a component of the earth's global albedo. A global surface albedo of 0.15 was calculated from averaging northern and southern hemispheric data with respect to season and latitude.

In general, the solar zenith angle varies significantly each hour of the day, and sunlight duration varies with season as well as latitude. In order to compensate for these variations, a set of weighting factors was obtained for the six discrete solar zenith angles used in the transfer program.

The annual radiation budget of the earth–atmosphere system is presented in Fig. 8.19. The radiation from the sun, averaged for the entire year, is represented by 100 units. Using a solar constant of 1.94 cal cm^{-2} min^{-1} (Thekaekara, 1976), the average insolation at the top of the atmosphere was calculated to be 0.485 cal cm^{-2} min^{-1}. The upward flux of infrared radiation at the surface was computed using the surface temperature in the Stefan–Boltzmann law.

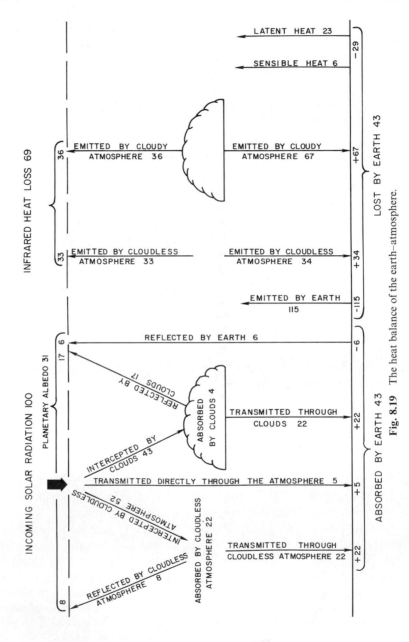

Fig. 8.19   The heat balance of the earth–atmosphere.

Basically, Fig. 8.19 consists of three sections: one dealing with solar radiation and the manner in which it is apportioned in the atmosphere, the second concerning infrared radiation and its distribution, and the third dealing with nonradiative processes. Of the 100 units of incoming solar flux, 26 are absorbed within the atmosphere, 22 by cloud-free air, and 4 by clouds. A total of 31 units are reflected back to space including 8 from cloudless atmospheres, 17 from cloudy atmospheres, and 6 directly from the earth's surface. The remaining 43 units are absorbed by the earth's surface. Meanwhile, the earth and the atmosphere emit thermal infrared according to the temperature and composition distributions. The upward flux from the warmer surface accounts for 115 units. The relatively colder troposphere emits both upward and downward fluxes with 69 and 101 units at the top and surface, respectively. The net upward flux at the surface, which is the difference between the flux emitted by the surface and the downward flux from the atmosphere reaching the surface, is 14 units. As a result of thermal emission, the atmosphere loses 55 units. With the absorption of only 26 units of incoming solar flux, the net radiative loss from the atmosphere amounts to 29 units. This deficit is balanced by an upward flux of latent and sensible heat. According to Sasamori *et al.*, since the average annual ratio of sensible to latent heat loss at the surface (Bowen ratio) has a global value of 0.27, the latent and sensible components should read 23 and 6 units, respectively, to produce an overall balance at the surface. It is apparent that the global atmosphere as a whole experiences a net radiative cooling that is balanced by the latent heat of condensation released in precipitation processes and by the conduction of sensible heat from the underlying surface. If there were no latent and sensible heat transfer, the earth's surface would have had a temperature higher than the present observed value of $288°$ K in order to achieve the balance requirement for radiative equilibrium.

In Table 8.1 comparisons between a number of models are presented for the various components of the radiation budget. It should be noted that the work of Houghton and London is for the northern hemisphere, and that of Sasamori *et al.* is only for the southern hemisphere, in which highly parameterized methods are used in the radiative transfer calculations.

The total absorption of solar radiation computed by Wittman exceeds the absorption computed by the previous investigators. This is due, in part, to the increased absorption from aerosols, the effects of scattering by clouds and aerosols, and reflection from the surface, all of which increase the amount of energy available for absorption by increasing the optical path length through the atmosphere. The comparisons of the global albedo show that Wittman's value (31%) is 3–4% lower than those of the earlier works. According to satellite observations, an albedo of 0.30 was calculated from TIROS, Nimbus, and ESSA satellite measurements by Vonder Haar and

TABLE 8.1 *Annual Radiation Budget of the Earth-Atmosphere System*

	Houghton (1954)	London (1957)	Sasamori et al. (1972)	Wittman (1978)
I. SOLAR RADIATION				
1. Insolation at top of atmosphere	100	100.0	100	100
2. Absorption in the atmosphere				
a. by the cloudless atmosphere	9	15.8	17	22
b. by clouds	10	1.6	4	4
Total Absorption	19	17.4	21	26
3. Reflection and scattering back to space				
a. by atmosphere	9	6.8	6	8
b. by clouds	25	24.2	29	17
c. by earth's surface	—	4.2	—	6
Total Reflection	34	35.2	35	31
4. Absorbed by earth's surface				
a. direct	24	22.4	24	5[a]
b. transmitted through clouds	17	14.4	—	22[a]
c. scattered	6	10.6	21	22[a]
Total Absorption at Earth's Surface	47	47.4	45	43

II. INFRARED RADIATION				
1. Net radiation from earth's surface				
a. total emission by earth	119	114.4	112	115
b. back radiation from cloudless atmosphere	—	—	—	34
c. back radiation from cloudy atmosphere	—	—	—	67
d. total back radiation	105	96.4	96	101
Net Radiation from Earth's Surface	14	18.0	16	14
2. Infrared radiation lost to space				
a. from cloudless atmosphere	—	—	—	36
b. from cloudy atmosphere	—	—	—	33
Total Lost to Space	66	64.8	66	69
3. Net radiation lost by atmosphere	52	46.8	51	55
III. TRANSPORT TO ATMOSPHERE				
1. Latent heat	23	18.6	23	23
2. Sensible heat	10	10.8	7	6
Total Heat Transport	33	29.4	30	29

[a] These values represent the radiation incident on the earth's surface; therefore, the radiation reflected by the earth's surface has been subtracted from the total.

Suomi. Moreover, from Nimbus II satellite observations a value of 0.29 and 0.31 was calculated for June and July of 1966 by Raschke and Bandeen. Using Nimbus III data, the global albedo was computed to be 0.284 by Raschke *et al.* (1973), and recently a value of 0.30 was derived from the Nimbus 6 earth radiation budget experiment data by Smith *et al.* (1977).

The infrared radiation emitted from the earth's surface and the atmosphere computed by Wittman is somewhat greater than those values calculated previously. Consequently, the net radiation from the earth's surface is smaller, while the net radiation lost by the atmosphere is greater than those of the earlier works. These differences result from the different water vapor, ozone, temperature, and cloud distributions, as well as the different values of the solar constant, and the mean temperature of the earth's surface utilized in the studies. The global infrared heat loss of 0.335 cal cm^{-2} min^{-1} from Wittman's study is in good agreement with the satellite observed upwelling infrared flux. The comprehensive study of Vonder Haar and Suomi derived a global infrared loss of 0.34 cal cm^{-2} min^{-1} for the five-year period, 1962–1966. Raschke *et al.* reported an infrared heat loss to space of 0.345 cal cm^{-2} min^{-1} based on Nimbus III measurements during 1969–1970. Using Nimbus 6 data for the months July and August 1975, Smith *et al.* calculated the long-wave radiation flux to be 0.344 cal cm^{-2} min^{-1}.

8.4.5   Radiation and General Circulation

As illustrated in Fig. 8.19, in the mean, the net incoming solar flux absorbed by the earth–atmosphere system must be equal to the outgoing thermal infrared flux emitted to space. However, as evident in Figs. 8.6 and 8.18c, the absorbed solar flux depends significantly on the latitude, having a maximum at the equator and minima at the poles. The strong latitudinal gradient is largely caused by the sharp decrease in insolation during the winter season and the high surface albedo in the polar region. On the other hand, the outgoing infrared flux is only slightly latitudinally dependent. This is owing to the larger burden of atmospheric water vapor and the higher and colder cloudtop temperatures in the tropics, which produce the greenhouse effect to reduce the thermal infrared emission loss. Thus, there is a radiation excess in the equatorial region, and a radiation deficit in the polar regions

As a result of radiative energy excess and deficit, the equator-to-pole temperature gradient is generated, and subsequently, a growing store of zonal mean available potential energy is produced. In the equatorial region, warm air expands upward and creates a poleward pressure gradient force at the upper altitudes. Because of the earth's rotation and the inhomogeneity of the surface, air flows poleward from the equator, while the upper levels cool and sink in the subtropical high pressure belts ($\sim 30°$) and return to the equator. Kinetic energy is generated as a result of work done by the horizontal pressure gradient force.

This thermally driven circulation between equator and subtropics is now called *Hadley circulation* or *Hadley cell*. Because of the earth's rotation (Coriolis force), air flowing toward the equator at the surface deflects to the west and creates the easterly trade winds. In the upper level of the Hadley cell, the Coriolis deflection of the poleward moving air generates westerly winds.

In the polar regions a similar thermally driven circulation is found. Cold air shrinks downward, producing a poleward directed pressure gradient force and motion in the upper altitudes. The sinking motion over the poles results in airflow in the lower level towards the equator and into the low pressure belts ($\sim 60°$). Thus, a Hadley cell develops between the poles and the subpolar low pressure regions. Here, the effect of the Coriolis force is the same, i.e., east winds are produced at the surface and westerly winds aloft. In the Hadley cell, the atmosphere may be regarded as an engine which absorbs net heat from a high temperature reservoir and releases heat to a low temperature reservoir. The temperature differences generate available potential energy which is in turn partly converted to kinetic energy to overcome the effect of friction.

The poleward zonal thermal winds at the upper altitudes become unstable (referred to as baroclinic instability) in middle latitudes and generate a reverse cell. Here warm air sinks in the subtropical highs, and cold air rises in the subpolar lows in which westerly winds prevail in all levels. The meridional cell in this region cannot be explained by the direct heating and cooling effects as in the Hadley cell, and cannot generate kinetic energy. The maintenance of the westerlies in middle latitudes is explained by the continuous transfer of angular momentum from the tropics, influenced by the large-scale wave disturbances. The baroclinic waves transport heat poleward and will intensify until heat transport is balanced by the radiation deficit in the polar regions.

The foregoing brief description presents a gross picture of the general circulation of the atmosphere in relation to the earth's radiation field.

8.5   SIMPLE RADIATION AND CLIMATE MODELS

8.5.1   Global Radiative Equilibrium Model

The simpliest climate model for the earth–atmosphere system is to consider the earth and the atmosphere as a whole and to evaluate the global radiative equilibrium temperature from the balance of the incoming solar flux and outgoing thermal infrared flux. Let the global albedo be $\bar{r}$, the solar constant be S, and the radius of the earth be a_e. Over a long period of time, say one year, there should be a balance between the energy absorbed and emitted so that a radiative equilibrium temperature is maintained. Thus we should have

$$\pi a_e^2 (1 - \bar{r})S = 4\pi a_e^2 \sigma T_e^4, \tag{8.29}$$

where πa_e^2 represents the cross sectional area of the earth–atmosphere which intercepts the incoming solar flux, and the spherical area $4\pi a_e^2$ denotes emission in all directions. It follows that the equilibrium temperature of the system is

$$T_e = [(1 - \bar{r})S/4\sigma]^{1/4}. \tag{8.30}$$

With this simple equation, we may study the effect of changes in the global albedo and/or the solar constant on the equilibrium temperature of the entire system. However, the surface temperature which is a fundamental parameter in climate studies cannot be related to the solar constant nor the global albedo change. The information of the surface temperature has to be related to the transparency and opacity of the atmosphere with respect to solar and thermal infrared radiation, respectively.

To include the surface temperature and the radiative properties of the atmosphere in the simplest radiative equilibrium model, we construct a two layer model and utilize the global radiative budget parameters depicted in Fig. 8.19. Let the mean solar absorptivity and the thermal infrared emissivity of the earth's atmosphere be $\bar{A}$ and $\bar{\varepsilon}$, respectively, and assume that the earth's surface is a blackbody with a temperature of T. In reference to Fig. 8.20, we may write down the energy balance equations at the top of the atmosphere and the surface, respectively, in the forms

$$Q(1 - \bar{r}) - \bar{\varepsilon}\sigma T_a^4 - (1 - \bar{\varepsilon})\sigma T^4 = 0, \tag{8.31}$$

$$Q(1 - \bar{r} - \bar{A}) + \bar{\varepsilon}\sigma T_a^4 - \sigma T^4 = 0, \tag{8.32}$$

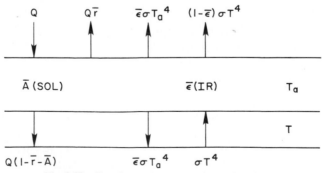

Fig. 8.20   Two-layer global radiative budget model.

where $Q = S/4$, and we note that the global albedo is prescribed without reference to the reflection properties of the atmosphere or the surface. The solutions for the surface and atmospheric temperatures are

$$T^4 = Q[2(1 - \bar{r}) - \bar{A}]/[\sigma(2 - \bar{\varepsilon})], \qquad (8.33)$$

$$T_a^4 = Q[\bar{A} + \bar{\varepsilon}(1 - \bar{r} - \bar{A})]/[\sigma\bar{\varepsilon}(2 - \bar{\varepsilon})]. \qquad (8.34)$$

These equations are highly nonlinear with many coupling terms. Thus, it is very difficult to carry out sensitivity analyses concerning the perturbation of the radiative parameters on temperature values. However, if the absorptivity and emissivity of the atmosphere are assumed to be constants, the effect of changes of the solar constant on the equilibrium surface and atmospheric temperatures may be studied.

8.5.2   One-Dimensional Radiative Equilibrium Model with Vertical Resolution

In Chapters 3 and 4, we introduced the concept of heating and cooling rates due to solar and thermal infrared radiative energy transfer. Moreover, the detailed heating and cooling rates of the earth's atmosphere based on climatological profiles also have been presented in the previous section. The heating and cooling rates are expressed in terms of the temperature change with respect to time. Thus, in the presence of a purely radiative energy exchange, the so-called *radiative equilibrium* may be achieved and the vertical temperature profile formed.

Let the total solar heating rate be $(\partial T/\partial t)_S$ and the total thermal infrared cooling rate be $(\partial T/\partial t)_{IR}$. Thus, the net heating or cooling for a given altitude or pressure may be written in the form

$$\left(\frac{\partial T}{\partial t}\right)_{rad} = \left(\frac{\partial T}{\partial t}\right)_{IR} + \left(\frac{\partial T}{\partial t}\right)_S. \qquad (8.35)$$

This equation is general and is applicable to both clear and cloudy atmospheres. The state of pure radiative equilibrium may be approached by the method of numerical integration. Let n be the time step of the integration, and Δt be the time interval; the temperature at a given pressure level P may be expressed by

$$T^{(n+1)}(p) = T^{(n)}(P) + \left(\frac{\partial T}{\partial t}\right)_{rad}^{(n)} \Delta t. \qquad (8.36)$$

An initial guess of the temperature profile is needed and numerical differencing schemes need to be employed. Radiative equilibrium is reached when the temperature difference $|T^{(n+1)} - T^{(n)}|$ is less than a small preset value.

Specific attention should be given to the radiative equilibrium conditions at the surface and the top of the atmosphere where the net solar flux (downward) must be equal to the net thermal infrared flux (upward). These conditions will ensure the radiative equilibrium of the earth's surface and the atmosphere, and the earth as a whole.

The heavy solid curve shown in the left-hand side of Fig. 8.21 is the vertical temperature profile derived from the radiative equilibrium calculations reported by Manabe and Strickler (1964) for a clear atmosphere. The gases considered are water vapor, ozone, and carbon dioxide whose profiles are for 35°N in April. The solar constant is assumed to be 2 cal cm^{-2} min^{-1}. The effective mean zenith angle of the sun is 60°, and the fractional daylight hour per day is 0.5. The surface albedo used is 0.102. The term Δt is set to be 8 hr, and the convergence criterion is 10^{-3} deg day^{-1}. The radiative equilibrium temperatures of the upper troposphere and the surface obtained from the numerical experiment are much lower and higher, respectively, than the observed values. These deviations apparently are due to the neglect of the upward heat transfer by atmospheric motion in the calculations. To overcome this shortcoming, Manabe and Strickler introduced a simple numerical procedure called *convective adjustment* to approximate the vertical heat transport. The procedure is to adjust the lapse rate to the critical lapse rate whenever the critical lapse rate is exceeded in the numerical iterations. The vertical temperature profile derived from this adjustment is said to be in *thermal equilibrium*. Due to the balance of the upward heat transport associated with moist and dry convection on small and large scales and to the transfer of radiative energy, the observed tropospheric temperature lapse rate is about 6.5 deg km^{-1}. This value may be used as a reference for the critical lapse rate.

In the thermal equilibrium computations, a number of requirements must be satisfied in the final state. These include: (1) the net incoming solar flux must be equal to the net outgoing thermal infrared flux at the top of the atmosphere; (2) at the surface, the difference of the net downward solar flux and net upward thermal infrared flux must be equal to the net integrated radiative cooling of the atmosphere so that a balance is maintained between the net gain of energy by radiation at the surface and the loss of heat by convection transfer into the atmosphere; and (3) the condition of local radiative equilibrium is satisfied when the computed lapse rate is less than the critical value.

On the basis of these physical principles, Manabe and Strickler (1964) and Manabe and Wetherald (1967) introduced the additional numerical procedures: For a convective layer which is in contact with the surface, the computed lapse rate is set to be equal to the critical lapse rate. Moreover, to assure that the earth's surface gives away as much heat as it receives,

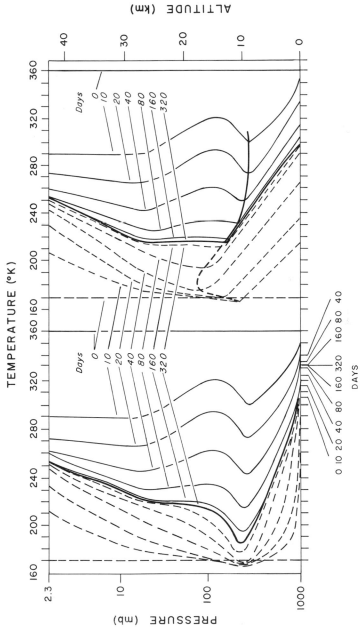

Fig. 8.21   The approach to states of pure radiative (left) and thermal (right) equilibrium. The solid and dashed lines show the approach from a warm and cold isothermal atmosphere (after Manabe and Strickler, 1964).

we must have

$$\frac{C_p}{g} \int_{P_t}^{P_s} \left(\frac{\partial T}{\partial t}\right)^{(n)}_{net} dP = \frac{C_p}{g} \int_{P_t}^{P_s} \left(\frac{\partial T}{\partial t}\right)^{(n)}_{rad} dP + [-F_{IR}^{(n)} + F_S], \quad (8.37)$$

where F_{IR} and F_S represent the net infrared and solar flux, respectively, at the surface, $(\partial T/\partial t)_{rad}$ is the radiative temperature change denoted earlier, and P_s and P_t are the pressures at the earth's surface and at the top of the convective layer. For a convective layer which is within the atmosphere, we also must have

$$\frac{C_p}{g} \int_{P_t}^{P_b} \left(\frac{\partial T}{\partial t}\right)^{(n)}_{net} dP = \frac{C_p}{g} \int_{P_t}^{P_b} \left(\frac{\partial T}{\partial t}\right)^{(n)}_{rad} dP \quad (8.38)$$

to ensure the energy continuity, where P_b is the atmospheric pressure at the bottom of the convective layer. For a nonconvective layer, however, we have simply

$$\left(\frac{\partial T}{\partial t}\right)^{(n)}_{net} = \left(\frac{\partial T}{\partial t}\right)^{(n)}_{rad}. \quad (8.39)$$

The temperature iteration equation under the local-thermal equilibrium condition then is given by

$$T^{(n+1)}(P) = T^{(n)}(P) + \left(\frac{\partial T}{\partial t}\right)^{(n)}_{net} \Delta t. \quad (8.40)$$

To obtain the vertical temperature profile under the local-thermal equilibrium, the convective adjustments denoted in Eqs. (8.37) and (8.38) must be satisfied at every time step for every predivided layer. The heavy solid curve in the right-hand side of Fig. 8.21 shows the temperature profile derived from the convective adjustment. It is seen that the surface temperature in local thermal equilibrium with a 6.5 deg km^{-1} adjustment is 300° K, which is more realistic than the surface temperature of 332.3°K obtained from pure radiative equilibrium. Furthermore, more realistic temperatures at the upper troposphere also are obtained under thermal equilibrium. The calculations shown in Fig. 8.21 were carried out for initial isothermal temperature distributions of warm (solid curves) and cold (dashed curves) cases. Note that in this case it takes almost one year to reach thermal equilibrium temperature.

The foregoing discussions regarding radiative and thermal equilibrium provide the fundamentals for the investigation of the variation of various gases, aerosol, and cloudiness on the globally averaged temperature profile. In the one-dimensional radiation and climate model, in which radiative transfer and parameterized vertical convection are included, the impact of the increase of carbon dioxide and aerosols distribution on the globally averaged temperature field may be examined.

8.5.3   Energy Balance Climate Model

The energy balance model is concerned with the computation of surface temperature from the balance between incoming solar and outgoing infrared flux. The model is primarily characterized by latitudinal variations in which global energy budgets are assumed to be expressible in terms of surface temperature. Moreover, planetary albedo is assumed to rely upon the ice or no ice cover, and the convergence of dynamic heat fluxes is represented by either a linear function of the surface temperature deviation or a simple diffusion law. Basically, the energy balance model may be described by an equation of the form

$$C \frac{\partial T(x, t)}{\partial t} = F_s(x) - F_{IR}(x) - R(x), \qquad (8.41)$$

where C denotes the thermal inertia coefficient for a zonal column of the atmosphere–ocean system, t the time, and $x = \sin \lambda$, where λ is the latitude. (It is more convenient to use x instead of λ in model calculations.) F_s and F_{IR} are the incoming solar and outgoing infrared flux densities at the top of the atmosphere, and $-R(x)$ denotes the divergence of the atmospheric and oceanic heat flux. Under a steady state condition, Eq. (8.41) reduces to

$$F_s(x) - F_{IR}(x) = R(x). \qquad (8.42)$$

We now introduce the basic physical principle encountered in the energy balance climate model.

8.5.3.1   *Linear Heating Law*   On the basis of the monthly mean values of radiation flux density at the top of the atmosphere for 260 stations, Budyko (1969) developed an empirical formula relating the outgoing infrared flux density, the surface temperature, and the fractional cloud cover in the form

$$F_{IR}(x) = a_1 + b_1 T(x) - [a_2 + b_2 T(x)]\eta, \qquad (8.43)$$

where T is the surface temperature, η the fractional cloud cover, and the empirical constants $a_1 = 0.324$, $b_1 = 0.00324$, $a_2 = 0.0694$, and $b_2 = 0.00232$. Using these coefficients, F_{IR} is in units of cal cm^{-2} min^{-1}, and T is in degrees Celsius. The influence of the deviation of cloudiness from its mean global value (50%) on the temperature is normally neglected because of the intricate interaction of clouds with the radiation field and surface albedo. With the fractional cloud cover $\eta = 0.5$, Eq. (8.43) may be rewritten in the form

$$F_{IR}(x) = a + bT(x), \qquad (8.44)$$

with $a = 0.286 \, \text{cal cm}^{-2} \, \text{min}^{-1}$ and $b = 0.00206 \, \text{cal cm}^{-2} \, \text{min}^{-1} \, \text{C}^{-1}$. The physical explanation for the linear relation between the outgoing infrared flux and the surface temperature is that since the temperature profiles have more or less the same shape at all latitudes, the infrared cooling, which depends on the temperature at all levels, may be expressed in terms of the surface temperature.

The incoming solar flux density may be expressed by

$$F_S(x) = Qs(x)[1 - r(x)] = Qs(x)A(x), \tag{8.45}$$

where $Q = S/4$, S is the solar constant, r the planetary albedo which is allowed to depend on temperature, A the solar flux density absorbed by the earth–atmosphere system, and $s(x)$ a normalized mean annual distribution of insolation at each latitude such that $\int_0^1 s(x) \, dx = 1$. $Qs(x)$ can be found in Fig. 2.10.

Moreover, to relate the surface temperature distribution and the horizontal heat transfer in the atmosphere and hydrosphere, Budyko derived a simple empirical equation by comparing the observed mean latitudinal values of $R(x)$ with the difference of the annual mean temperature at a given latitude and the global mean temperature $\bar{T}$. It is given by

$$R(x) = c[T(x) - \bar{T}], \tag{8.46}$$

with the empirical constant $c = 0.00538 \, \text{cal cm}^{-2} \, \text{min}^{-1} \, \text{C}^{-1}$.

Upon substituting Eqs. (8.44) and (8.45) into Eq. (8.42), we have

$$Qs(x)A(x) - [a + bT(x)] = c[T(x) - \bar{T}]. \tag{8.47}$$

The surface temperature is then given by

$$T(x) = \frac{Qs(x)A(x) - a + c\bar{T}}{c + b}. \tag{8.48}$$

Furthermore, over a climatological time scale, the earth–atmosphere as a whole should be in radiative equilibrium so that

$$Q\bar{A} - (a + b\bar{T}) = 0, \tag{8.49}$$

where the global surface temperature, global absorptivity, and global albedo are defined by

$$\bar{T} = \int_0^1 T(x) \, dx, \qquad \bar{A} = 1 - \bar{r} = \int_0^1 s(x)A(x) \, dx. \tag{8.50}$$

Consequently,

$$\bar{T} = (Q\bar{A} - a)/b. \tag{8.51}$$

At this point, the latitudinally dependent surface temperature may be computed as a function of x for given $s(x)$ and $A(x)$. The planetary albedo usually is given by a simple step function depending on whether or not there is an ice sheet. By letting x_s represent the sine of latitude λ_s of the ice line, the absorptivity or albedo may be expressed by

$$A(x, x_s) = 1 - r(x, x_s) = \begin{cases} A_1, & x > x_s \\ A_2, & x < x_s. \end{cases} \tag{8.52}$$

Computations involving the temperature as a function of x are given in Excercise 8.4.

Using a linear perturbation analysis on the global surface temperature, solar constant, and global albedo, i.e., letting $\bar{T} = T_0 + \Delta \bar{T}$ and so on, it can be shown that the temperature at a particular latitude may be given by

$$T(x) = \frac{1}{c+b}\left\{ Q_0 s(x)[1 - r(x)]\left(1 + \frac{\Delta Q}{Q_0}\right) - a + c\bar{T}_0 \right.$$
$$\left. + \frac{cQ_0}{b}\left[\frac{\Delta Q}{Q_0}(1 - \bar{r}_0 - \Delta\bar{r}) - \Delta\bar{r} \right] \right\}, \tag{8.53}$$

where $\bar{T}_0$, Q_0, and $\bar{r}_0$ are mean values for the present conditions. This equation will allow us to study effects of the change of the solar constant on the earth's mean annual surface temperature, such that the coupling effect of the changing global albedo, which is an indication of the change in a glaciated area, is included. Budyko expressed $\Delta\bar{r}$ in terms of an empirical function in the form $0.3l(x, \bar{x}_s)s'(x_s)$, where $\bar{x}_s$ represents the present ice-line position ($\bar{x}_s = 0.95$), l the ratio of the change in the ice-covered area to the total area of the northern hemisphere, and s' the ratio of the mean solar flux in the zone of the additional ice area to the mean solar flux for the entire hemisphere.

Using Eq. (8.53), calculations can be carried out to investigate the positions of ice line for different values of $\Delta Q/Q_0$. Shown in Fig. 8.22 are the latitude λ_s corresponding to the ice-line position, and the global temperature $\bar{T}$ as a function of $\Delta Q/Q_0$. In these calculations, the existing mean value for the solar constant is 1.92 cal cm^{-2} min^{-1}, and for the global albdo, it is 0.33. The ice-line temperature $T(x_s)$ is assumed to be $-10°C$ based on climatological data. The step function for absorption is $A_1 = 0.38$, $A_2 = 0.68$, and at the ice line it is 0.5. With a 1% change for the incoming solar flux the global surface temperature is reduced by about 5°C. Further, a 1.5% decrease of the incoming solar flux decreases the global surface temperature by about 9°C. The response to these decreases in temperature is a southward advance of glaciation by 8 to 18° of latitude, which corresponds to the advance of Quarternary glaciation. Based on these calculations, when $\Delta Q/Q_0$ is reduced by about 1.6%, the ice line reaches a latitude of about 50°N. At this latitude the global surface temperature decreases to

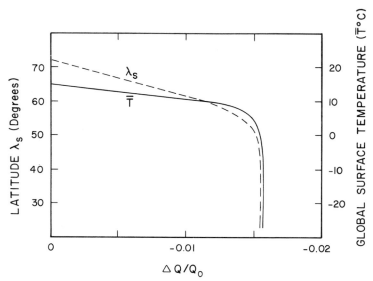

Fig. 8.22 Dependence of the global surface temperature and the latitude of glaciation on the change of incoming solar flux (after Budyko, 1969, with modification).

several tens of degrees below zero. As a result, the ice sheet begins to advance southward all the way to the equator with no further reduction in solar radiation required. We note here that with a constant global mean albedo (i.e., $\Delta \bar{r} = 0$), a 1% decrease in $\Delta Q / Q_0$ lowers the global surface temperature by only 1.2–1.5°C. The significance of the ice-albedo coupling is quite evident.

8.5.3.2 *Simple Diffusion Law* Instead of expressing the horizontal heat transport in terms of a linear function of the surface temperature, Sellers (1969) employed the energy balance equation for the earth–atmosphere system denoted in Eq. (8.28)

$$R = L\Delta C_v + \Delta C + \Delta F = F_S - F_{IR}. \qquad (8.54)$$

Parameterized equations are proposed for the net fluxes from the transport of water vapor by atmospheric currents C_v, and from the transport of sensible heat either by atmospheric currents C or by ocean currents F. Symbolically, we may write

$$C_v = f(\Delta T), \qquad C = g(\Delta T), \qquad F = h(\Delta T).$$

That is, these transport parameters are expressed in terms of the linear function of the surface temperature differences. Consequently, a second-order equation in ΔT may be obtained. We note that Sellers also expressed the outgoing infrared flux density in a more sophisticated form

$$F_{IR} = T^4(x)\{1 - m\tanh[19T^6(x) \times 10^{-16}]\},$$

where m denotes the atmospheric attenuation coefficient. According to Sellers, it is equal to 0.5 for present conditions.

The second-order equation in ΔT for the horizontal heat transport is in essence the diffusion approximation. On the basis of the diffusion law, North (1975a,b), and Held and Suarez (1974) developed more rigorous mathematic models. A thermal diffusion form $-D\nabla^2 T$ for the horizontal heat transport was adopted with D an empirical coefficient to be determined by fitting the present climate. Thus, all the transport processes are parameterized within the single coefficient. It is similar to an eddy diffusion approach to dispersion by macroturbulence in the entire geofluid system. Using the spherical coordinates for the Laplace operator [see Eq. (5.29)], and noting that only a one-dimensional latitudinal variation is considered, we find

$$-D\nabla^2 T = \frac{-D}{a_e^2 \sin\theta} \frac{d}{d\lambda}\left(\sin\theta \frac{dT}{d\theta}\right) = \frac{-D}{a_e^2}\frac{d}{dx}(1-x^2)\frac{d}{dx}T(x), \quad (8.55)$$

where the polar angle $\theta = 90° - \lambda$, and a_e is the radius of the earth. Let $D' = D/a_e^2$, we find from Eq. (8.42)

$$D'\frac{d}{dx}(1-x^2)\frac{d}{dx}T(x) = F_{IR}(x) - Qs(x)A(x, x_s). \quad (8.56)$$

Since F_{IR} and T are linearly related through Eq. (8.44), we may rewrite Eq. (8.56) in the form

$$\left[\frac{d}{dx}(1-x^2)\frac{d}{dx} - \frac{1}{D''}\right]F(x) = -\frac{Q}{D''}s(x)A(x, x_s), \quad (8.57)$$

with $D'' = D'/b$, and we let $F_{IR} = F$ for convenience. We must now specify the ice-sheet edge x_s. It is generally assumed that if $T(x) < T_s$, ice will be present, whereas if $T(x) > T_s$ there will be no ice. In terms of infrared flux density $F(x_s) = F_s$. As mentioned earlier, T_s is normally assumed to be $-10°C$. From Eq. (8.44), this corresponds to $F_s = 0.2680$ cal cm^{-2} min^{-1}.

For a mean annual model with symmetric hemispheres, the boundary condition must be that there is no heat flux transport at the poles or across the equator, i.e., $\nabla F(x)$ or $\nabla T(x) = 0$ at $x = 1$ and 0, respectively. Thus,

$$(1-x^2)^{1/2}\frac{d}{dx}F(x)\bigg|_{x=0} = (1-x^2)^{1/2}\frac{d}{dx}F(x)\bigg|_{x=1} = 0. \quad (8.58)$$

The solution for $F(x)$ may be obtained by expanding it in Legendre polynomials in the form

$$F(x) = \sum_{n=\text{even}} F_n P_n(x), \quad (8.59)$$

where only even terms are taken because $F(x)$ is an even function of x in a mean annual case, i.e., symmetric between hemispheres, and F_n represent the unknown coefficients to be determined. Since the Legendre polynomials are the eigenfunction of the spherical diffusion equation as previously described in Eq. (5.43) (for $l = 0$), we have

$$\frac{d}{dx}(1 - x^2)\frac{d}{dx}P_n(x) = -n(n + 1)P_n(x). \tag{8.60}$$

Moreover, $(1 - x^2)^{1/2}\,dP_n(x)/dx = 0$ for $x = 0, 1$ when $n =$ even. It follows that the imposed boundary conditions described in Eq. (8.58) are satisfied by the expansion.

Upon substituting Eq. (8.59) into Eq. (8.57) and making use of the orthogonal property of P_n (see Appendix E), we find

$$F_n = QH_n(x_s)/[1 + n(n + 1)D''], \tag{8.61}$$

where

$$H_n(x_s) = (2n + 1)\int_0^1 s(x)A(x, x_s)P_n(x)\,dx, \tag{8.62}$$

which may be evaluated from the known values for $s(x)$ and $A(x, x_s)$. The final procedure to complete the solution is to determine the diffusion transport coefficient D''. This may be done empirically by varying D'' in Eq. (8.59) with F_n given in Eq. (8.61) until the present climate conditions are fitted $[\bar{x}_s = 0.95, F(\bar{x}_s) = F_s, Q = Q(\bar{x}_s) = Q_0]$. The solution now may be employed to investigate the ice-line position as a function of Q. Upon utilizing Eqs. (8.59) and (8.61) and letting $x = x_s$, we obtain

$$Q(x_s) = F_s\left[\sum_{n=\text{even}}\frac{H_n(x_s)P_n(x_s)}{1 + n(n + 1)D''}\right]^{-1}. \tag{8.63}$$

The normalized mean annual distribution of insolation $s(x)$ may be fitted by Legendre polynomial expansions. To a good approximation, within about 2% accuracy, it is given by

$$s(x) = \sum_{n=\text{even}} s_nP_n(x) \approx 1 + s_2P_2(x), \tag{8.64}$$

where $s_2 = -0.482$. Based on observed data, the absorptivity for ice-free latitudes also may be fitted by polynomial expansions in the form

$$A_2(x, x_s) = d_0 + d_2P_2(x), \qquad x < x_s, \tag{8.65}$$

with $d_0 = 0.697$, $d_2 = -0.0779$. The absorptivity over ice or snow having 50% cloud cover is assumed to be 0.38 mentioned previously. Figure 8.23a illustrates the surface temperature distribution as a function of x based on a

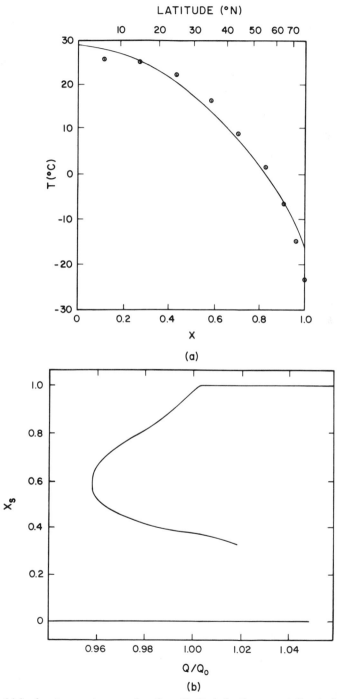

Fig. 8.23 (a) Surface temperature as a function of latitude for the present climate derived from the two-mode approximation. Circles are observed values (after North, 1975b, with modification). (b) The ice-line position as a function of the incoming solar flux in units of its present value (after North, 1975b, with modification).

two-mode expansion in Eq. (8.59). Also shown are observed values. The agreement appears quite reasonable in view of the simplicity of the diffusion model. Dependence of the ice-line position on the change of incoming solar flux is shown in Fig. 8.23b. The multiple-branch nature of the solution in the simple diffusion model is apparent. The upper branch indicates a southward advance of glaciation caused by the decrease of the incoming solar flux. After the ice line reaches about 45–50°N, its southward advance continues even when the incoming solar flux increases. This conclusion is essentially consistent with the results shown in Fig. 8.22.

In a recent paper by Lindzen and Farrell (1977), it was pointed out that the simple climate models described in the foregoing are not in reasonable agreement with the nearly isothermal surface temperatures observed within 30° of latitude of the equator. To introduce the tropical transport (referred to as Hadley cell transport) into the simple climate models, an empirical adjustment is made in which a heat flux is assumed to exist. This heat flux goes to zero for latitudes greater than some latitude λ_h ($\sim 25°$N). For the Budyko linear heating model, $Qs(x)A(x, x_s)$ is replaced by its average over the region $0 \leq \lambda \leq \lambda_h$. Shown in Fig. 8.24 is a schematic illustration suggested by Lindzen and Farrell for the dependence of the ice-line position on varying incoming solar flux. At 25° latitude, identified as the Hadley stability ledge, a reduction in the incoming solar flux does not significantly alter the ice-line position. It is not until the reduction of solar radiation is on the order of about 15 to 20% that the glaciation will advance continuously southward despite the increase of the incoming solar flux. From 25° to about 60°,

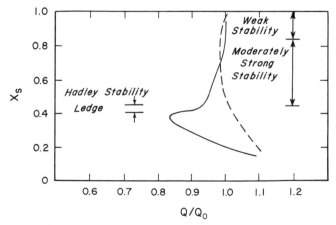

Fig. 8.24 A schematic illustration of the ice-line position as a function of the normalized incoming solar flux with (solid lines) and without (dashed line) a Hadley adjustment (after Lindzen and Farrell, 1977, with modification).

identified as the moderately strong stability region, a southward advance of glaciation is related to the decrease of incoming solar radiation. Moreover, for $\lambda \geq 60°$, referred to as a weak stability region, it was suggested that the ice-line position may be related to the orbital parameters. Comparing this plot with those shown in Figs. 8.22 and 8.23b, it is seen that substantial reductions in incoming solar flux ($\gg 1.6\%$) are necessary for an ice-covered earth, and that the large portions of the globe are stable.

EXERCISES

8.1   Assume that the atmosphere acts as a single isothermal layer with a temperature T_a which transmits solar radiation but absorbs all thermal infrared radiation. Show that the global surface temperature $T = \sqrt[4]{2}T_a$. Let the global albedo be 30%, and the solar constant be 1.94 cal cm^{-2} min^{-1}. What would be the global surface temperature?

8.2   The mean global surface temperature is only about 10°C. The mean global absorptivity of solar radiation by the atmosphere is about 0.2. Use the global albedo and solar constant given in Exercise (8.1) and compute the mean global emissivity and temperature of the atmosphere. Repeat the calculation if the solar constant decreases by 1%.

8.3   Let the global reflectivities of the atmosphere and surface be $\bar{r}$ and r_s, respectively. Consider the multiple reflection between the surface and the atmosphere and show that the global albedo of the earth–atmosphere system is given by

$$\bar{r} + (1 - \bar{A} - \bar{r})^2 r_s/(1 - r_s\bar{r}).$$

8.4   (a)   The normalized mean annual distribution of insolation is approximately given by Eq. (8.64), and the albedo is given by the step function

$$r(x, x_s) = \begin{cases} 0.62, & x > 0.95 \\ 0.32, & x < 0.95. \end{cases}$$

Compute and plot the latitudinal surface temperature as a function of x from Eq. (8.48).
   (b)   Also compute the temperature at the ice line assuming an albedo of 0.5. Show that the solution of the ice-line position x_s is quadratic. Plot x_s as a function of Q/Q_0 from 0.97 to 1.2.

8.5   (a)   Derive Eq. (8.53) by means of a linear perturbation analysis.
   (b)   Show that the area covering the earth poleward of latitude λ is given by $2\pi a_e^2(1 - x)$, where a_e is the radius.

(c)*   Let $\Delta \bar{r} = 0.3 s'(x) l(x, \bar{x}_s)$, where l is the ratio of the change in the ice-covered area to the total area of the northern hemisphere, and

$$s'(x) = \int_{x_s}^{0.95} s(x) \, dx.$$

Show that the solution of x_s is given by a fourth order polynomial equation. Compute and plot x_s as a function of $\Delta Q / Q_0$, and compare your result with those depicted in Fig. 8.22.

8.6   From Eq. (8.61), we find

$$F_0 = Q H_0(x_s), \quad F_2 = Q H_2(x_s)/(6D'' + 1).$$

Based on the two-mode approximation, we also have from Eq. (8.59)

$$F(x) = F_0 + F_2 P_2(x).$$

By fitting the present climate conditions, i.e. $\bar{x}_s = 0.95$, $4Q_0 = 1.94$ cal cm^{-2} min^{-1}, find the empirical coefficient D''.

    * Simple computer programming is required.

SUGGESTED REFERENCES

Budyko, M. I. (1974). *Climate and Life.* Academic Press, New York. Chapter 5 gives a brief description of climatic change and the influence of solar radiation on climate.

Freeman, K. P., and Liou, K. N. (1979). Climate effects of cirrus clouds. *Adv. Geophys.* **21**, 231–287. This paper presents a comprehensive and updated discussion on the radiation balance based on theoretical calculations and investigates the effects of varying cirrus cover on the radiation balance of the earth–atmosphere system.

Kondratyev, K. Ya. (1972). *Radiation Processes in the Atmosphere.* World Meteorological Organization, WMO-No. 309. The entire book contains authoritative discussions on radiation balance components derived from satellites and radiation climatology.

Schneider, S. H., and Dickinson, R. E. (1974). Climate modeling. *Rev. Geophys. Space Phys.* **12**, 447–793. This paper gives a comprehensive and delightful review on hierarchy of climate models of varying complexities including the simple models discussed in this chapter.

Sellers, W. D. (1965). *Physical Climatology.* Univ. of Chicago Press, Chicago, Illinois. Chapters 3, 4, and 8 contain materials on insolation, radiation, and energy balance.

Appendix A
PRINCIPAL SYMBOLS

Only the principal symbols used in this book are listed here. Some symbols formed by adding subscripts or superscripts or primes to principal symbols are not listed. Constants and notations that appear only once and that are transient in nature also are not listed. Bold letters denote vectors or matrices.

a	Radius of a spherical particle
a_e	Radius of the earth
a_s	Radius of the sun
a_j	Gaussian weights
a_l	Amplitude of the wave in l direction
a_r	Amplitude of the wave in r direction
$\mathbf{a}_x, \mathbf{a}_y, \mathbf{a}_z$	Unit vectors in Cartesian coordinates
$\mathbf{a}_r, \mathbf{a}_\theta, \mathbf{a}_\phi$	Unit vectors in spherical coordinates
A	Area
A	Absorptivity
A_λ, A_ν	Monochromatic absorptivity
$A_{\tilde{\nu}}$	Mean absorptivity
$B_{\tilde{\nu}}, B_\lambda, B_\nu$	Planck function
$\mathbf{B}$	Magnetic induction
c	Velocity of light
C_p	Specific heat at constant pressure
$\mathbf{C}$	Covariance matrix
d	Distance between the earth and the sun
d_m	Mean distance between the earth and the sun
$\mathbf{D}$	Electric displacement
e	Charge of an electron
e	Partial pressure of absorbing gas
E	Energy

E_n	Exponential integral
$\mathbf{E}$	Electric field vector
$\mathbf{E}_l, \mathbf{E}_r$	Electric field vectors parallel (l) and perpendicular (r) to a plane through the direction of propagation
$\mathbf{E}_r, \mathbf{E}_\theta, \mathbf{E}_\phi$	Electric field vectors in spherical coordinates
f	Flux, power
f	Coriolis parameter
F	Flux density (irradiance, emittance)
$F_{\tilde{v}}, F_\lambda, F_v$	Monochromatic flux density
F_0	Incident solar flux density
F_S	Flux density of solar radiation
F_{IR}	Flux density of infrared radiation
g	Acceleration of gravity
g	Asymmetry factor
G	Gain relative to an isotropic scatterer
h	Planck's constant
h	height, length
$H(\mu)$	Chandrasehkar's H function
$H_n^{(2)}$	Hankel function of the second kind
$\mathbf{H}$	Magnetic vector
$\mathbf{H}_r, \mathbf{H}_\theta, \mathbf{H}_\phi$	Magnetic vectors in spherical coordinates
i	Square root of -1
i_1, i_2	Spherical angles
i_1, i_2, i_3, i_4	Intensity functions in the Mie theory
I	Intensity (radiance)
$I_{\tilde{v}}, I_\lambda, I_v$	Monochromatic intensity (radiance)
I_0	Incident intensity
I_n	Modified Bessel function of the first kind
I_n	Intensity component for each order of scattering
I_l, I_r	Intensity components parallel (l) and perpendicular (r) to a plane through the direction of propagation
I_S	Intensity of solar radiation
I_{IR}	Intensity of infrared radiation
j_λ	Source function coefficient
$\mathbf{j}$	Electric current density
$J_\lambda, J_{\tilde{v}}, J$	Source function
J_n	Bessel function of the first kind
J_2	Number of quanta absorbed by O_2 per unit volume per unit time per molecule
J_3	Number of quanta absorbed by O_3 per unit volume per unit time per molecule
k	Wave number ($2\pi/\lambda$)
k, k_j	Eigenvalues
$k, k_{\tilde{v}}, k_v$	Absorption coefficient
k_λ	Absorption coefficient, mass extinction cross section
K	Boltzmann's constant
K	Weighting function
l	parallel direction
L	Latent heat of vaporization
L	Length of a ice crystal
$L(x)$	Ladenberg and Reiche function
L_j	Coefficients in the solution of the discrete-ordinates method

L	Transformation (rotational) matrix
LP	Degree of linear polarization
m	Air mass ($\sec \theta_0$)
m	Complex index of refraction
m_r	Real part of the refractive index
m_i	Imaginary part of the refractive index
m_e	Mass of electron
M	Molecular weight of dry air
M	Transformation matrix in the Mie theory
M_{ij}	Matrix elements in the transformation matrix
n	Quantum number
$n(x)$	Particle size distribution
N	Number density
N_0	Avogadro's number
N_n	Neumann function
$\not\hspace{-1pt}\rho$	Number of refraction and reflection within a particle
p	Pressure
p_0	Standard pressure
p_s	Surface pressure
$p(v)$, $p(S)$	Probability function
P	Power
P	Degree of polarization
$P(\cos \Theta)$	Phase function
P_1, P_2	Perpendicular (1) and parallel (2) components of the phase function
P	Induced dipole moment
P	Phase matrix
P_{ij}	Elements of the phase matrix
P_n	Legendre polynomials
P_n^l	Associated Legendre polynomial
q	Specific humidity
Q	The second element of the Stokes parameter
Q	Daily insolation (also used as $S/4$)
Q_e	Extinction efficiency
Q_s	Scattering efficiency
Q_a	Absorption efficiency
r	Perpendicular direction
r	Distance (also used as radius)
$r(\mu)$	Reflection (or local or planetary albedo)
r_s	Surface albedo
$\bar{r}$	Spherical albedo
R	Net flux density divergence due to solar and infrared radiation
R	Radius of a hexagonal crystal
$R(\mu, \phi; \mu_0, \phi_0)$	Reflection function
R_1, R_2	Perpendicular (1) and parallel (2) components of the amplitude coefficients in Fresnel formula
R_λ	Reflectivity (in Chapter 1)
s	Distance
$s(x)$	Normalized mean annual distribution of insolation
S	Solar constant
S	Line strength

$S(\mu, \phi; \mu_0, \phi_0)$	Chandrasehkar's scattering function
S_1, S_2	Perpendicular (1) and parallel (2) components of the scattering function in the Mie theory
$\mathbf{S}$	Poynting vector
t	Time
t_{dif}	Diffuse transmission
t_{dir}	Direct transmission
$\bar{t}$	Global diffuse transmission
t^{f}	Broadband flux transmissivity
T	Temperature (also used as surface temperature in Section 8.5)
T_0	Standard temperature
T_{s}	Surface temperature
T_{e}	Effective equilibrium temperature of the earth
T_{B}	Brightness temperature
$\mathcal{T}_{\bar{\nu}}, \mathcal{T}_{\lambda}, \mathcal{T}_{\nu}$	Monochromatic transmission function (transmittance, or transmissivity)
$\mathcal{T}_{\bar{\nu}}$	Mean transmission function
$\mathcal{T}_{\nu}^{\text{f}}$	Monochromatic slab (diffuse) transmission function
$\mathcal{T}_{\bar{\nu}}^{\text{f}}$	Mean slab (diffuse) transmission function
$T(\mu, \phi; \mu_0, \phi_0)$	Transmission function (in Chapter 6)
$T_{\text{c}}(\mu, \phi; \mu_0, \phi_0)$	Chandrasehkar's transmission function
u	Path length
u_0, u_p	Amplitude of the wave disturbance in the diffraction theory
$u_{\bar{\nu}}$	Energy density
U	The third element of the Stokes parameters
v	Velocity
V	Fourth element of the Stokes parameters
x	Size parameter (ka)
x	Sine of the latitude $(\sin \lambda)$
X	One of the Cartesian coordinates
$X(\mu, \phi; \mu_0, \phi_0)$	Chandrasehkar's X function
Y	One of the Cartesian coordinates
$Y(\mu, \phi; \mu_0, \phi_0)$	Chandrasehkar's Y function
z	height
Z	One of the Cartesian coordinates
W	Equivalent width
α	Lorentz half width
α	Polarizability (in Chapter 3)
α_{D}	Doppler half width
α_{N}	Natural half width
β	Angle denoting the ellipticity
β_{e}	Extinction coefficient
β_{s}	Scattering coefficient
β_{π}	Backscattering coefficient
δ	Absorption line spacing
δ	Inclination angle (in Chapter 2)
$\delta, \delta_r, \delta_l$	Phase of the wave
δ	Dirac's δ function
ε	Permittivity
ε	Pyranometer excess (in Chapter 2)
ε	Measurement error (in Chapter 7)

ε_0	Permittivity of a vacuum
$\varepsilon_\lambda, \varepsilon_{\tilde{v}}$	Monochromatic emissivity
ε^f	Broadband flux emissivity
θ	Zenith angle
θ	Scattering angle in the Mie theory
Θ	Scattering angle
θ_0	Solar zenith angle
λ	Wavelength
λ	Latitude
μ	Cosine of the zenith angle
μ	Permeability
μ_i	Gaussian points
μ_0	Cosine of the solar zenith angle
μ_0	Permeability of a vacuum
v	Wave number (cm^{-1})
$\tilde{v}$	Frequency
v^*	Shape factor in the Junge size distribution
ξ	ξ function in the discrete-ordinates method
ρ	Density
ρ_w	Water vapor density
ρ_3	Ozone density
ρ_a	Density of air
σ	Stefan–Boltzmann constant
σ	Cross section area
σ_e	Extinction cross section
σ_s	Scattering cross section
σ_a	Absorption cross section
σ_π	Backscattering cross section
ω	Circular frequency
$\tilde{\omega}$	Single-scattering albedo
$\tilde{\omega}_l$	Coefficients in the Legendre polynomial expansion for the phase function
τ, τ_λ	Optical depth
Ω	Solid angle
Ω	Total ozone concentration
χ	Orientation angle
η	Fraction cloud cover
ϕ	Azimuthal angle
ϕ_j^m	Eigenfunction in the discrete-ordinates method
$\phi(\tilde{v}, v)$	Instrumental response function
ψ	Longitude
γ	Direct plus diffuse transmission
$\bar{\gamma}$	Total global transmission
$\sum$	Summation operator
π	3.1415926
$\prod$	Multiplication operator

Appendix B
SOME USEFUL CONSTANTS

Velocity of light	$c = 2.99793 \pm 1 \times 10^{10}$ cm sec^{-1}
Planck's constant	$h = 6.62620 \times 10^{-27}$ erg sec
Boltzmann's constant	$K = 1.38062 \times 10^{-16}$ erg deg^{-1}
Stefan–Boltzmann constant	$\sigma = 5.66961 \times 10^{-5}$ erg cm^{-2} sec^{-1} deg^{-4}
Solar constant	$S = 1.35300 \pm 0.021 \times 10^{6}$ erg cm^{-2} sec^{-1}
	$= 1.94 \pm 0.03$ cal[a] cm^{-2} min^{-1}
Mean radius of the earth	$a_e = 6.37120 \times 10^{8}$ cm
Mean radius of the sun (visible disk)	$a_s = 6.96000 \times 10^{10}$ cm
Mean distance between the earth and the sun	$d_m = 1.49598 \times 10^{13}$ cm
Angular velocity of rotation of the earth	$\omega = 7.29221 \times 10^{-5}$ rad sec^{-1}
Standard temperature	$T_0 = 273.16°$K
Standard pressure	$P_0 = 1.01325 \times 10^{6}$ dyn cm^{-2} = 1013.25 mb
Density of air at standard pressure and temperature	$\rho = 1.273 \times 10^{-3}$ g cm^{-3}
Acceleration of gravity (at sea level and 45° latitude)	$g = 9.80616 \times 10^{2}$ cm sec^{-2}
Molecular weight of dry air	$M = 28.97$ g mole^{-1}
Specific heat at constant pressure	$C_p = 1.004 \times 10^{6}$ cm^2 sec^{-2} deg^{-1}
Specific heat at constant volume	$C_v = 7.17 \times 10^{6}$ cm^2 sec^{-2} deg^{-1}
Universal gas constant	$R^* = 8.31432 \times 10^{7}$ erg mole^{-1} deg^{-1}
Avogadro's number	$N_0 = 6.02297 \times 10^{23}$ mole^{-1}
Loschmidt's number (at standard temperature and pressure)	$n_0 = 2.68719 \times 10^{19}$ molecule cm^{-3}
Mass of an electron	$m_e = 9.10956 \times 10^{-28}$ g

[a] 1 cal = 4.1855×10^{7} erg

Electronic charge

$e = 1.60219 \times 10^{-19}$ C (coulomb, mks)

$= 4.803 \times 10^{-10}$ statcoulomb[b] (also esu, cgs)

Permittivity of a vacuum

$\varepsilon_0 = 8.85419 \times 10^{-12}$ C kg^{-1} m^{-3} sec^2 (mks)

$= 1$ Gaussian unit (cgs)

Permeability of a vacuum

$\mu_0 = 12.56637 \times 10^{-7}$ kg m C^{-2} (mks)

$= 1$ Gaussian unit (cgs)

[b] 1 statcoulomb $= 1 \sqrt{\text{erg cm}}$

Appendix C
DERIVATION OF THE PLANCK FUNCTION

In accordance with Boltzmann statistics, if N_0 denotes the number of oscillators in any given energy state, then the number N in a state having energy higher by an amount ε is given by

$$N = N_0 e^{-\varepsilon/KT}, \tag{C.1}$$

where K is Boltzmann's constant, and T the absolute temperature. On the basis of Planck's first postulation, an oscillator cannot have any energy but only energies given by Eq. (1.16). Thus, the possible values of ε must be 0, $h\tilde{\nu}$, $2h\tilde{\nu}$, and so on. If the number of oscillators with zero energy is N_0, then by virtue of Eq. (C. 1), the number with energy $h\tilde{\nu}$ is $N_0 e^{-h\nu/KT}$, the number with energy $2h\tilde{\nu}$ is $N_0 e^{-2h\tilde{\nu}/KT}$, and so on. The total number of oscillators with frequency $\tilde{\nu}$ for all states is therefore

$$
\begin{aligned}
N &= N_0 + N_0 e^{-h\tilde{\nu}/KT} + N_0 e^{-2h\tilde{\nu}/KT} + \cdots \\
&= N_0[1 + e^{-h\tilde{\nu}/KT} + (e^{-h\tilde{\nu}/KT})^2 + \cdots] \\
&\cong N_0/(1 - e^{-h\tilde{\nu}/KT}).
\end{aligned}
\tag{C.2}
$$

The total energy of these oscillators may be obtained by multiplying each term in Eq. (C.2) by the appropriate energy:

$$
\begin{aligned}
E &= 0 \cdot N_0 + h\tilde{\nu} \cdot N_0 e^{-h\tilde{\nu}/KT} + 2h\tilde{\nu} \cdot N_0 e^{-2h\tilde{\nu}/KT} + 3h\tilde{\nu} \cdot N_0 e^{-3h\tilde{\nu}/KT} + \cdots \\
&= h\tilde{\nu}N_0 e^{-h\tilde{\nu}/KT}[1 + 2e^{-h\tilde{\nu}/KT} + 3(e^{-h\tilde{\nu}/KT})^2 + \cdots] \\
&\cong h\tilde{\nu}N_0 e^{-h\tilde{\nu}/KT}/(1 - e^{-h\tilde{\nu}/KT})^2.
\end{aligned}
\tag{C.3}
$$

The average energy per oscillator then is given by

$$\frac{E}{N} = \frac{N_0 h\tilde{v} e^{-h\tilde{v}/KT}/(1 - e^{-h\tilde{v}/KT})^2}{N_0/(1 - e^{-h\tilde{v}/KT})} = h\tilde{v}/(e^{h\tilde{v}/KT} - 1). \tag{C.4}$$

According to Planck's second postulation, the quanta of energy are emitted only when an oscillator changes from one to another of its quantized energy states. The average emitted energy of a group of oscillators therefore is given by Eq. (C.4), which is the factor appearing in Planck's formula.

To obtain the Planck function, we let $u_{\tilde{v}}$ denote the monochromatic energy density, i.e., the energy per unit volume per unit frequency interval in a cavity with temperature T. With this definition, we write

$$u_{\tilde{v}} = A h\tilde{v}/(e^{h\tilde{v}/KT} - 1), \tag{C.5}$$

where A is a constant to be determined. In accordance with the principle of equipartition of energy, the energy density in a cavity is given by the classical Rayleigh–Jeans formula

$$u_{\tilde{v}} = (8\pi\tilde{v}^2/c^3) KT. \tag{C.6}$$

This formula is valid when the temperature T is high and the frequency $\tilde{v}$ small. So letting $h\tilde{v}/KT \to 0$ in Eq. (C.5), we find $A = 8\pi\tilde{v}^2/c^3$. Thus, the monochromatic energy density is

$$u_{\tilde{v}} = \frac{8\pi h\tilde{v}^3}{c^3(e^{h\tilde{v}/KT} - 1)}. \tag{C.7}$$

For blackbody radiation, the emitted photons travel in all directions (4π solid angle) with the speed of light c. Thus, the emitted intensity (or radiance) in a cavity with a temperature T in units of energy/area/time/sr/frequency may be expressed by

$$B_{\tilde{v}}(T) = u_{\tilde{v}} c/(4\pi). \tag{C.8}$$

Upon substituting Eq. (C.7) into (C.8), we obtain the Planck function given by Eq. (1.18) in the form

$$B_{\tilde{v}}(T) = \frac{2h\tilde{v}^3}{c^2(e^{h\tilde{v}/KT} - 1)}.$$

Appendix D

COMPLEX INDEX OF REFRACTION, DISPERSION OF LIGHT, AND LORENTZ–LORENZ FORMULA

Within a dielectric, positive and negative charges are impelled to move in opposite directions by an applied electric field. As a result, electric dipoles are generated. The product of charges and the separation distance of positive and negative charges is called the dipole moment, which when divided by the unit volume is referred to as polarization $\mathbf{P}$. The displacement vector $\mathbf{D}$ (charge per area) within a dielectric is defined in cgs units by

$$\mathbf{D} = \varepsilon\mathbf{E} = \mathbf{E} + 4\pi\mathbf{P}, \qquad (D.1)$$

when ε is the permittivity of the medium. Thus,

$$\varepsilon = (1 + 4\pi\mathbf{P} \cdot \mathbf{E}/E^2). \qquad (D.2)$$

The velocity of light in terms of ε and the permeability μ is given by

$$c = \sqrt{\frac{1}{\mu\varepsilon}}. \qquad (D.3)$$

The permeability μ in air or water is nearly equal to the permeability μ_0 is vacuum, i.e., $\mu \approx \mu_0$. The index of refraction is defined as the ratio of the velocity of light in vacuum and in the medium, and may be expressed by

$$m = \frac{c_0}{c} \approx \sqrt{\varepsilon} = \sqrt{1 + \frac{4\pi\mathbf{P} \cdot \mathbf{E}}{E^2}}. \qquad (D.4)$$

But the polarization vector for N dipoles is [see Eq. (3.54)]

$$\mathbf{P} = N\alpha\mathbf{E}. \tag{D.5}$$

Inserting Eq. (D.5) into Eq. (D.4) leads to

$$m^2 = 1 + 4\pi N\alpha. \tag{D.6}$$

Now, we have to find the polarizibility in terms of frequency. On the basis of the definition of a polarization vector, we have

$$\mathbf{P} = N e\mathbf{r}. \tag{D.7}$$

where e is the charge of an electron, and $\mathbf{r}$ represents the vector distance. Combining Eqs. (D.5) and (D.7), we find

$$\alpha\mathbf{E} = e\mathbf{r}. \tag{D.8}$$

Further, from the Lorentz force equation, the force generated by the electric and magnetic fields are given by

$$\mathbf{F} = e[\mathbf{E} + (\mu/c)\mathbf{v} \times \mathbf{H}], \tag{D.9}$$

where v denotes the velocity of an electron, which is very small compared to the velocity of light. Hence, the force produced by the magnetic field may be neglected. The force in the vibrating system in terms of the displacement r is due to (1) the acceleration of the electron; (2) the damping force, which carries away energy when the vibrating electrons emit electromagnetic waves, and which is proportional to the velocity of the electrons; and (3) the restoring force of the vibration, which is proportional to the distance r. From Newton's second law we find

$$\frac{\mathbf{F}}{m_e} = \frac{e\mathbf{E}}{m_e} = \frac{d^2\mathbf{r}}{dt^2} + \gamma\frac{d\mathbf{r}}{dt} + \xi\mathbf{r}, \tag{D.10}$$

where γ and ξ are the damping and restoring coefficient, respectively, and m_e is the mass of the electron. In scalar form we write

$$\frac{d^2r}{dt^2} + \gamma\frac{dr}{dt} + \xi r = \frac{eE}{m_e}. \tag{D.11}$$

The homogeneous solution of this second-order differential equation simply is given by

$$r = r_0 e^{-i\omega t} = r_0 e^{-i2\pi\tilde{v}t}. \tag{D.12}$$

Substituting Eq. (D.12) into Eq. (D.11), we obtain

$$[(\xi - 4\pi^2\tilde{v}^2) - i2\pi\tilde{v}\gamma]r = eE/m_e. \tag{D.13}$$

The natural (or resonant) frequency is defined by $\tilde{v}_0 = \sqrt{\xi}/2\pi$. Thus, we find

$$\alpha = \frac{er}{E} = \frac{e^2}{m_e} \frac{1}{4\pi^2(\tilde{v}_0^2 - \tilde{v}^2) - i2\pi\gamma\tilde{v}}$$

$$= \frac{e^2}{m_e}\left[\frac{\tilde{v}_0^2 - \tilde{v}^2}{4\pi^2(\tilde{v}_0^2 - \tilde{v}^2)^2 + \gamma^2\tilde{v}^2} + \frac{i}{2\pi}\frac{\gamma\tilde{v}}{4\pi^2(\tilde{v}_0^2 - \tilde{v}^2)^2 + \gamma^2\tilde{v}^2}\right]. \quad (D.14)$$

Let the real and imaginary parts of the index of refraction be m_r and m_i, respectively. Then

$$m = m_r + im_i. \quad (D.15)$$

It follows from Eq. (D.6) that

$$m_r^2 - m_i^2 = 1 + \frac{4\pi Ne^4}{m_e}\frac{\tilde{v}_0^2 - \tilde{v}^2}{4\pi^2(\tilde{v}_0^2 - \tilde{v}^2)^2 + \gamma^2\tilde{v}^2},$$

$$2m_r m_i = \frac{2Ne^2}{m_e}\frac{\gamma\tilde{v}}{4\pi^2(\tilde{v}_0^2 - \tilde{v}^2)^2 + \gamma^2\tilde{v}^2}. \quad (D.16)$$

For air, $m_r \approx 1$ and $m_i \ll m_r - 1$. Also, in the neighborhood of the resonant frequency, $\tilde{v}^2 - \tilde{v}_0^2 = (\tilde{v}_0 + \tilde{v})(\tilde{v} - \tilde{v}_0) \approx -2\tilde{v}_0(\tilde{v} - \tilde{v}_0)$. Further, the half width of the natural broadening depends on the damping and is given in the form $\alpha_N = \gamma/4\pi$, while the line strength S is $\pi Ne^2/(m_e c)$. Thus, we obtain the real part

$$m_r - 1 = -\frac{Ne^2}{4\pi m_e \tilde{v}_0}\frac{\tilde{v} - \tilde{v}_0}{(\tilde{v} - \tilde{v}_0)^2 + \alpha_N^2}, \quad (D.17)$$

and the absorption coefficient (Born and Wolf, 1975, p. 614)

$$k_{\tilde{v}} = \frac{4\pi\tilde{v}_0 m_i}{c} = \frac{S}{\pi}\frac{\alpha_N}{(\tilde{v} - \tilde{v}_0)^2 + \alpha_N^2}. \quad (D.18)$$

Equation (D.18) is simply the Lorentz profile discussed in Section 1.3.

Shown in Fig. D.1 is the dependence of $(m_r - 1)$ and $k_{\tilde{v}}$ on the frequency. The value of $(m_r - 1)$ increases as the frequency increases when $\tilde{v}_0 - \alpha > \tilde{v}$. This mode is referred to as *normal dispersion* under which the light is dispersed by a prism into component colors. For the region $\tilde{v}_0 + \alpha > \tilde{v} > \tilde{v}_0 - \alpha$, $(m_r - 1)$ decreases with increasing frequency, and it is called *anomalous dispersion*. For the range $\tilde{v} > \tilde{v}_0 + \alpha$, normal dispersion takes place again, but $(m_r - 1)$ is smaller than unity.

In this appendix we also wish to prove Eq. (3.69). We consider a dielectric placed between the plates of a parallel plate condenser without the end effect. Moreover, we consider an individual molecule constituting this di-

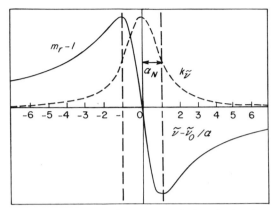

Fig. D.1   The real and imaginary parts of the complex index of refraction as functions of the frequency.

electric and draw a sphere with radius a about this molecule. The molecule therefore is affected by the fields caused by (1) the charges on the surfaces of the condenser plates, (2) the surface charge on the dielectric facing the condenser plates, (3) the surface charge on the spherical boundary of radius a, and (4) the charges of molecules (other than the one under consideration) contained within the sphere. For (1) and (2), the electric field produced by these charges are

$$\mathbf{E}_1 + \mathbf{E}_2 = (\mathbf{E} + 4\pi\mathbf{P}) - 4\pi\mathbf{P} = \mathbf{E}. \tag{D.19}$$

For (3), the electric field, which is produced by the polarization charge presented on the inside of the sphere, is given by

$$dE_3 = \frac{4\pi P \cos\theta\, dA}{4\pi a^2}, \tag{D.20}$$

where $P\cos\theta$ represents the component of the polarization vector in the direction of the electric field vector, and the differential area $dA = a^2 \sin\theta\, d\theta\, d\phi \times \cos\theta$. Thus,

$$E_3 = \int_0^{2\pi}\int_0^{\pi} \frac{4\pi P \cos\theta}{4\pi a^2} a^2 \sin\theta \cos\theta\, d\theta\, d\phi = \frac{4\pi P}{3}. \tag{D.21}$$

For (4), it turns out that $E_4 = 0$. Thus, the effective electric field is

$$\mathbf{E}' = \mathbf{E} + 4\pi\mathbf{P}/3. \tag{D.22}$$

But according to Eq. (D.5), we have

$$\mathbf{P} = \alpha N \mathbf{E}' = \alpha N(\mathbf{E} + 4\pi\mathbf{P}/3). \tag{D.23}$$

It follows that

$$\mathbf{P} = \alpha N \mathbf{E}/(1 - 4\pi\alpha N/3). \qquad \text{(D.24)}$$

Thus, from the definition of the index of refraction in Eq. (D.4), we find

$$m^2 = 1 + 4\pi\alpha N/(1 - 4\pi\alpha N/3). \qquad \text{(D.25)}$$

Rearranging the terms, we obtain the Lorentz–Lorenz formula

$$\alpha = \frac{3}{4\pi N}\frac{m^2 - 1}{m^2 + 2}. \qquad \text{(D.26)}$$

Appendix E
PROPERTIES OF THE LEGENDRE POLYNOMIALS

As indicated in Eqs. (5.42) and (5.43), the solution of the second-order differential equation

$$(1 - \mu^2)\frac{d^2 y}{d\mu^2} - 2\mu\frac{dy}{d\mu} + \left[l(l+1) - \frac{m^2}{1 - \mu^2}\right]y = 0 \qquad (E.1)$$

is given by

$$y(\mu) = P_l^m(\mu) = \frac{(1 - \mu^2)^{m/2}}{2^l l!}\frac{d^{m+l}}{d\mu^{m+l}}(\mu^2 - 1), \qquad (E.2)$$

where $\mu = \cos\theta$. When $m = 0, P_l^0(\mu) = P_l(\mu)$ are the Legendre polynomials. It is clear from Eq. (E.1) that

$$P_l^m(\mu) = (1 - \mu^2)^{m/2}\frac{d^m P_l(\mu)}{d\mu^m}. \qquad (E.3)$$

The associated Legendre polynomials satisfy the orthogonal properties

$$\int_{-1}^{1} P_l^m(\mu)P_k^m(\mu)\,d\mu = \begin{cases} 0, & l \neq k, \\ \dfrac{2}{2l+1}\dfrac{(l+m)!}{(l-m)!}, & l = k, \end{cases} \qquad (E.4)$$

$$\int_{-1}^{1} P_l^m(\mu)P_l^n(\mu)\,\frac{d\mu}{1 - \mu^2} = \begin{cases} 0, & m \neq n, \\ \dfrac{1}{m}\dfrac{(l+m)!}{(l-m)!}, & m = n. \end{cases} \qquad (E.5)$$

Some useful recurrence relations in conjunction with light scattering and radiative transfer are

$$\frac{dP_l^m}{d\theta} = -\sqrt{1-\mu^2}\,\frac{dP_l^m}{d\mu} = \tfrac{1}{2}\big[(l-m+1)(l+m)P_l^{m-1} - P_l^{m+1}\big], \quad \text{(E.6)}$$

$$(2l+1)\mu P_l^m = (l+m)P_{l-1}^m + (l-m+1)P_{l+1}^m, \quad \text{(E.7)}$$

$$(2l+1)(1-\mu^2)^{1/2}P_l^m = (P_{l+1}^{m+1} - P_{l-1}^{m+1}). \quad \text{(E.8)}$$

A number of low-order associated Legendre and Legendre polynomials are

$$
\begin{aligned}
& P_1^1(\mu) = (1-\mu^2)^{1/2}, && P_2^1(\mu) = 3\mu(1-\mu^2)^{1/2}, \\
& P_3^1(\mu) = \tfrac{3}{2}(5\mu^2-1)(1-\mu^2)^{1/2}, && P_2^2(\mu) = 3(1-\mu^2), && \text{(E.9)} \\
& P_3^2(\mu) = 15\mu(1-\mu^2), && P_3^3(\mu) = 15(1-\mu^2)^{3/2},
\end{aligned}
$$

$$
\begin{aligned}
& P_0(\mu) = 1, && P_1(\mu) = \mu, \\
& P_2(\mu) = \tfrac{1}{2}(3\mu^2-1), && P_3(\mu) = \tfrac{1}{2}(5\mu^3-3\mu), && \text{(E.10)} \\
& P_4(\mu) = \tfrac{1}{8}(35\mu^4-30\mu^2+3).
\end{aligned}
$$

Appendix F
THE SCATTERING GEOMETRY

We would like to prove that

$$\cos \Theta = \cos \theta \cos \theta' + \sin \theta \sin \theta' \cos(\phi - \phi').$$

In reference to Fig. F.1, we let

$$\overline{CD} = \overline{CO} \tan \theta', \qquad \overline{OD} = \overline{CO} \sec \theta',$$
$$\overline{CE} = \overline{CO} \tan \theta, \qquad \overline{OE} = \overline{CO} \sec \theta, \tag{F.1}$$

where $\overline{CD}$ and $\overline{CE}$ are the tangent lines of the arcs CA and CB. For the

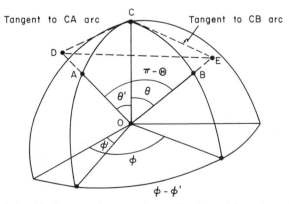

Fig. F.1 The relationship between the scattering angle, the zenith angle, and the azimuthal angle in spherical coordinates.

triangle $\triangle CDE$ we find

$$\overline{DE}^2 = \overline{CD}^2 + \overline{CE}^2 - 2\overline{CE}\,\overline{CD}\cos DCE. \tag{F.2}$$

For the triangle $\triangle ODE$ we find

$$\overline{DE}^2 = \overline{OD}^2 + \overline{OE}^2 - 2\overline{OD}\,\overline{OE}\cos DOE. \tag{F.3}$$

Upon substituting Eq. (F.1) into Eqs. (F.2) and (F.3), we obtain

$$\overline{DE}^2 = \overline{CO}^2[\tan^2\theta' + \tan^2\theta - 2\tan\theta'\tan\theta\cos(\phi - \phi')], \tag{F.4}$$

$$\overline{DE}^2 = \overline{CO}^2[\sec^2\theta' + \sec^2\theta - 2\sec\theta'\sec\theta\cos\Theta]. \tag{F.5}$$

It follows that

$$\begin{aligned}
\tan^2\theta' + \tan^2\theta - 2\tan\theta'\tan\theta\cos(\phi - \phi') \\
= \sec^2\theta' + \sec^2\theta - 2\sec\theta'\sec\theta\cos\Theta.
\end{aligned} \tag{F.6}$$

But $\sec^2\theta - \tan^2\theta = 1$, Eq. (F.6) becomes

$$2 - 2\sec\theta'\sec\theta\cos\Theta = -2\tan\theta'\tan\theta\cos(\phi - \phi'). \tag{F.7}$$

Thus,

$$\begin{aligned}
\cos\Theta &= \cos\theta\cos\theta' + \sin\theta\sin\theta'\cos(\phi - \phi') \\
&= \mu\mu' + (1 - \mu^2)^{1/2}(1 - \mu'^2)^{1/2}\cos(\phi - \phi').
\end{aligned} \tag{F.8}$$

Appendix G

ADDITION THEOREM FOR THE LEGENDRE POLYNOMIALS

Let $g(\mu, \phi)$ be an arbitrary function on the surface of a sphere where it and all of its first and second derivatives are continuous. Then $g(\mu, \phi)$ may be represented by an absolutely convergent series of surface harmonics as

$$g(\mu, \phi) = \sum_{l=0}^{\infty} \left[a_{l0} P_l(\mu) + \sum_{m=1}^{l} (a_{lm} \cos m\phi + b_{lm} \sin m\phi) P_l^m(\mu) \right]. \quad \text{(G.1)}$$

The coefficients are determined by

$$a_{l0} = \frac{2l+1}{4\pi} \int_0^{2\pi} \int_{-1}^{1} g(\mu, \phi) P_l(\mu) \, d\mu \, d\phi, \quad \text{(G.2)}$$

$$a_{lm} = \frac{(2l+1)(l-m)!}{2\pi(l+m)!} \int_0^{2\pi} \int_{-1}^{1} g(\mu, \phi) P_l^m(\mu) \cos m\phi \, d\mu \, d\phi, \quad \text{(G.3)}$$

$$b_{lm} = \frac{(2l+1)(l-m)!}{2\pi(l+m)!} \int_0^{2\pi} \int_{-1}^{1} g(\mu, \phi) P_l^m(\mu) \sin m\phi \, d\mu \, d\phi. \quad \text{(G.4)}$$

We note that

$$\int_{-1}^{1} P_l^m(\mu) P_k^m(\mu) \, d\mu = \begin{cases} 0, & l \neq k \\ \dfrac{2(l+m)!}{(2l+1)(l-m)!}, & l = k, \end{cases} \quad \text{(G.5)}$$

$$\int_0^{2\pi} \cos m\phi \cos n\phi \, d\phi = \begin{cases} 0, & m \neq n \\ \pi, & m = n, \end{cases} \quad \text{(G.6)}$$

and also that $P_l(1) = 1$, $P_l^m(1) = 0$. Thus, we write

$$[g(\mu, \phi)]_{\mu=1} = \sum_{l=0}^{\infty} a_{l0} = \frac{1}{4\pi} \sum_{l=0}^{\infty} (2l + 1) \int_0^{2\pi} \int_{-1}^1 g(\mu, \phi) P_l(\mu) \, d\mu \, d\phi. \quad \text{(G.7)}$$

We now define the surface harmonic function in the form

$$Y_l(\mu, \phi) = \sum_{m=0}^{l} (a_{lm} \cos m\phi + b_{lm} \sin m\phi) P_l^m(\mu). \quad \text{(G.8)}$$

Let $Y_l(\mu, \phi)$ of order l be $g(\mu, \phi)$, and by virtue of Eq. (G. 7), we find

$$[Y_l(\mu, \phi)]_{\mu=1} = \frac{2l + 1}{4\pi} \int_0^{2\pi} \int_{-1}^1 Y_l(\mu, \phi) P_l(\mu) \, d\mu \, d\phi. \quad \text{(G.9)}$$

On the basis of the scattering geometry, we have

$$\cos \Theta = \mu\mu' + (1 - \mu^2)^{1/2}(1 - \mu'^2)^{1/2} \cos(\phi - \phi'). \quad \text{(G.10)}$$

Thus, we may let

$$P_l(\cos \Theta) = \sum_{m=0}^{l} (c_m \cos m\phi + d_m \sin m\phi) P_l^m(\mu)$$

$$= \frac{c_0}{2} P_l(\mu) + \sum_{m=1}^{l} (c_m \cos m\phi + d_m \sin m\phi) P_l^m(\mu). \quad \text{(G.11)}$$

Upon utilizing the orthogonal properties denoted in Eqs. (G.5) and (G.6), we find

$$\int_0^{2\pi} \int_{-1}^1 P_l(\cos \Theta) P_l^m(\mu) \cos m\phi \, d\mu \, d\phi = \frac{2\pi(l + m)!}{(2l + 1)(l - m)!} c_m. \quad \text{(G.12)}$$

We let $P_l^m(\mu) \cos m\phi = Y_l(\mu, \phi)$, and by virtue of Eq. (G.9), Eq. (G.12) becomes

$$\int_0^{2\pi} \int_{-1}^1 P_l(\cos \Theta)[P_l^m(\mu) \cos m\phi] \, d\mu \, d\phi = \frac{4\pi}{2l + 1} [P_l^m(\mu) \cos m\phi]_{\cos\Theta = 1}$$

$$= \frac{4\pi}{2l + 1} P_l^m(\mu') \cos m\phi'. \quad \text{(G.13)}$$

Note that $\cos \Theta = 1$ and $\Theta = 0$, so we have $\mu = \mu'$, and $\phi = \phi'$. It follows from Eq. (G.11) that

$$c_m = \frac{2(l + m)!}{(l - m)!} P_l^m(\mu') \cos m\phi'. \quad \text{(G.14)}$$

In a similar manner, we find

$$d_m = \frac{2(l + m)!}{(l - m)!} P_l^m(\mu') \sin m\phi'. \tag{G.15}$$

Thus, from Eqs. (G.14), (G.15), and (G.11), we obtain

$$P_l(\cos \Theta) = P_l(\mu)P_l(\mu') + 2 \sum_{m=1}^{l} \frac{(l - m)!}{(l + m)!} P_l^m(\mu)P_l^m(\mu') \cos m(\phi' - \phi). \tag{G.16}$$

Appendix H

UNITED STATES METEOROLOGICAL SATELLITE PROGRAMS[a]

Name	Launched	Period (min)	Perigee (km)	Apogee (km)	Incli-nation (deg)	Instrument[b]
TIROS I	01 Apr 60	99.2	796	867	48.3	1 TV-WA and 1 TV-NA
TIROS II	23 Nov 60	98.3	717	837	48.5	1 TV-WA, 1 TV-NA, passive and active IR scan
TIROS III	12 Jul 61	100.4	854	937	47.8	2 TV-WA, HB, IR, IRP
TIROS IV	08 Feb 62	100.4	817	972	48.3	1 TV-WA, IR, IRP, HB
TIROS V	19 Jun 62	100.5	680	1119	58.1	1 TV-WA, 1 TV-MA
TIROS VI	18 Sep 62	98.7	783	822	58.2	1 TV-WA, 1 TV-MA
TIROS VII	19 Jun 63	97.4	713	743	58.2	2 TV-WA, IR, ion probe, HB
TIROS VIII	21 Dec 63	99.3	796	878	58.5	1st APT TV direct readout and 1 TV-WA
Nimbus I	28 Aug 64	98.3	487	1106	98.6	3 AVCS, 1 APT, HRIR
TIROS IX	22 Jan 65	119.2	806	2967	96.4	First "wheel"; 2 TV-WA global coverage
TIROS X	02 Jul 65	100.6	848	957	98.6	Sun synchronous, 2 TV-WA
ESSA 1	03 Feb 66	100.2	800	965	97.9	1st operational system, 2 TV-WA, FPR
ESSA 2	28 Feb 66	113.3	1561	1639	101.0	2 APT, global operational APT
Nimbus II	15 May 66	108.1	1248	1354	100.3	3 AVCS, HRIR, MRIR
ESSA 3	02 Oct 66	114.5	1593	1709	101.0	2 AVCS, FPR
ATS 1	06 Dec 66	24 hr	41,257	42,447	0.2	Spin scan camera
ESSA 4	26 Jan 67	113.4	1522	1656	102.0	2 APT
ESSA 5	20 Apr 67	113.5	1556	1635	101.9	2 AVCS, FPR
ATS III	05 Nov 67	24 hr	41,166	41,222	0.4	Color spin scan camera
ESSA 6	10 Nov 67	114.8	1622	1713	102.1	2 APT TV
ESSA 7	16 Aug 68	114.9	1646	1691	101.7	2 AVCS, FPR, S Band
ESSA 8	15 Dec 68	114.7	1622	1682	101.8	2 APT TV
ESSA 9	26 Feb 69	115.3	1637	1730	101.9	2 AVCS, FPR, S Band
Nimbus III	14 Apr 69	107.3	1232	1302	101.1	SIRS A, IRIS, MRIR, IDCS, MUSE, IRLS
ITOS 1	23 Jan 70	115.1	1648	1700	102.0	2 APT, 2 AVCS, 2 SR, FPR
Nimbus IV	15 Apr 70	107.1	1200	1280	99.9	SIRS B, IRIS, SCR, THIR, BUV, FWS, IDCS, IRLS, MUSE

Name	Launched	Period (min)	Perigee (km)	Apogee (km)	Incli-nation (deg)	Instrument[b]
NOAA 1	11 Dec 70	114.8	1422	1472	102.0	2 APT, 2 AVCS, 2 SR, FPR
NOAA 2	15 Oct 72	114.9	1451	1458	98.6	2 VHRR, 2 VTPR, 2 SR, SPM
Nimbus 5	11 Dec 72	107.1	1093	1105	99.9	SCMR, ITPR, NEMS, ESMR, THIR
NOAA 3	06 Nov 73	1161.1	1502	1512	101.9	2 VHRR, 2 VTPR, NEMS, ESMR, SPM
SMS 1	17 May 74	1436.4	35,605	35,975	0.6	VISSR, DCS, WEFAX, SEM
NOAA 4	15 Nov 74	101.6	1447	1461	114.9	2 VHRR, 2 VTPR, 2 SR, SPM
SMS 2	06 Feb 75	1436.5	35,482	36,103	0.4	VISSR, DCS, WEFAX, SEM
Nimbus 6	12 Jun 75	107.4	1101	1115	99.9	ERB, ESMR, HIRS, LRIR, T&DR, SCAMS, TWERLE, PMR
GOES 1	16 Oct 75	1436.2	35,728	35,847	0.8	VISSR, DCS, WEFAX, SEM
NOAA 5	29 Jul 76	116.2	1504	1518	102.1	2 VHRR, 2 VTPR, 2 SR, SPM
GOES 2	16 Jun 77	1436.1	35,600	36,200	0.5	VISSR, DCS, WEFAX, SEM
GOES 3	15 Jun 78	1436.1	35,600	36,200	0.5	VISSR, DCS, WEFAX, SEM
TIROS-N	13 Oct 78	98.92	849	864	102.3	AVHRR, HIRS-2, SSU, MSU, HEPAD, MEPED
Nimbus 7	24 Oct 78	99.28	943	956	104.09	LIMS, SAMS, SAM-II, SBUV/TOMS, ERB, SMMR, THIR, CZCS
NOAA 6	27 Jun 79	101.26	807.5	823	98.74	AVHRR, HIRS-2, SSU, MSU, HEPAD, MEPED

[a] Courtesy A. Schnapf, RCA Corporation, Astro-Electronics, Princeton, New Jersey, in a paper, "Evolution of the Operational Satellite Service 1958–1984" presented at NOAA A Colloquium, May 1979, Washington, D.C.

[b] APT   Automatic Picture Transmission TV   NEMS   Nimbus E Microwave Spectrometer

AVCS   Advanced Vidicon Camera System (1-in. Vidicon)   PMR   Pressure Modulated Radiometer

AVHRR   Advanced Very High Resolution Radiometer   SAM-II   Stratospheric Aerosol Measurement-II

BUV   Backscatter Ultraviolet Spectrometer   SAMS   Stratospheric and Mesospheric Sounder

CZCS	Coastal Zone Color Scanner
DCS	Data Collection System
ERB	Earth Radiation Budget
ESMR	Electronic Scanning Microwave Radiometer
FPR	Flat Plate Radiometer
FWS	Filter Wedge Spectrometer
HB	Heat Budget Instrument
HEPAD	High Energy Proton and Alpha Particle Detector
HIRS	High Resolution Infrared Sounder
HRIR	High Resolution Infrared Radiometer
IDCS	Image Dissector Camera System
IR	Infrared—5 Channel Scanner
IRIS	Infrared Interferometer Spectrometer
IRLS	Interrogation, Recording, and Location Subsystem
IRP	Infrared Passive
ITPR	Infrared Temperature Profile Radiometer
LIMS	Limb Infrared Monitoring of the Stratosphere
LRIR	Limb Radiance Infrared Radiometer
MEPED	Medium Energy Proton and Electron Detector
MRIR	Medium Resolution Infrared Radiometer
MSU	Microwave Scanner Unit
MUSE	Monitor of Ultraviolet Solar Energy
SBUV	Solar Backscatter Ultraviolet Spectrometer
SCAMS	Scanning Microwave Spectrometer
SCMR	Surface Composition Mapping Radiometer
SCR	Selective Chopper Radiometer
SEM	Solar Environmental Monitor
SIRS	Satellite Infrared Spectrometer
SMMR	Scanning Multichannel Microwave Radiometer
SPM	Solar Proton Monitor
SR	Scanning Radiometer
SSU	Stratospheric Sounding Unit
T&DR	Tracking and Data Relay
THIR	Temperature Humidity Infrared Radiometer
TOMS	Total Ozone Mapping Spectrometer
TV	Television Cameras (1/2-in. Vidicon)
	NA, narrow angle—12°
	MA, medium angle—78°
	WA, wide angle—104°
TWERLE	Tropical Wind Energy Reference Equipment
VHRR	Very High Resolution Radiometer
VISSR	Visible Infrared Spin-Scan Radiometer
VTPR	Vertical Temperature Profile Radiometer
WEFAX	Weather Facsimile

Appendix I
ANSWERS TO SELECTED EXERCISES

CHAPTER 1

1.3 $B_v(T) = 2hv^3c^2/(e^{hcv/KT} - 1)$

1.4   Show that $(5 - x) = 5 - e^{-x}$, where $x = hc/(K\lambda T)$, and find x.

1.5   Insert $\lambda_m = \alpha/T$ into Eq. (1.19).

1.6 $\sim 300°K$

1.7 7.42×10^{-20} erg sec^{-1} cm^{-2} sr^{-1}/μm, 81.3 erg sec^{-1} cm^{-2} sr^{-1}/(cm^{-1}), 8.57×10^{-14} erg sec^{-1} cm^{-2} sr^{-1}/Hz

1.8 5.22×10^5 erg sec^{-1} cm^{-2}, 4.69×10^5 erg sec^{-1} cm^{-2}, $9.36\ \mu$m

1.9(b) $n = 1, 2, \lambda_{12} = 1216$ Å

1.13 $\beta_\lambda = 0.1, 0.5$ m^{-1}, $\tau = 1, 5$

1.15 $I_\lambda(s) = I_\lambda(0)(1 - R_\lambda)^2 \mathcal{T}_\lambda/(1 - R_\lambda^2 \mathcal{T}_\lambda^2)$

1.17 $F_v(\tau = 0) = \pi B_v(T_s)2E_3(\tau_1) + \pi B_v(T)[1 - 2E_3(\tau_1)]$. See Eq. (4.8) for the definition of the exponential integral E_3.

CHAPTER 2

2.1   Solar irradiance $F_\lambda = \pi B_\lambda(T)(a_s/d_m)^2$

2.2 $5754°K$

2.3 1.48×10^{29} erg day^{-1}

2.4 4.53×10^{-10}

2.5 1.14×10^{16} erg sec^{-1} (from the cloud), 1.82×10^5 erg sec^{-1} (at the surface)

2.6 $F_{\lambda 0} \approx 0.032$ cal cm^{-2} min^{-1}, $\mathcal{T}_\lambda \approx 0.68$

2.8   254°K (Earth)

2.9   70.8°K

2.10   4.2°K

2.11(a) 1114 cal cm^{-2} day^{-1}; (b) 889 cal cm^{-2} day^{-1}

2.12   Elevation angle $\varepsilon = \delta$ (poles), 15.44 hr at 45°N in solstice

CHAPTER 3

3.1 $[O]\left\{[O]^2 - \dfrac{[O_2]J_2}{K_{11}[M]}\right\} + \dfrac{K_{11}J_3[M] + K_{12}K_{13}[O_2][M]}{K_{11}K_{13}[M]}\left\{[O]^2 - \right.$

$\left.\dfrac{[O_2]J_2}{K_{11}[M] + K_{12}K_{13}[O_2][M]/J_3}\right\} = 0$, to a good approximation

$[O]^2 \approx [O_2]J_2/\{K_{11}[M] + K_{12}K_{13}[O_2][M]/J_3\}$

3.2 $[O]^2 = [O_2]J_2/\{K_{11}[M] + K'_{11}\}$

3.4 Let $I = 1, [1 \quad 0 \quad 0 \quad 0], [1 \quad -1 \quad 0 \quad 0], [1 \quad 0 \quad 0 \quad \pm 1]$

3.6 Let $I_0 = 1$. (c) $[2 \quad -\frac{1}{2} \quad 0 \quad \frac{1}{2}], I = 2, P = \dfrac{1}{2\sqrt{2}}\%, I_r = \frac{5}{4}$; (d) $\chi = 0°$,

$\beta = 22.5°$; (e) $\frac{1}{4}[(4 - \sqrt{2})(2\sqrt{2} - 1)0(2\sqrt{2} - 1)], \frac{1}{4}[(4 - \sqrt{2})(1 - 2\sqrt{2})0(1 - 2\sqrt{2})]$; (f) $\frac{1}{4}[(4 + \sqrt{2})(2\sqrt{2} - 3)0(1 + 2\sqrt{2})]$ for right-hand polarization.

3.7 $\sigma_s(0.7 \ \mu m) = 1.71 \times 10^{-27}$ cm^2

3.8 $\tau(0.7 \ \mu m) \approx 0.033$

3.9 At $z = 10$ km, $mr \approx 1.000099$

3.10 Let the laser beam width be ℓ cm; $F \approx F_0\ell \ 1.37 \times 10^{-11}$ (at 10 km)

3.11 For $\lambda = 10$ cm, $\beta_\pi \approx 9.76 \times 10^{-10}$ km^{-1}, 8.66×10^{-10} km^{-1}

CHAPTER 4

4.1 Start from Eqs. (4.23) and (4.24), and use Eq. (4.67).

4.2 *Note:* $\displaystyle\int_0^\infty \dfrac{dx}{1 + x^2} = \dfrac{\pi}{2}, \int_0^\infty x^{-1/2}e^{-x}\,dx = \Gamma(\tfrac{1}{2}) = \sqrt{\pi}$

4.7 $\mathcal{T}_{\Delta\nu}(u) = F_{\Delta\lambda}/F_{0,\,\Delta\lambda}$

4.8 $F^\uparrow \approx 0.5$ cal cm^{-2} min^{-1}, $F^\downarrow \approx 0.2$ cal cm^{-2} min^{-1}, $\partial T/\partial t \approx -1.5°C$ day^{-1}

CHAPTER 5

5.1 *Note:* $\nabla^2 \mathbf{A} = \nabla(\nabla \cdot \mathbf{A}) - \nabla \times (\nabla \times \mathbf{A})$

5.4 Let $\psi = e^{i\omega t}R(r)Z(z)\Phi(\phi)$.

5.5   See Eqs. (3.42) and (5.104).

5.6 $\theta_i = \cos^{-1}\sqrt{1/(1 + m^2)}$

5.7(a)   For white corona, use $\lambda = 0.55\ \mu m$; $a \approx 3.5\ \mu m$; (b) $\not{p} = 2$ (primary), θ (red) $= 137.78°$; (c) $A = 60°$, θ' (red) $= 21.61°$; (d) *Note*: $\phi_t = A/2$, $\phi_i = (A + \theta_h')/2$, where θ_h' is the minimum deviation angle projected on the horizontal plane, and $\sin\dfrac{\theta'}{2} = \cos\varepsilon_i \sin\dfrac{\theta_h'}{2}$ (prove), θ' (red) $= 24.54°$.

CHAPTER 6

6.1   Let $F_0 = 1$ and $\phi_0 = 0°$. For $\tau = 0.1$, $I(\mu = 1, \phi = 0°) = 0.028$; for $\tau = 1$, $I(\mu = 0.4, \phi = 90°) = 0.135$

6.2 $I_2(0, \mu, \phi) = \dfrac{\mu_0 F_0 \tilde{\omega}^2}{4\pi} \displaystyle\int_0^{2\pi}\int_{-1}^{1} P(\mu, \phi; \mu', \phi')P(\mu', \phi'; -\mu_0, \phi_0)g(\mu, \mu_0,$

$\mu')\,d\mu'\,d\phi'$, where $g(\mu, \mu_0, \mu') = \dfrac{1}{4(\mu_0 + \mu')}\left[\dfrac{\mu_0}{\mu_0 + \mu}\left\{1 - \exp\left[-\tau_1\right.\right.\right.$

$\left.\left.\left(\dfrac{1}{\mu_0} + \dfrac{1}{\mu}\right)\right]\right\} + \dfrac{\mu'}{\mu - \mu'}\left\{\exp\left[-\tau_1\left(\dfrac{1}{\mu_0} + \dfrac{1}{\mu'}\right)\right] - \exp\left[-\tau_1\left(\dfrac{1}{\mu_0} + \dfrac{1}{\mu}\right)\right]\right\}\right]$,

$\mu \neq \mu'$

6.3 $I^+(0, \mu_1) = \dfrac{1}{2}\left\{\dfrac{\mu_0}{\mu_1}\left[2S^- - 2b\dfrac{\mu_0}{\mu_1}(S^+ + S^-)\right] + \dfrac{K\mu_1}{2b} + H\right\}$, where $K =$

$-\dfrac{2\mu_0}{\mu_1}\left[\left(S^- - \dfrac{b\mu_0}{\mu_1}\right)e^{-\tau_1/\mu_0} + S^+ + \dfrac{b\mu_0}{\mu_1}(S^+ + S^-)\right]\bigg/\left(\dfrac{\tau_1 + \mu_1}{b}\right)$, $H =$

$\dfrac{2\mu_0}{\mu_1}\left[S^+ + \dfrac{b\mu_0}{\mu_1}(S^+ + S^-)\right] + \dfrac{K\mu_1}{2b}$

6.4 $I_0(\tau) = Ke^{k\tau} + He^{-k\tau} + \beta e^{-\tau/\mu_0}$, $\beta = \eta/(k^2 - 1/\mu_0^2)$

6.6 $I(\tau; \mu_i) = \displaystyle\sum_{\alpha = -n}^{n} \dfrac{L_\alpha}{1 + \mu_i k_\alpha} e^{-k_\alpha\tau} + B_\nu(T)$

6.7 $\tilde{\omega}(\tilde{\nu}) = (1 + x^2)/(2 + x^2)$, where $x = (\nu - \nu_0)/\alpha$, $R(\mu, \mu_0) = \dfrac{\tilde{\omega}}{4(\mu + \mu_0)}$

$\dfrac{(1 + \sqrt{3}\mu)(1 + \sqrt{3}\mu_0)}{[1 + \mu\sqrt{3(1 - \tilde{\omega})}][1 + \mu_0\sqrt{3(1 - \tilde{\omega})}]}$

6.8 $\tilde{\omega} = 0.8$, $r(\mu_0 = 1) \approx 0.37$, $\bar{r} \approx 0.34$

CHAPTER 7

7.1(b) $v^* = 3.13$

7.2

$$\text{First difference, } \mathbf{H} = \begin{bmatrix} 1 & -1 & & & & & \\ -1 & 2 & -1 & & & & 0 \\ & -1 & 2 & -1 & & & \\ & & -1 & 2 & -1 & & \\ & & & & \ddots & & \\ & & 0 & & & \ddots & \\ & & & & & -1 & 1 \end{bmatrix}$$

7.3(a) $g(k) = -\dfrac{12}{k^2} e^{-k} + \left(\dfrac{24}{k^2} - \dfrac{8}{k} + 2\right)\left[\dfrac{1}{k^2} - \left(\dfrac{1}{k} + \dfrac{1}{k^2}\right)e^{-k}\right],\ i = 10,\ k = 5,$

$g(k) = 0.0490$; (b) $i = 10,\ k = 5,\ g(k) = 0.0489$; (d) $\gamma = 1,\ j = 15,$
$f(x_j) = 0.5251(0.9375),\ \gamma = 10^{-7}, j = 15,\ f(x_j) = 1.005(0.9375)$

CHAPTER 8

8.1 $302°\text{K}$
8.2 $\bar{\varepsilon} = 0.88,\ T_a = 250°\text{K}$
8.4   Temperature at the ice line $T(x_s = 0.95) = -6.96°\text{C}$ ($S = 1.92$ cal cm^{-2} min^{-1})
8.6 $D'' \approx 0.26$

BIBLIOGRAPHY

Ambartsumyan, V. A. (1942). A new method for computing light scattering in turbid media. *Izv. Akad. Nauk SSSR, Ser. Geogr. i Geofiz.* **3**, 97–104.

Ambartsumyan, V. A. (1958). "Theoretical Astrophysics." Pergamon Press, New York.

Backus, G. E., and Gilbert, J. F. (1970). Uniqueness in the inversion of inaccurate gross earth data. *Philos. Trans. Roy. Soc. London, Ser. A.* **266**, 123–192.

Bellman, R. E., Kalaba, R. E., and Prestrud, M. C. (1963). "Invariant Imbedding and Radiative Transfer in Slabs of Finite Thickness." Elsevier, New York.

Born, M., and Wolf, E., (1975). "Principles of Optics," 5th ed. Pergamon Press, New York.

Budyko, M. I. (1969). The effect of solar radiation variations on the climate of the earth. *Tellus* **21**, 611–619.

Budyko, M. I. (1974). "Climate and Life." Academic Press, New York.

Bullrich, K. (1964). Scattered radiation in the atmosphere and the natural aerosol. *Adv. Geophys.* **10**, 99–260.

Chahine, M. (1970). Inverse problems in radiative transfer: Determination of atmospheric parameters. *J. Atmos. Sci.* **27**, 960–967.

Chandrasekhar, S. (1950). "Radiative Transfer." Dover, New York.

Chapman, S. (1930). A theory of upper-atmospheric ozone. *Mem. Roy. Meteorol. Soc.* **3**, 103–125.

Charlock, T., and Herman, B. M. (1976). Discussion of the Elsasser formulation for infrared fluxes. *J. Appl. Meteorol.* **15**, 657–661.

Coleman, R. F. (1979). Light scattering by hexagonal ice crystals. M.S. thesis, Dept. of Meteorology, Univ. of Utah, Salt Lake City, Utah.

Collis, R. T. H. (1969). Lidar. *Adv. Geophys.* **13**, 113–139.

Coulson, K. L. (1975). "Solar and Terrestrial Radiation." Academic Press, New York.

Coulson, K. L., Dave, J. V., and Sekera, Z. (1960). "Tables Related to Radiation Emerging from a Planetary Atmosphere with Rayleigh Scattering." Univ. of California Press, Berkeley.

Craig, R. A. (1965). "The Upper Atmosphere: Meteorology and Physics." Academic Press, New York.

Dave, J. V. (1978). Effect of aerosols on the estimate of total ozone in an atmospheric column from the measurements of its ultraviolet radiance. *J. Atmos. Sci.* **35**, 899–911.

Dave, J. V., and Mateer, C. L. (1967). A preliminary study on the possibility of estimating total atmospheric ozone from satellite measurements. *J. Atmos. Sci.* **24**, 414–427.

Davis, P. A. (1963). An analysis of the atmospheric heat budget. *J. Atmos. Sci.* **20**, 5–22.

Davison, B. (1957). "Neutron Transport Theory." Oxford Univ. Press (Clarendon), London and New York.

Deirmendjian, D. (1969). "Electromagnetic Scattering on Spherical Polydispersions." Elsevier, New York.

Derr, V. E., Ed. (1972). "Remote Sensing of the Troposphere." Supt. of Documents, US Govt. Printing Office, Washington, D.C.

Dobson, G. M. B. (1957). Observers' handbook for the ozone spectrometer. *Ann. Int. Geophys. Year* **5**, 46–89.

Dopplick, T. G. (1972). Radiative heating of the global atmosphere. *J. Atmos. Sci.* **29**, 1278–1294.

Eddy, J. A. (1977). Historical evidence for the existence of the solar cycle. *In* "The Solar Output and Its Variation" (O. R. White, ed.), pp. 51–71. Colorado Assoc. Univ. Press, Boulder, Colorado.

Elsasser, W. M. (1938). Mean absorption and equivalent absorption coefficient of a band spectrum. *Phys. Rev.* **54**, 126–129.

Elsasser, W. M. (1942). Heat transfer by infrared radiation in the atmosphere. *Harvard Meteorol. Studies* No. 6, 107 pp. Harvard Univ. Press, Cambridge, Massachusetts.

Elsasser, W. M., and Culbertson, M. F. (1960). Atmospheric radiation tables. *Meteorol. Monogr.* **4**, 1–43.

Fleagle, R. G., and Businger, J. A. (1963). "An Introduction to Atmospheric Physics." Academic Press, New York.

Freeman, K. P., and Liou, K. N. (1979). Climatic effects of cirrus clouds. *Adv. Geophys.* **21**, 231–287.

Friedman, H. (1960). The sun's ionizing radiations. *In* "Physics of Upper Atmosphere" (J. R. Ratcliffe, ed.), pp. 133–218. Academic Press, New York.

Fymat, A. L., and Zuev, V. E., Eds. (1978). "Remote Sensing of the Atmosphere: Inversion Methods and Applications." Elsevier, New York.

Gille, J. C., and House, F. B. (1971). On the inversion of limb radiance measurements. I: Temperature and thickness. *J. Atmos. Sci.* **28**, 1427–1442.

Goldman, A., and Kyle, T. G. (1968). A comparison between statistical model and line calculation with application to the 9.6 μm ozone and the 2.7 μm water vapor bands. *Appl. Opt.* **7**, 1167–1177.

Goody, R. M. (1952). A statistical model for water vapor absorption. *Quart. J. Roy. Meteorol. Soc.* **78**, 165–169.

Goody, R. M. (1964). "Atmospheric Radiation. I: Theoretical Basis." Oxford Univ. Press (Clarendon), London and New York.

Grody, N. C. (1976). Remote sensing of atmospheric water content from satellites using microwave radiometry. *IEEE Trans. Antennas Propag.* **AP 24**, 155–162.

Halliday, D., and Resnick R. (1974). "Fundamentals of Physics," rev. ptg. Wiley, New York.

Hansen, J. E. (1971). Multiple scattering of polarized light in planetary atmospheres. Part II. Sunlight reflected by terrestrial water clouds. *J. Atmos. Sci.* **28**, 1400–1426.

Hansen, J. E., and Arking, A. (1971). Clouds of Venus: Evidence for their nature. *Science* **171**, 669–672.

Hansen, J. E., and Hovenier, J. W. (1974). Interpretation of the polarization of Venus. *J. Atmos. Sci.* **31**, 1137–1160.

Hansen, J. E., and Travis, L. D. (1974). Light scattering in planetary atmospheres. *Space Sci. Rev.* **16**, 527–610.

Hays, J. D., Imbrie, J., and Shackleton, N. J. (1976). Variations in the earth's orbit: Pacemaker of the ice ages. *Science* **194**, 1121–1132.

Held, I. M., and Suarez, M. J. (1974). Simple albedo feedback models of the icecaps. *Tellus* **26**, 613–629.

Herzberg, G. (1945). "Molecular Spectra and Molecular Structure." Van Nostrand Reinhold, Princeton, New Jersey.

Hinkley, E. D., ed. (1976). "Laser Monitoring of the Atmosphere." Springer-Verlag, Berlin and New York.

Houghton, H. G. (1954). On the annual heat balance of the Northern Hemisphere. *J. Meteorol.* **11**, 1–9.

Houghton, J. T., and Hunt, G. E. (1971). The detection of ice clouds from remote measurements of their emission in the far infrared. *Quart. J. Roy. Meteorol. Soc.* **97**, 1–7.

Howard, J. N., Burch, D. L., and Williams, D. (1956). Near-infrared transmission through synthetic atmospheres. *J. Opt. Soc. Am.* **46**, 186–190.

Humphreys, W. J. (1954). "Physics of the Air." Dover, New York.

Hunt, G. E. (1977). Studies of the sensitivity of the components of the earth's radiation balance to changes in cloud properties using a zonally averaged model. *J. Quant. Spectrosc. Radiat. Transfer* **18**, 295–307.

Jacobowitz, H. (1971). A method for computing the transfer of solar radiation through clouds of hexagonal ice crystals. *J. Quant. Spectrosc. Radiat. Transfer* **11**, 691–695.

Jacobowitz, H., Smith, W. L., Howell, H. B., and Nagle, F. W. (1979). The first 18 months of planetary radiation budget measurements from the Nimbus 6 ERB experiment. *J. Atmos. Sci.* **36**, 501–507.

Jastrow, R., and Thompson, M. H. (1974). "Astronomy: Fundamentals and Frontiers," 2nd ed. Wiley, New York.

Joseph, J. H., Wiscombe, W. J., and Weinman, J. A. (1976). The Delta–Eddington approximation for radiative flux transfer. *J. Atmos. Sci.* **33**, 2452–2459.

Kaplan, L. D. (1959). Inference of atmospheric structure from remote radiation measurements. *J. Opt. Soc. Am.* **49**, 1004–1007.

Katayama, A. (1966). On the radiation budget of the troposphere over the northern hemisphere (I). *J. Meteorol. Soc. Japan* **44**, 381–401.

Kattawar, G. W., Plass, G. N., and Adams, C. N. (1971). Flux and polarization calculations of the radiation reflected from the clouds of Venus. *Astrophys. J.* **170**, 371–386.

Kerker, M. (1969). "The Scattering of Light and Other Electromagnetic Radiation." Academic Press, New York.

King, J. I. F. (1956). The radiative heat transfer of planet Earth. *In* "Scientific Uses of Earth Satellites" pp. 133–136. Univ. of Michigan Press, Ann Arbor.

Kondratyev, K. Ya. (1969). "Radiation in the Atmosphere." Academic Press, New York.

Kondratyev, K. Ya. (1972). "Radiation Processes in the Atmosphere." World Meteorol. Organization, WMO-No. 309.

Kourganoff, V. (1952). "Basic Methods in Transfer Problems." Oxford Univ. Press (Clarendon), London and New York.

Kunde, V. G., Conrath, B. J., Hanel, R. A., Maguire, W. C., Prabhakara, C., and Salomonson, V. V. (1974). The Nimbus IV infrared spectroscopy experiment. 2. Comparison of observed and theoretical radiances from 425–1450 cm^{-1}. *J. Geophys. Res.* **79**, 777–784.

Lenoble, J., Ed. (1977). "Standard Procedures to Compute Atmospheric Radiative Transfer in a Scattering Atmosphere." Radiation Commission, IAMAP, published by National Center for Atmospheric Research, Boulder, Colorado.

Leovy, C. B. (1969). Energetics of the middle atmosphere. *Adv. Geophys.* **13**, 191–221.

Lindzen, R. S., and Farrell, B. (1977). Some realistic modification of simple climate models. *J. Atmos. Sci.* **34**, 1487–1501.

Liou, K. N. (1973). A numerical experiment on Chandraskhar's discrete-ordinate method for radiative transfer: Applications to cloudy and hazy atmospheres. *J. Atmos. Sci.* **30**, 1303–1326.

Liou, K. N. (1974). Analytic two-stream and four-stream solutions for radiative transfer. *J. Atmos. Sci.* **31**, 1473–1475.

Liou, K. N. (1977). Remote sensing of the thickness and composition of cirrus clouds from satellites. *J. Appl. Meteorol.* **16**, 91–99.

Liou, K. N. (1977). A complementary theory of light scattering by homogeneous spheres. *Appl. Math. Comp.* **3**, 331–358.

Liou, K. N., and Duff, A. D. (1979). Atmospheric liquid water content derived from parameterization of Nimbus 6 scanning microwave spectrometer data. *J. Appl. Meteorol.* **18**, 99–103.

Liou, K. N., and Hansen, J. E. (1971). Intensity and polarization for single scattering by polydisperse spheres: A comparison of ray optics and Mie theory. *J. Atmos. Sci.* **28**, 995–1004.

Liou, K. N., and Sasamori, T. (1975). On the transfer of solar radiation in aerosol atmospheres. *J. Atmos. Sci.* **32**, 2166–2177.

List, R. J. (1958). "Smithsonian Meteorological Tables," 6th rev. ed. Random House (Smithsonian Inst. Press), New York.

London, J. (1975). A study of the atmospheric heat balance. 99 pp. College of Engineering, New York Univ., Final Rept. Contract AF19(122)–165.

Manabe, S., and Strickler, R. F. (1964). Thermal equilibrium of the atmosphere with a convective adjustment. *J. Atmos. Sci.* **21**, 361–385.

Manabe, S., and Wetherald, R. T. (1967). Thermal equilibrium of the atmosphere with a given distribution of relative humidity. *J. Atmos. Sci.* **24**, 241–259.

Manabe, S., and Wetherald, R. T. (1975). The effects of doubling the CO_2 concentration on the climate of a general circulation model. *J. Atmos. Sci.* **32**, 3–15.

Mateer, C. L., Heath, D. F., and Krueger, A. J. (1971). Estimation of total ozone from satellite measurements of backscattered ultraviolet earth radiance. *J. Atmos. Sci.* **28**, 1307–1311.

McCartney, E. J. (1976). "Optics of the Atmosphere." Wiley, New York.

McClatchey, R. A., and Selby, J. E. A. (1972). Atmospheric transmittance, $7–30$ μm: Attenuation of CO_2 laser radiation. *Environ. Res. Pap.* No. 419, AFCRL-72-0611.

McClatchey, R. A., Fenn, R. W., Selby, J. E. A., Volz, F. E., and Garing, J. S. (1971). Optical properties of the atmosphere. *Environ. Res. Pap.* No. 354, AFCRL-71-0279.

McEwan, M. J., and Phillips, L. F. (1975). "Chemistry of the Atmosphere." Wiley, New York.

Meeks, M. L., and Lilley, A. E. (1963). The microwave spectrum of oxygen in the Earth's atmosphere. *J. Geophys. Res.* **68**, 1683–1703.

Mie, G. (1908). Beigrade zur Optik trüber Medien, speziell kolloidaler Metallösungen. *Ann. Physik.* (4), **25**, 377–445.

Milankovitch, M. (1941). Kanon der Erdbestrahlung und seine Anwendung auf des Eiszeitproblem. Roy. Serbian Acad., Belgrade.

Möller, F. (1943). "Das strahlungsdiagramm." Reichsamt für Wetterdienst, Berlin.

North, J. R. (1975a). Analytical solution to a simple climate model with diffusive heat transport. *J. Atmos. Sci.* **32**, 1301–1307.

North, J. R. (1975b). Theory of energy-balance climate models. *J. Atmos. Sci.* **32**, 2033–2043.

Penndorf, R. (1962). Scattering and extinction coefficients for small spherical aerosols. *J. Atmos. Sci.* **19**, 193.

Penner, S. S. (1959). "Quantitative Molecular Spectroscopy and Gas Emissivities." Addison-Wesley, Reading, Massachusetts.

Perrin, F. (1942). Polarization of light scattered by isotropic opalescent media. *J. Chem. Phys.* **10**, 415–527.

Phillips, D. L. (1962). A technique for the numerical solution of certain integral equations of the first kind. *J. Assoc. Comput. Mach.* **9**, 84–97.

Raschke, E., and Bandeen, W. R. (1970). The radiation balance of the planet Earth from radiation measurements of the satellite Nimbus II. *J. Appl. Meteorol.* **9**, 215–238.

Raschke, E., Vonder Haar, T. H., Bandeen, W. R., and Pasternak, M. (1973). The annual radiation balance of the earth–atmosphere system during 1969–1970 from Nimbus III measurements. *J. Atmos. Sci.* **30**, 341–364.

Rigone, J. L., and Strogryn, A. P. (1977). Data processing for the DMSP Microwave Radiometer System. *11th Inter. Symp. Remote Sensing of the Environment.* Univ. of Michigan, pp. 1599–1608.

Roberts, E., Selby, J. E. A., and Biberman, I. M. (1976). Infrared continuum absorption by atmospheric water vapor in the 8–12μm window. *Appl. Opt.* **15**, 2085–2090.

Robinson, N., ed. (1966). "Solar Radiation." Elsevier, New York.

Rodgers, C. D. (1967). The radiative heat budget of the troposphere and lower stratosphere. Res. Rep. No. A2, Dept. of Meteorology, MIT, Cambridge, Massachusetts.

Rodgers, C. D. (1976). Retrieval of atmospheric temperature and composition from remote measurements of thermal radiation. *Rev. Geophys. Space Phys.* **14**, 609–624.

Rodgers, C. D., and Walshaw, C. D. (1966). The computation of infrared cooling rate in planetary atmospheres. *Quart. J. Roy. Meteorol. Soc.* **93**, 43–54.

Roewe, D., and Liou, K. N. (1978). Influence of cirrus clouds on the infrared cooling rate in the troposphere and lower stratosphere. *J. Appl. Meteorol.* **17**, 92–105.

Sasamori, T. (1968). The radiative cooling calculation for application to general circulation experiments. *J. Appl. Meteorol.* **7**, 721–729.

Sasamori, T., London, J., and Hoyt, D. V. (1972). Radiation budget of the southern hemisphere. *Meteorol. Monogr.* **13**, No. 35, Chapter 2.

Sassen K., and Liou, K. N. (1979). Scattering of polarized laser light by water droplet, mixed-phase, and ice crystal clouds. Part I: Angular scattering patterns. *J. Atmos. Sci.* **36**, 838–851.

Schneider, S. H., and Dickinson, R. E. (1974). Climate modeling. *Rev. Geophys. Space Phys.* **12**, 447–493.

Schotland, R. M. (1969). Some aspects of remote atmospheric sensing by laser radar. *In* "Atmospheric Exploration by Remote Probes," pp. 179–200, Committee on Atmospheric Sciences, Nat. Acad. Sci.–Nat. Res. Council, Washington, D.C.

Sellers, W. D. (1965). "Physical Climatology." Univ. of Chicago Press, Chicago, Illinois.

Sellers, W. D. (1969). A global climatic model based on the energy balance of the earth–atmosphere system. *J. Appl. Meteorol.* **8**, 392–400.

Shettle, E. P., and Weinman, J. A. (1970). The transfer of solar irradiance through inhomogeneous turbid atmosphere evaluated by Eddington's approximation. *J. Atmos. Sci.* **27**, 1048–1055.

Smith, W. L. (1970). Iterative solution of the radiative transfer equation for the temperature and absorbing gas profile of an atmosphere. *Appl. Opt.* **9**, 1993–1999.

Smith, W. L., Hickey, J., Howell, H. B., Jacobowitz, H., Hilleary, D. T., and Drummond, A. J. (1977). Nimbus-6 earth radiation budget experiment. *Appl. Opt.* **16**, 306–318.

Sobolev, V. V. (1975). "Light Scattering in Planetary Atmospheres." Pergamon Press, New York.

Staelin, D. H., Kunzi, K. F., Pettyjohn, R. L., Poon, R. K. L., and Wilcox, R. W. (1976). Remote sensing of atmospheric water vapor and liquid water with the Nimbus 5 microwave spectrometer. *J. Appl. Meteorol.* **15**, 1204–1214.

Staley, D. O., and Jurica, G. M. (1970). Flux emissivity tables for water vapor, carbon dioxide, and ozone. *J. Appl. Meteorol.* **9**, 365–372.

Study of Man's Impact on Climate (SMIC) (1971). "Inadvertent Climate Modification". MIT Press, Cambridge, Massachusetts.

Stratton, J. A. (1941). "Electromagnetic Theory." McGraw-Hill, New York.

Suomi, V. E., Hanson, K. J., and Vonder Haar, T. H. (1967). The theoretical basis for low-resolution radiometer measurements from a satellite. Ann. Rep., Dept. of Meteorology, Univ. of Wisconsin, 79–100.

Thekaekara, M. P., (1974). Extraterrestrial solar spectrum, 3000–6100 Å at 1 Å intervals. *Appl. Opt.* **13**, 518–522.

Thekaekara, M. P. (1976). Solar irradiance: Total and spectral and its possible variations. *Appl. Opt.* **15**, 915–920.

Tiwari, S. N. (1978). Models for infrared atmospheric radiation. *Adv. Geophys.* **20**, 1–80.

Tricker, R. A. R. (1970). "Introduction to Meteorological Optics." Elsevier, New York.

Twomey, S. (1963). On the numerical solution of Fredholm integral equations of the first kind by the inversion of the linear system produced by quadrature. *J. Assoc. Comput. Mach.* **10**, 97–101.

Twomey, S. (1977). "Introduction to the Mathematics of Inversion in Remote Sensing and Indirect Measurements." Elsevier, New York.

US Standard Atmosphere (1976). NOAA-S/T76–1562. Supt. of Documents, US Govt. Printing Office, Washington, D.C.

van de Hulst, H. C. (1957). "Light Scattering by Small Particles." Wiley, New York.

van de Hulst, H. C. (1963). A New Look at Multiple Scattering. Sci. Rep., 81 pp. NASA, Goddard Inst. Space Studies, New York.

van de Hulst, H. C., and Grossman, K. (1968). Multiple light scattering in planetary atmospheres. *In* "The Atmospheres of Venus and Mars," (J. C. Brandt and M. B. McElroy, eds.), pp. 35–55. Gordon & Breach, New York.

Volz, F. E. (1974). Economical multispectral sun photometer for measurements of aerosol extinction from 0.44 to 1.6 μm and precipitable water. *Appl. Opt.* **13**, 1732–1733.

Vonder Haar, T. H., and Suomi, V. E. (1971). Measurements of the Earth's radiation budget from satellites during a five-year period. Part I: Extended time and space means. *J. Atmos. Sci.* **28**, 305–314.

Waters, J. W., Kunzi, K. F., Pettyjohn, R. L., Poon, R. K. L., and Staelin, D. H. (1975). Remote sensing of atmospheric temperature profiles with the Nimbus 5 microwave spectrometer. *J. Atmos. Sci.* **32**, 1953–1969.

Watson, G. N. (1944). "A Treatise on the Theory of Bessel Functions," 2nd ed. Cambridge Univ. Press, London and New York.

Wendling, P., Wendling, R., and Weickmann, H. K. (1979). Scattering of solar radiation by hexagonal ice crystals. *Appl. Opt.* **18**, 2663–2671.

Whittacker, E. T., and Watson, G. N. (1940). "Modern Analysis," 4th ed. Cambridge Univ. Press, London and New York.

Winston, J. S. (1969). Temporal and meridional variations in zonal mean radiative heating measured by satellites and related variations in atmospheric energetics. Ph.D. dissertation. Dept. of Meteorology and Oceanography, New York Univ., New York.

Wittman, G. D. (1978). Parameterization of the solar and infrared radiative properties of clouds. M.S. thesis, Dept. of Meteorology, Univ. of Utah, Salt Lake City, Utah.

Yamamoto, G. (1952). On a radiation chart. *Sci. Rep. Tohoku Univ. Ser. 5 Geophys.* **4**, 9–23.

Zdunkowski, W. G., Barth, R. E., and Lombardo, F. A. (1966). Discussion on the atmospheric radiation tables by Elsasser and Culbertson. *Pure Appl. Geophys.* **63**, 211–219.

INDEX